ENVIRONMENTAL
HYDROLOGY

Edited by
Andy D. Ward
William J. Elliot

LEWIS PUBLISHERS

Boca Raton   New York   London   Tokyo

Library of Congress Cataloging-in-Publication Data

Ward, Andrew D.
  Environmental hydrology / Andrew Ward, William J. Elliot.
    p.   cm.
  Includes bibliographical references and index.
  ISBN 0-87371-886-0 (alk. paper)
    1. Hydrology.   2. Hydrology--Environmental aspects.   I. Elliot, William J.   II. Title.
  GB665.W28 1995
  551.48--dc20   95-19757
                                                             CIP

No claim to original U.S. Government works
International Standard Book Number 0-87371-886-0
Library of Congress Card Number 95-19757
Printed in the United States of America   2 3 4 5 6 7 8 9 0
Printed on acid-free paper

Acknowledgments

Preparation of this book would not have been possible without the outstanding collective contributions of many people. We regret it is not possible for us to fully express our gratitude for these efforts or to adequately recognize every contribution. Development of the book started as a project to update the book *Agricultural and Forest Hydrology* by L.L. Harrold, G.O. Schwab, and B.L. Bondurant. We would like to thank Glenn Schwab for providing permission to use materials from that book. Particular thanks are extended to Jan Sauris for her extraordinary efforts in coordinating development activities, word processing several drafts of the text, reviewing each chapter, and providing many valuable suggestions. Illustrations for Chapters 1, 3, 4, 5, 7, 9, and 10 were prepared by Beth Daye. Two photos on the cover were taken by Keith Weller, provided courtesy of the USDA-Agricultural Research Service. Other illustrations were prepared by Jeff Blatt (Chapter 8) and Chase Langford (Chapter 12) or by the authors. Larry Brown made substantial contributions to the scientific content of Chapters 1 and 5. Comprehensive reviews of one or more chapters were provided by Russ Congalton, Jeff de Roche, Jay Dorsey, Mike Lichtensteiger, Nancy Shaffer, Stan Trimble, Qiong Wu, David Hall, and Bruce Wilson. Other important contributions were provided by Jean Boenish, Ann Chin, Annette Elliot, George Elliot, Chris Finton, Satish Gupta, John Hewlett, Lisa Lindenmann, Norman Meek, Alex Mendel, Myron Molnau, Linda O'Hirok, Paul Price, Manuel Reyes, Elizabeth Ridder, Abe Springer, Earl Swales, and Steve Workman. The corrections, suggestions, and patience of the many students who have used parts of the book during the past few years are greatly appreciated. The authors would also like to recognize the valuable support and many sacrifices that were made by their families during the development of the book. The permission which was granted by many people and organizations to use a diversity of copyright materials helped us to produce a book of this quality. We apologize if we have inadvertently failed to recognize anyone for his or her contribution. Preparation of this book was made possible through the support of the Agricultural Engineering Department at The Ohio State University, the Ohio Agricultural Research and Development Center, and the multi-agency regional Management Systems Evaluation Areas Program, which is primarily funded by the United States Department of Agriculture.

Andrew David Ward, Ph.D., is a Professor of Agricultural Engineering at The Ohio State University.

In 1971, Dr. Ward graduated from Imperial College, London, England, with a B.S. degree in Civil Engineering. He then worked as an engineer in England and as a school teacher in New York City before pursuing graduate studies in Agricultural Engineering. He obtained his M.S. and Ph.D. degrees in 1977 and 1981, respectively, from the Department of Agricultural Engineering, University of Kentucky, Lexington. He then worked in South Africa and the United States for an international consulting group prior to joining The Ohio State University in 1986.

Dr. Ward is a registered professional engineer and a member of the American Society of Agricultural Engineering, the Soil & Water Conservation Society, and the American Institute of Hydrology.

He has taught courses and workshops on hydrology, hydrogeology, water quality, soil erosion, water management, irrigation, drainage, surface reclamation, and surveying. He is the recipient of numerous competitively awarded research grants and conducts research on modeling hydrologic systems, drainage, water quality, remote sensing, soil erosion, reservoir sedimentation, and the development and implementation of techniques to prevent or control water quality impacts on the environment. He has provided leadership to the development of several hydrologic computer programs including the WASHMO watershed storm hydrograph program, the DEPOSITS reservoir sedimentation program, the ADAPT agricultural water quality model, and the BESTAQUA agricultural expert system.

Dr. Ward is the author of more than 140 papers, manuals, and book contributions. He is the coauthor (with Dr. Lyon et al.) of a paper which received the 1994 Autometric Award from the American Society of Photogrammetry and Remote Sensing for the best interpretation of remote sensing data.

William J. Elliot, Ph.D., is the Project Leader of the Engineering Technology for Improved Forest Access Research Work Unit at the Intermountain Research Station, USDA Forest Service, Moscow, ID.

In 1971, Dr. Elliot received his B.S. in agricultural engineering from Iowa State University. He then worked for the Peace Corps at Embu Institute of Agriculture, Kenya, Africa until 1973. After briefly returning to the United States he accepted a position as the Resident Engineer for the Bahamas Agricultural Research, Development, and Training Project on Andros Island. In 1980, he obtained an M.S. degree in agricultural engineering from the University of Aberdeen, Scotland. He then worked in Liberia, Scotland, and England before pursuing a Ph.D. in agricultural engineering, which he received from Iowa State University in 1988. Prior to his current position with the Forest Service he was a member of the faculty of the Department of Agricultural Engineering, The Ohio State University, for a period of four years.

Dr. Elliot is a registered professional engineer and a member of the American Society of Agricultural Engineering, the Soil & Water Conservation Society, and the British Institute of Agricultural Engineers. He has taught courses and workshops on hydrology, water quality, soil erosion, water management, irrigation, drainage, crop handling and storage, machinery management, and surveying. He is the recipient of several competitively awarded research grants and currently conducts research on soil erosion mechanics, and the application of the Water Erosion Prediction Project (WEPP) model to forest condition. During the past decade, Dr. Elliot has played a significant role in the development of the WEPP model and received an award from the USDA-Agricultural Research Service for his contribution.

Dr. Elliot is the author of more than 100 manuscripts. He is an invited author or editor of several book chapters and books.

Contributors

E. Scott Bair
Associate Professor
Department of Geological
        Sciences
The Ohio State University
Columbus, Ohio

Jay Dorsey
Post-Doctoral Scientist
Department of Soil Science
University of Minnesota
St. Paul, Minnesota

Terry J. Logan
Professor of Soil Chemistry
School of Natural Resources
The Ohio State University
Columbus, Ohio

Charles H. Luce
Research Hydrologist
USDA-Forest Service
Moscow, Idaho

John Grimson Lyon
Associate Professor
Department of Civil Engineering
The Ohio State University
Columbus, Ohio

Sue E. Nokes
Assistant Professor
Department of Agricultural
        Engineering
University of Kentucky
Lexington, Kentucky

Stanley W. Trimble
Professor
Department of Geography
University of California
Los Angeles, California

Preface

A Personal Perspective and Introduction

Andy D. Ward

I was only a few years old and the heat and humidity were unbearable as we bounced along the dirt road in a cloud of dust. Finally, the vehicle came to a halt. We started walking stealthily through the parched scrub land and thorn bushes. We stopped by a group of trees that provided some shade. A short distance below us was a muddy river. Antelope were cautiously drinking and a group of women and children walked along a narrow well-trodden path. Suddenly, our guide signaled us to be quiet. Not 50 feet away, an enormous grey body appeared and then another and still another. I held my breath in awe as a herd of African elephants moved toward the river. The antelope were Impala and the reason for their caution soon became apparent. There was a sudden commotion and loud splash as a young impala was dragged into the river by an enormous crocodile.

We were in the Luangwa River Valley in Zambia, a small country in the southern part of Africa, a continent where I spent much of my youth; and there are only two seasons—dry or rainy. The women and children were walking to the river to bathe, wash their meager clothes, and take water back to their village ten miles away. On their return, the women balanced huge open containers of water on their heads—a remarkable feat, as one false step and the precious contents would be spilled. Amazingly, hundreds of thousands of other women throughout Africa repeat a similar journey daily. Many hours of walking and the perils of snakes, lions, elephants, and crocodiles have to be endured to obtain relatively small amounts of muddy polluted water.

I lived in several small towns, and the lifestyle was spartan compared to living conditions of most of the population in the United States. Electrical power failures occurred almost daily and when water flowed from our taps it always seemed to be a new shade of brown, yellow, or orange. All drinking water was boiled.

Things I remember from my childhood in Africa, in the early 1960s, were concerns about the population explosion and the need for birth control; how the world might run out of food before the year 2000; how a nuclear war or some dreadful disease like the great plague would help to balance nature and wipe out most of humanity; and how, by the end of the century, civil war and anarchy might occur throughout the world because of unstable economies and limited food and water. Fortunately, we have not had a nuclear war, but perhaps AIDS is the dreadful disease which was imagined in my childhood. We have not yet reached the year 2000 but we have seen severe famine in many parts of the world, particularly in Africa. Droughts, famine, and anarchy have decimated some developing countries. There are severe food and water shortages even in developed parts of the world such as eastern Europe and the former Soviet Union.

The population of the world is growing by 92 million people annually. This increase is equivalent to an increase of nearly 3 people every second or a monthly increase equal to the population of New York City. Within 60 years the population of the world could double. About 95% of the growth is in developing countries. Worldwide, one in three children is malnourished and 1.2 billion people (five times the U.S. population) lack safe water to drink.

My interest in hydrology was not stirred just by childhood images of drought and food shortages. Following thunderstorms I often saw huge raging rivers, such as the mighty Zambezi, running the color of chocolate with sediment. I was amazed that any soil remained on the land. Yet a review of the literature suggests that erosion rates in Africa might be lower than in Asia, South America, Central America, and many parts of North America. In the United States, sediment is the main pollutant in surface water systems and agricultural activities contribute more than 50% of the nation's sediment. In the early 1960s, the U.S. Soil Conservation Service (SCS)[1] established guidelines that identify a maximum soil loss tolerance of 5 tons/acre annually, approximately the rate of soil formation by natural processes. It takes about 10,000 years for one foot of soil to be formed due to natural processes. Even today, soil losses from agriculture in the U.S. can be ten times this rate. World-wide, even higher rates can be attributed to surface mining, urban development, and deforestation activities.

In the mid-1960s I spent two years in Switzerland, replacing the relatively flat landscape of Africa with huge mountains, lush vegetation, frequent rainfall, and cold snowy winters. In midsummer the rivers ran full and were icy cold for they were fed by snowmelt from the Alps. Snowmelt is the main source of water for great rivers such as the Rhine which runs through Germany, France, Belgium and Holland. I lived near the head waters of the Rhine where it forms the border between Switzerland and Austria. Farming was a year-round activity in this region and cows seemed to be everywhere. "Honey" wagons deposited endless loads of fresh manure on pastures and crops already darkened from earlier loads. The Rhine was already being polluted by high discharges of nitrates from the manure.

In the 1980s and 1990s, lack of snow in the Alps, salt production works on the banks of the Rhine in France, and pollution problems associated with PCBs and heavy metals from paper, mining and industrial activities in Germany have caused serious water supply and pollution problems in Holland. These activities are polluting the Issel River which discharges into Issel Lake, a source of water for Amsterdam. Holland is a hydrologic wonder because more than 50% of the country is below sea level and much of the land has been reclaimed from the sea through the formation of polders.

Water and hydrologic phenomena have always been a part of my life. I was born in Southend-on-Sea, Essex, England, which has the longest pier in the world (about 1.2 miles long) because when the tide goes out, mud banks stretch from the shore more than a mile. While working as a counselor for a camp in Maine, I discovered there is nothing more exhilarating than white-water canoeing in rivers like the St.

---

[1] In 1994, the SCS changed its name to the Natural Resources Conservation Service. However, most material references in this book bear the name SCS, and for convenience we have retained the old name in the book.

Croix. The occasional capsize quickly established a realization of the enormous power of water. One of my first professional jobs was as a site engineer on a new residence for nurses at St. Thomas hospital, located across the River Thames from the Houses of Parliament in London. During the daily walks across the bridge I realized I was more interested in the river than the building I was helping to build.

While studying for a degree in Agricultural Engineering at the University of Kentucky, I came face to face with the environmental debate on how to maintain the high standard of living many Americans enjoy, while preserving our environment. The focus of my research was on developing a method to predict how sediment deposits accumulate in sediment ponds. These ponds are constructed downstream from surface mining operations to trap soil and spoil materials that rainfall washes away from the mining operations. I worked in the Appalachian Mountains of Eastern Kentucky and Tennessee where enormous machinery reshape mountains to reach the rich, underlying coal seams. The benefits of this environmental disturbance are affordable electricity, transportation, food, and manufactured commodities.

Surface mining activities and environmental pollution problems are not unique to one part of the world or to the United States. An example of problems associated with mining can be found in Tasmania, a small island the size of West Virginia, located about 140 miles southeast of Australia. The interior of Tasmania is very mountainous, and to the west of these mountains there is heavy annual rainfall and there are impenetrable rain forests; home of the notorious Tasmanian Devil, a small carnivorous bear. Much of Tasmania's mineral wealth of silver, zinc, and gold is located in the northwest. Trees have been removed from the mountain sides to provide fuel for deep underground mining operations and the mining families. The removal of trees, combined with acid rain from the processing of ore, has denuded the area of all vegetation. High rainfall and steep mountain slopes have resulted in the removal of all soil and thwarted efforts to revegetate the area. This is perhaps the bleakest, most desolate landscape I have ever seen—it is comparable to scenes from the moon.

During visits to the Kingdom of Swaziland, located on the eastern side of Southern Africa, I have seen stark granite mountains and rivers and reservoirs choked with sediment. Farming is conducted on terraces constructed on the sides of steep mountains. However, the need to cut wood to provide fuel for domestic use and overgrazing by cattle has resulted in a gradual denuding of many areas and the formation of huge gullies called dongas. Throughout the world, forests are lost to clear cutting at a rate of 43 million acres annually, an area about the size of Oklahoma or Washington State.

In the United States we have many sources of surface and ground-water contamination including landfills, surface impoundments, underground storage tanks, waste disposal wells, agricultural practices, military facilities, mining, septic systems, and transportation systems. We have more than four million acres of land disturbed by coal mining; 67,000 inactive or abandoned mines; 75,000 industrial landfills; 18,500 municipal landfills; more than 25,000 closed or abandoned municipal landfills; 180,000 waste impoundments; more than 7,000 hazardous waste sites; 10 million underground storage tanks; 50,000 pesticide products composed of 600 active ingredients; and 22 million septic systems that serve one-third of the nation's population (Jorgensen, 1989).

For most people in developed countries, the weather is a topic of conversation and little thought is given to water resources or electric power generation. This complacency is only occasionally interrupted by droughts, floods, temporary disruptions in water supplies, and short-term pollution problems. It would be nice to think that one day, all of humanity could share this complacency regarding water resources. Unfortunately, there is increasing evidence that water-related issues are rapidly becoming some of the most critical issues in all countries, regardless of their state of development. In this century, worldwide water use has increased tenfold. The total annual withdrawal of water by humans is currently near 4,000 billion cubic meters — about one-tenth of the total renewable supply of water, and a quarter of the stable supply available throughout the year (Hillel, 1992).

In the last two decades the United States has experienced cycles of extreme droughts and floods. In 1988, most of the nation experienced one of the worst droughts in history. Drought conditions peaked in mid-July when severe or extreme drought gripped 45% of the country. The exact cause of the drought is uncertain but is widely believed to have been caused by unusual Pacific Ocean temperatures. Although severe, the 1988 drought was not a unique event. Despite the record high temperatures and record low rainfall experienced by some parts of the country that year, the droughts of the 1930s and 1950s were more extreme, and it is likely that other more severe droughts occurred in times before accurate weather records were maintained (Heim, 1988). Droughts continue to occur in several parts of the country. For example, periodically low snowfall in the mountains has caused serious water supply problems in the fertile valleys of California and has lead to water rationing in some urban communities. In 1993, a severe drought in the southeast, and particularly South Carolina almost resulted in a total crop loss.

Part of the reason for these cycles of dry, hot weather has been attributed to global warming. Water vapor is the most important greenhouse gas in the atmosphere and the major obstacle to sunlight reaching the earth. The following statement is cited from Bilger (1992):

"The 1980s was the hottest decade in history and included seven of the 10 hottest years ever recorded. By the summer of 1988, the hottest months in living memory throughout much of the Northern Hemisphere, the timing was perfect for a global warming panic.... Thinking as concerned human beings, however, most scientists know that a global temperature rise of 2 to 90°F is a real possibility and that the public should be warned that such an increase could cause drought, famine, rising sea levels, and a host of other environmental disasters over the next 100 years."

Catastrophic flooding in the Mississippi River Basin has occurred on several occasions during this century. In 1927, thousands of square miles of farmland in the lower Mississippi flooded, and more than 650,000 people were left homeless. In response to this flooding, the U.S. Corp of Engineers constructed an enormous network of earth and concrete levees, dams, floodgates, locks, and spillways that cost more than eight billion dollars. Much of this effort focused on potential flood flows entering the lower Mississippi River basin from the Missouri and Ohio Rivers. In 1973 and 1983 there was severe flooding in the upper Mississippi basin. The damage associated with these events was small compared to the flooding which occurred during the summer of 1993. Rainfall events of 4–10 inches in a few hours occurred repeatedly during June, July, and August in Nebraska, Iowa, Minnesota, and part of Missouri. The ground quickly became saturated and flows in the Missouri and upper Mississippi Rivers were eight to ten times normal levels. Rivers remained at record levels for days, and sometimes weeks. Eventually most levees were breached. Nearly 90% of all the land in flood plains of the Mississippi River was flooded, and damage exceeded $10 billion.

In 1985, Tom Haan (an author of several prominent hydrology books) and I found ourselves drifting along a small chain of lakes on the northeast coast of South Africa. We were evaluating the feasibility of growing rice in a nearby wetland area that was virtually untouched by civilization. If constructed, this project would drain much of the area and convert it into crop land. Suddenly there was commotion all around us and we discovered that we had drifted into a family of hippopotami. Fortunately, they had little interest in us and we continued on our way. As we slowly moved from one lake into the next, we saw a wide variety of birds and an occasional crocodile. We stopped on a small mound of soggy land and watched several native fishermen catching freshwater fish in crude handmade wooden traps. As we continued toward sea, the water became more brackish and the current stronger. We stopped again just a few hundred feet from the outlet to the ocean. Here the fishermen used poles and lines to catch saltwater fish. On the north side of the estuary, in Mozambique, we noticed a lookout post manned by a soldier with a machine gun. The countryside in Mozambique had been ravaged by years of civil war. It was sad to think that this beautiful chain of lakes, the nearby wetland area, swamps to the south, and the birds and wildlife might also be destroyed due to political uncertainties and the need to feed a nation.

Interest in water and related problems has grown markedly throughout all segments of global society in the last half of the 20th century. An understanding of the occurrences, distribution, and movement of water is essential in agriculture, forestry, botany, soil science, geography, ecology, geology, and geomorphology. In short, water is an important element of the physical environment. We should all seek knowledge as an aid in understanding the physical environment in which humanity has developed and in which we now live.

The purpose of this book is to provide a qualitative understanding of hydrologic processes and an introduction to methods for quantifying hydrologic parameters and processes. It has been prepared for use in introductory hydrology courses taught at universities to environmental science, natural resource, geography, agricultural engineering, and environmental engineering students. A comprehensive understanding of the presented topics and problems should provide sufficient knowledge for students to make an assessment of hydrologic processes associated with environmental systems and to develop initial conceptual evaluations that are part of most assessments.

Topics at the beginning of each chapter are intended for students who have not taken university level basic science and mathematics courses. Topics that require prerequisites in the basic sciences, advanced knowledge of mathematics, or are not fully described because of their complexity, are marked with an asterisk.

Readers will note that we have frequently used English units of measure in this book. Although the rest of the world has converted to SI, système international, (metric) units, most field-level work in the U.S. is still done in English Units. Also, several U.S. scientific journals are switching back to English units so that published papers will have more impact. In any case, in the U.S. we need to know how to use and convert between both systems. This may be an unfortunate imposition, but it is a reality. A table of unit conversion factors is presented in Appendix A.

The topic of hydrology contains many different terms which may not be familiar to the reader. Therefore, we have presented a glossary of terms in Appendix B. Publications cited in the preface are included in the list of references in Chapter 1.

Table of Contents

Chapter 1
The Hydrologic Cycle, Water Resources, and Society
Andy D. Ward and *William J. Elliot*

Chapter 2
Precipitation
William J. Elliot

Chapter 3
Infiltration and Soil Water Processes
Andy D. Ward and *Jay Dorsey*

Chapter 4
Evapotranspiration
Sue E. Nokes

Chapter 5
Surface Runoff and Subsurface Drainage
Andy D. Ward

Chapter 6
Soil Erosion and Control Practices
William J. Elliot and *Andy D. Ward*

Chapter 7
Flow in Channels, Rivers, and Impoundments
Andy D. Ward

Chapter 8
Forests and Wetlands
Charles H. Luce

Chapter 9
Hydrogeology
E. Scott Bair

Chapter 10
Water Quality
Terry J. Logan

Chapter 11
Remote Sensing and Geographic Information Systems in Hydrology
John Grimson Lyon

Chapter 12
**Practical Exercises on Conducting and
Reporting Hydrologic Studies**
Stanley W. Trimble and *Andy D. Ward*

CHAPTER 1

The Hydrologic Cycle, Water Resources, and Society

Andy D. Ward and William J. Elliot

1.1. THE HYDROLOGIC CYCLE

The earth holds more than 300 million cubic miles of water beneath the land surface, on the surface, and in the atmosphere. This vast amount of water is in constant motion, known as the hydrologic cycle. Hydrology is concerned with the transport of water through the air, over the ground surface, and through the strata of the earth. A knowledge of hydrology is important in practically all problems that involve the use and supply of water. Therefore, hydrology is of value not only in the field of engineering but also in forestry, agriculture, and other branches of the environmental sciences. The term "hydrology" can be broken down into two terms: "hydro," relating to water, and "loge," a Greek word meaning knowledge. Thus, hydrology is the study, or knowledge, of water.

The hydrologic cycle, illustrated in Figure 1.1, shows the pathways where water travels as it circulates throughout global systems by various processes. The visible components of this cycle are precipitation and runoff. However, other components, such as evaporation, infiltration, transpiration, percolation, groundwater recharge, interflow, and groundwater discharge, are equally important. A summary of the world water balance by continent is presented in Table 1.1. In the following sections, we will discuss each of these components and their relation to each other.

1.1.1. Precipitation

Water that evaporates from the earth is temporarily stored as water vapor in the atmosphere. While in the atmosphere, this vapor and small water droplets form clouds. As the atmosphere becomes saturated, water is released back to earth as some form of precipitation (rain, snow, sleet, or hail). Some of the precipitation

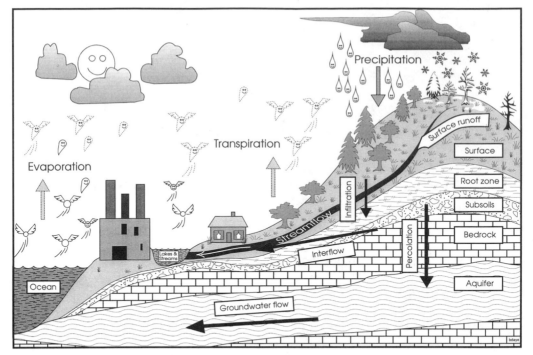

Figure 1.1. The Hydrologic Cycle.

might evaporate before it reaches the ground. Precipitation reaching the ground can evaporate from anywhere including bare soil surfaces, plant surfaces, and the surfaces of ponds, lakes, and streams. Precipitation is a natural phenomenon that humans can do very little to control.

1.1.2. Evaporation

Evaporation occurs when water is changed from a liquid to a vapor. Increases in air and water temperatures, wind, and solar radiation all increase evaporation rates while a high water vapor percentage in the air (high relative humidity) decreases the potential for evaporation. Through the process of evaporation, water moves back to the atmosphere in the form of vapor.

1.1.3. Transpiration

Water can take several paths after it enters the soil. Some water becomes part of the soil storage. This water is not stationary and moves downward at a rate that depends on various soil properties such as hydraulic conductivity and porosity. While in storage near the surface, some of this water is used by plants and is eventually returned to the atmosphere as water vapor. The process by which plants release water vapor to the atmosphere is called transpiration. This water vapor is a natural by-product of photosynthesis.

Table 1.1.   Annual world water balance, by continent expressed as inches of water.   (Summarized from van der Leeden et al., 1991).

Water Balance Elements[1]	Europe	Asia	Africa	North America	South America	Australia	Weighted Average
Area, millions of mi^2	3.8	17.6	11.8	8.1	4.0	3.4	
Precipitation	28.9	28.6	27.0	26.4	64.9	29.0	34.1
Total river runoff	12.6	11.5	5.5	11.3	23.0	8.9	12.1
Groundwater runoff	4.3	3.0	1.9	3.3	8.3	2.1	3.8
Surface water runoff	8.3	8.5	3.6	8.0	14.7	2.8	7.6
Infiltration and soil water	20.6	20.0	23.4	18.4	50.2	22.2	25.8
Evaporation	16.3	17.0	21.5	15.1	41.9	20.1	22.0

1. Total river runoff is the sum of the groundwater runoff and surface water runoff. The sum of the total river runoff and evaporation equals the precipitation.

1.1.4. Evapotranspiration

Because of the difficulty in separating the processes of evaporation and plant transpiration, we usually view these two processes as one process called evapotranspiration. This term includes both the water that evaporates from soil and plant surfaces and the water that moves out of the soil profile by plant transpiration. More than half of the water that enters the soil is returned to the atmosphere through evapotranspiration.

1.1.5. Infiltration

Infiltration is the entry of water into the soil. The amount of water that infiltrates into the ground varies widely from place to place. The rate at which water infiltrates depends on soil properties such as soil water content, texture, density, organic matter content, hydraulic conductivity (permeability), and porosity. Hydraulic conductivity is a measure of how fast water flows through certain soils or rock layers. Infiltration and hydraulic conductivity are greater in porous materials, such as sands, gravels, or fractured rock than in clay soils or solid rock. Porosity is a measure of the amount of open space in soil or rock which may contain water.

Conditions at the soil's surface also influence infiltration. For example, a compacted soil surface restricts the movement of water into the soil profile. Vegetation can play a prominent role in infiltration. The surface soil layer in a forest or a pasture will generally have far greater infiltration rates than a paved parking area or a compacted soil surface. Topography, slope, and the roughness of the surface also affect infiltration as do human activities in urban and agricultural areas where alteration of soil properties and surface conditions have occurred.

1.1.6. Percolation and Groundwater Recharge

Another path that water can take after it enters the soil surface is that of percolation. Percolation is water moving downward through the soil profile by gravity after it has entered the soil. Water that moves downward through the soil below the plant root zone towards the underlying geologic formation is called deep percolation. For the most part, deep percolation is beyond the reach of plant roots and this water contributes to replenishing the groundwater supply. The process of replenishing or refilling the groundwater supply is called groundwater recharge.

1.1.7. Runoff and Overland Flow

Runoff is the portion of precipitation, snow melt, or irrigation water that flows over and through the soils, eventually making its way to surface water systems. Contributions to runoff might include overland flow (surface runoff), interflow and groundwater flows. Once the precipitation rate exceeds the infiltration rate of the soil, depressions on the soil surface begin to fill. The water held in these depressions is called surface storage. If surface storage is filled and precipitation continues to exceed infiltration, water begins to move down slope as overland flow or in defined channels. This process is called surface runoff or overland flow. A large percentage of surface runoff enters stream channels. Surface runoff can also occur when the soil is saturated (soil storage is filled). In this case, all the voids, cracks, and crevices of the soil profile are filled with water and the excess begins to flow over the soil surface.

1.1.8. Interflow

As water percolates, some of it may reach a layer of soil or rock material that restricts downward movement. Restrictive layers can be formed naturally (clay pan or solid bedrock) or as a result of human activities. Once water reaches a restrictive layer, it may move laterally along this layer and eventually discharge to a surface-water body such as a stream or lake. The lateral movement of water is called interflow.

1.1.9. Groundwater Flow

Groundwater comprises approximately 4% of the water contained in the hydrologic cycle and can flow to surface-water bodies such as oceans, lakes, and rivers. This process creates a baseflow for a surface-water body which is an important contribution to groundwaters and surface waters. More than 50% of the population depends upon groundwater as its primary source of drinking water. Approximately 75% of American cities derive their supplies totally or partially from groundwater. In 1980, 88 billion gallons/day of groundwater were used in the United States, and 68% of this total was used for irrigation.

1.2. WATER SUPPLY

As a society, we face increasing demands for water and increasing threats to water quality. As water circulates between earth and the atmosphere, chemicals and particles contained and transported by it are modified as a result of natural processes and human activities. Chemicals and particles from dust, smoke, and smog in the atmosphere eventually fall back to the earth with precipitation. Runoff from roadways and parking lots wash grit and metal particles directly into storm sewers and streams. Water moving across the soil surface as runoff can detach soil particles and transport them to a stream or lake. Some chemicals attach to soil particles and also are transported to receiving waters. Runoff from lawns, pastures, and agricultural fields can also carry dissolved nutrients and pesticides. An assessment of the impact of nonpoint pollution on our surface water resources is summarized in Table 1.2 (USGS, 1986). The percentage of rivers, lakes, and estuaries which is threatened or already impaired is very high.

Water that flows to a groundwater system can be polluted by the leaching of chemicals, nutrients, and/or organic wastes from the land surface or from materials buried in landfills. Groundwater close to the surface or in porous sands and gravels

Table 1.2.   Surface waters impacted by nonpoint pollution sources in the United States, excluding Alaska, 1985 (assessed waters only; USGS, 1986).

Status	Rivers (Thousands of miles)	Lakes (Millions of acres)	Estuaries (Thousands of square miles)	Oceans (thousands of shore-line miles)
No use impairment	230	7.20	11.1	6.30
Threatened	48	3.70	3.0	0.10
Moderately impaired	87	3.50	1.6	0.34
Severely impaired	30	0.88	0.8	0.01
Undistinguished	9	0.09	2.5	0.07
Total assessed	404	15.37	19.0	6.82

is vulnerable to pollution. Deep groundwater is also vulnerable, especially if connected to the surface by fissures or sinkholes in underlying formations as in limestone rock areas.

Hillel (1992) provides the following insight on the escalation in water use: *"Historical records show that in 1900, the World's annual water use was about 400 billion cubic meters, or 242 cubic meters per person. By 1940, the total water use had doubled, while the per capita use had grown by some 40% to about 340 cubic meters.... By 1970, it had reached 700 cubic meters per capita. Both agricultural and industrial water use grew twice as much in the 20 years between 1950 and 1970 as they had during the first half of the century."*

Worldwide, agriculture is the major user of water, accounting for about 70% of all withdrawals. More than 17% of the world's cropland is irrigated and produces over a third of the harvest. The amount of irrigated land per capita in various parts of the world is presented in Table 1.3 (ICID, 1987). Between 1960 and 1980, the majority of the world's increased food production was achieved through additional irrigation. Unfortunately, irrigation can also cause damage to the land through waterlogging and salinization. For example, in India and Pakistan, 13% and 22% of the irrigated land, respectively, is unusable because of salinity problems (Tarrant, 1991).

The hydrologic cycle for the United States is summarized in Table 1.4 (Federal Council for Science and Technology, 1962). The water available in surface water systems is somewhat misleading as more than 90% of the total is located in the Great Lakes. However, we are fortunate to have a vast network of rivers. A map showing the relative size and location of the largest rivers in the United States is presented in Figure 1.2 (Iseri and Langbein, 1974). Water undergoes repeated cycles of being withdrawn, treated, partially consumed, and the remaining amount returned back to the river. Some portion of the flow might be used 15–20 times before it discharges into the ocean. Rivers such as the Mississippi and St. Lawrence are also important transportation systems linking inland agricultural and commercial areas with ocean ports.

Average annual water use in selected countries is shown in Table 1.5 (World Resources Institute, 1986). It can be seen that the United States has the highest per capita use, primarily due to the high level of development, the standard of living

Table 1.3. Amount of irrigated land per capita by world region (based on ICID, 1987).

World Region	1984 Irrigated Area (million acres)	1987 Population (millions)	Irrigated Area (acres/capita)
Africa	25.68	589	0.04
Americas	87.43	691	0.13
Asia and Oceania	342.83	2,938	0.12
Europe and USSR	86.69	779	0.11
World	542.65	4,997	0.11

Table 1.4. Distribution of water in the continental United States (Source, Federal Council for Science and Technology, 1962).

Water Supply	Volume ($\times 10^9$ m^3)	Volume (%)	Annual Circulation ($\times 10^9$ m^3/year)	Replacement Period (in years)
Groundwater				
Shallow (<800 m deep)	63,000	43.2	310	>200
Deep (>800 m deep)	63,000	43.2	6.2	>10,000
Freshwater lakes	19,000	13.0	190	100
Soil moisture (1-m root zone)	630	0.43	3,100	0.2
Salt lakes	58	0.04	5.7	>10
Average in stream channels	50	0.03	1,900	<0.03
Water vapor in atmosphere	190	0.13	6,200	>0.03
Frozen water, glaciers	67	0.05	1.6	>40

enjoyed by most Americans, and the need to use water for electric power cooling and agriculture. Let's look more closely at how water is used in the United States.

Approximately two-thirds of the precipitation in the U.S. is used to satisfy evaporation and evapotranspiration requirements (Figure 1.3). The remaining third enters surface and groundwater systems. Only a small portion of this water supply (8.1% of the precipitation or about 25% of the surface flow) is withdrawn for use in thermoelectric power generation, industrial and mining purposes, commercial and domestic water use, and for agricultural purposes. Depending on location, water will cost the consumer less than one dollar per 1000 gallons to more than two dollars per 1000 gallons.

Average urban domestic water use is presented in Table 1.6 (van der Leeden et al., 1991). Typically, families in many European countries use less than half the amount of water used in the U.S., while in developing countries, the rate of domestic water use is sometimes less than 10% of the average domestic amount used in U.S. households. It can be seen that a 33% savings could be obtained by simply eliminating lawn watering, swimming pools, automobile washing, and garbage disposal units. In European countries, other savings are obtained by using more efficient toilet systems which reduce water use for this purpose to less than one-third of U.S. consumption. In 1994, federal legislation was passed which specifies the maximum flow rates of new shower and toilet installations. These maximum rates are similar to current European levels. In developing countries, only small amounts

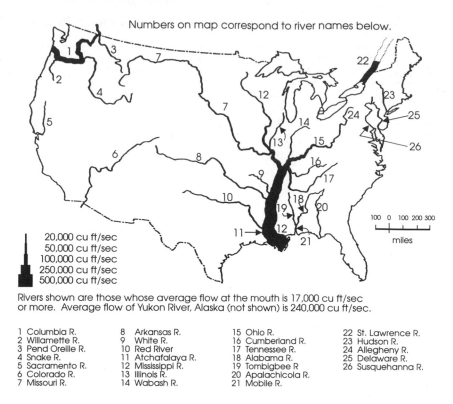

Numbers on map correspond to river names below.

20,000 cu ft/sec
50,000 cu ft/sec
100,000 cu ft/sec
250,000 cu ft/sec
500,000 cu ft/sec

100 0 100 200 300

miles

Rivers shown are those whose average flow at the mouth is 17,000 cu ft/sec
or more. Average flow of Yukon River, Alaska (not shown) is 240,000 cu ft/sec.

1 Columbia R.	8 Arkansas R.	15 Ohio R.	22 St. Lawrence R.
2 Willamette R.	9 White R.	16 Cumberland R.	23 Hudson R.
3 Pend Oreille R.	10 Red River	17 Tennessee R.	24 Allegheny R.
4 Snake R.	11 Atchafalaya R.	18 Alabama R.	25 Delaware R.
5 Sacramento R.	12 Mississippi R.	19 Tombigbee R	26 Susquehanna R.
6 Colorado R.	13 Illinois R.	20 Apalachicola R.	
7 Missouri R.	14 Wabash R.	21 Mobile R.	

Figure 1.2. Relative size and location of the largest rivers in the United States.

of water are used for bathing and laundering, due primarily to limited access to running water, hand bathing, and bathing and laundering in rivers by part of the population.

A summary of water used to produce various commodities is presented in Table 1.7 (Kollar and MacAuley, 1980). The quantities of water needed to manufacture most commodities is large and water is one of the main limiting factors in furthering worldwide industrial development. It can be seen from Table 1.7 that although a lot of water is needed to manufacture a commodity, only a small amount of this water is consumed. The rest is discharged or recycled, treated and then used again for this or another purpose. Improved technologies are required to reduce the amount of water needed and consumed.

1.3. THE IMPORTANCE OF HYDROLOGY TO SOCIETY

To the beginning student of hydrology, the natural questions that first arise are "Why must we be concerned with the complexity of the many relationships that affect various phases of the hydrologic cycle?" and "Hydrology wasn't as important 20 years ago, so why is it now?" Water has been central to the history of humanity. Civilizations have persisted or perished as they experienced situations of too much or too little water. Abandoned irrigation projects the world over, including native American works in the Southwestern United States (1100 AD) illustrate there was a

Table 1.5.   Average annual water use in selected countries (World Resources Institute, 1986).

Country	Water Withdrawal Total (km³)	Per Capita (m³)	Share Withdrawn by Sector (percent) Public	Industry	Electric Cooling	Agricultural/ Irrigation
United States	472,000	1,986	10	11	38	41
Canada	30,000	1,172	13	39	39	10
Egypt	45,000	962	1	0	0	98
Finland	4,610	946	7	85	0	8
Belgium	8,260	836	6	37	47	10
Former USSR	226,000	812	8	15	14	63
Panama	1,300	596	12	11	0	77
India	380,000	499	3	1	3	93
China	460,000	460	6	7	0	87
Poland	15,900	423	14	21	40	25
South Africa	9,200	284	17[*]	0	0	83

*   Public and Industry

woeful lack of knowledge of hydrology on which to base the development of water-supply systems. Salinity problems in agriculture of semiarid and arid regions of the world are also evidence of the lack of hydrologic principles of water management. Locating costly developments in flood plains of large river systems indicates a lack of understanding of river hydrology or direct ignorance of the hydrologist's forecast of flooding. In the past, there were signs of deficiencies in the science of hydrology but they affected only a few people and went almost unnoticed. It has been quite different since the mid-20th century.

Water is still central to our life today but we have grown more aware of the accelerated growth of industry and population and their use of water as related to the earth's fixed supply of water. We are coming to understand that our prosperity and prospects for survival vary with the amount and distribution of fresh, unpolluted water, and that each year there are millions more of us, but no more water than before. Multiple demands for the use of the same gallon of water and prospects for even greater demands tomorrow indicate that more and more people of different disciplines need to have a knowledge of hydrology. Today's student will be tomorrow's water management decision maker.

As the science of hydrology becomes more generally understood, parts of the hydrologic cycle are examined to determine how and to what extent the cycle can be modified by human activity in practical ways. Attempts to modify the weather to increase rainfall in specific areas, for example, have been made through research, trial, and demonstration. However, success in this effort has been limited. Extensive drainage projects in low-lying swamps opens up land for food production, thus lowering groundwater levels, reducing evaporation, changing rainfall-runoff relationships, and affecting wildlife habitats. In other areas, large-scale irrigation programs increase soil water content, evaporation, and crop use of water. In watersheds where the vegetative cover can be considerably modified by afforestation or deforestation

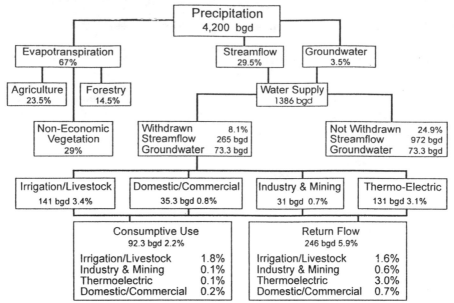

Figure 1.3. Distribution of annual precipitation in the United States (bgd = billion gallons per day).

Table 1.6. Typical urban water use by a family of four (van der Leeden et al., 1991, based on U.S. Water Resources Council, 1975; percentages added).

Type of Household Use	Daily Family Use		Daily Per Capita Use Gallons/day
	Gallons/day	%	
Drinking and water used in kitchen	8	2	2.00
Dishwasher (3 loads per day)	15	4	3.75
Toilet (16 flushes per day)	96	28	24.00
Bathing (4 baths or showers per day)	80	23	20.00
Laundering (6 loads per week)	34	10	8.50
Automobile washing (2 car washes per month)	10	3	2.50
Lawn watering and swimming pools (180 hours per year)	100	29	25.00
Garbage disposal unit (1 percent of all other uses)	3	1	0.75
Total	346	100	86.50

and where huge tracts of grassland can be plowed under for grain production, the movement of water into and over the land surface may be drastically altered. Numerous dams and reservoirs provide water, recreation and flood protection but modify natural stream flow regimes.

A practical knowledge of the science of hydrology will help the decision-maker and general public understand the overall effect of humanity's influences on the hydrologic cycle and the side effects of projects on other people, their activities, and the environment. The informed decision-maker will be able to weigh the advantages of each proposed change in the hydrologic cycle against the disadvantages.

Table 1.7.   Water use vs. industrial units of production in the United States (based on Kollar, and MacAuley, 1980).

Industry	Parameters of Water Use	Intake by Unit of Production	Consumption by Unit of Production	Discharge by Unit of Production
Meatpacking	gal/lb carcass weight	2.2	0.1	2.1
Dairy products	gal/lb milk processed	0.52	0.03	0.48
Canned fruits and vegetables	gal/case 24-303 cans e/case	107	10	98
Frozen fruits and vegetables	gal/lb frozen product	7.1	0.2	6.9
Wet corn milling	gal/bu corn grind	223	18	205
Cane sugar	gal/ton cane sugar	18,250	950	17,300
Malt beverages	gal/barrel malt beverage	420	90	330
Textile mills	gal/lb fiber consumption	14	1.4	12.8
Sawmills	gal/bd. ft lumber	3.3	0.6	2.7
Pulp and paper mills	gal/ton pulp and paper	38,000	1,800	36,200
Paper converting	gal/ton paper converted	3,9000	270	3,600
Industrial gases	gal/1000 ft³ industrial gases	226	31	193
Industrial inorganic chemicals	gal/ton chemicals	4,570	470	4,300
Plastic materials and resins	gal/lb plastic	6.7	0.6	6.1
Synthetic rubber	gal/lb synthetic rubber	6.5	1.4	5.1
Cellulosic man-made fibers	gal/lb fiber	68	4.6	63
Organic fibers, non cellulosic	gal/lb fiber	38	1.1	37
Paints and pigments	gal/gal paint	7.8	0.4	7.4
Industrial organic chemicals	gal/ton chem. building	54,500	2,800	51,700
Nitrogenous fertilizers	gal/ton fertilizer	4,001	701	3,299
Petroleum refining	gal/barrel crude oil input	289	28	261
Tires and inner tubes	gal/tire car and truck tires	153	14	139
Hydraulic cement	gal/ton cement	830	150	680
Steel	gal/ton steel net production	38,200	1,400	36,800
Iron and steel foundries	gal/ton ferrous castings	3,030	260	2,760
Primary copper	gal/lb copper	17	4.1	13
Primary aluminum	gal/lb aluminum	12	0.2	11.8
Automobiles	gal/car domestic automobiles	11,464	649	10,814

1.4. MODELING THE HYDROLOGIC CYCLE

Traditional methods for estimating hydrologic responses using tables, charts, and graphs are currently being replaced with computer models in research, in the class-room, and in the field. Since the mid-1960s, engineers and scientists have been developing hydrologic applications for computers. Government and university scientists with access to main frame computers began to develop models to describe climate, infiltration, runoff, erosion, soil water movement, drainage, groundwater flow, and water quality. With the advent of the personal computer in the mid-1980s, the use of models spread from laboratories to agencies concerned with hydrologic systems. As portable computers develop, scientists and technicians with regulatory agencies and consulting companies will be taking models directly to the field to assist in analyzing hydrologic problems with managers of farms, forests, mines, land development projects, and other hydrologic units. Computer applications feature strongly in current research and future development in all areas of hydrologic science. An understanding of the importance and applications of computer modeling of hydrologic processes is important for all who are interested in hydrology.

1.4.1. Types of Models

Models may be empirical, deterministic, or stochastic. Empirically-based models are developed by analyzing a large set of data, and developing statistical relationships between the inputs and the outputs (Woolhiser, 1982). An example of an empirical model is the Universal Soil Loss Equation or USLE (Wischmeier and Smith, 1978) and its computer counterpart, the Revised Universal Soil Loss Equation or RUSLE (Renard et al., 1991). Empirical models are not easily transferable between geographic regions.

Deterministic models, sometimes described as theoretical (Woolhiser, 1982) or process-based models, mathematically describe the processes being modeled, such as runoff. As the processes are independent of geographic variations, deterministic models can be applied to a wider range of conditions than empirical. In some instances, however, it may not be possible to adequately describe a process. An example is the gain or loss of dissolved chemicals as water percolates through a soil. In some cases, the amount of data required to describe a process may restrict the use of the model. Excessive input data may be necessary to provide all the information necessary to fully describe a process. This restriction was particularly apparent in the earlier modeling attempts when computer memory was limited.

Stochastic models seek to identify statistical probabilities of hydrologic events (Woolhiser, 1982), like rainfall or flood flows, and to predict the probability of a given outcome. They also consider the natural variability that might occur in some model input parameters. As users become more acquainted with the statistical nature of hydrology, stochastic modeling will increase. Recent developments in fractal theory and spatial distribution statistics will lead to further developments in stochastic modeling.

1.4.2. Advantages and Disadvantages

Computers allow managers to consider many more options than would be possible with hand calculations. For example, a recreational park planner can compare the effects of subsurface drainage depth and spacing on the number of days that a playing field may be available for use. Many combinations of depth and spacing can be considered, and the benefits of improved drainage can be compared to the cost of each system. Computer models are particularly helpful in determining which erosion management practices may best suit a given farm.

Research scientists can identify critical areas for further research and study and the sensitivity of hydrologic systems to the various parameters that are needed for a given model. Researchers are aware of which parameters they require for a model to operate, but the importance of each of the parameters on the model predictions is not known. If a model shows that the rate of plant canopy growth is not very important in the operation of a computer model, then research resources can be directed to other areas where a model may be highly sensitive to changes in inputs. The development of computer models requires a sound understanding of the system being modeled and the dominant processes in the system. The need to understand such processes is increasingly dictating the direction of much research (e.g., Laflen et al., 1991).

Scientists and engineers can better understand key parameters in a hydrologic system by using a computer model of that system. By developing the necessary input files to run a model, users gain a much greater appreciation of the importance of considering the entire system and not simply concentrating on one or two aspects of that system.

Computer models can lead to wrong conclusions if not properly applied. Numerous models have been developed not to provide absolute answers, but rather to give relative results to allow the user to compare the effect of different management systems on some hydrologic response. The water quality model GLEAMS is such a model (Leonard et al., 1987) although numerous users have found that the absolute predictions by the model have been good. A user may also input unrealistic combinations of values into a model, which can lead to misleading results. Altering the clay content in an input file without altering a property closely related to clay, like the cation exchange capacity, can lead to misleading results with an erosion model.

Considerable time may be required to build the input files necessary to run a computer model. Sometimes the necessary data are not available for the conditions under which the model is being applied. Many models have been developed for specific geographic areas, and when applied outside that area, have not been successful. In other cases, users may not fully understand the importance of some of the inputs, and poor estimates of some parameters may lead to poor model performance.

1.4.3. Typical Model Architecture

Most hydrologic computer models consist of three main components: (1) the input file(s); (2) the output file(s); and (3) the computer model (Woolhiser, 1982). The user will either develop the input file with a word processor or text editor, or the model will include a file builder to assist in assembling the input file. More sophisticated models have developed user friendly interfaces to assist with building the input files and interpreting the output files.

In some cases, the output file from one part of the model may serve as the input file for another. This practice is more common with complex models like CREAMS

(Knisel, 1980). Users may wish to access these intermediate files to get a better grasp on the hydrologic system or modify them to study the performance of a particular component of the model.

Typical input files for most hydrologic models include climate, soil, topographic or structure, and management files. With some models, some of the files may be combined. Some models may require additional files describing chemical properties for water quality.

The climate file may describe a single storm or runoff hydrograph. Some models have the option of daily or hourly rainfall distributions. Models with crop components will require information to estimate evapotranspiration or plant growth rate, like daily temperatures, wind speeds, and/or solar radiation.

The soil files generally include the necessary information to describe infiltration and water movement within the soil profile. Specific infiltration and permeability values or properties which are more easily measured and which will predict those values may be required. In programs where water holding characteristics are important, like drainage and irrigation programs, a method of describing water content is required.

The topographic or structure file is necessary to describe slope lengths and steepnesses, spacing of drains, or the relationship between model elements. In watershed models, the structure file can become quite complex, and users need to be extremely methodical in their approach. A structure file for the WEPP (Water Erosion Prediction Project) water erosion model (Nearing et al., 1989) describing a 40 ha- (100 acre-) watershed was found to require 35 elements to fully describe the combination of soils, crops, and drainageways.

The management files contain the necessary information to describe surface roughness as it affects runoff and water surface storage. It may also contain information describing plant growth rates and residue conditions as affected by tillage and decomposition. Plant water use is frequently required for models that predict soil water conditions for drainage or erosion prediction.

Generally, the better the quality of the data in the input files, the better the model will perform. Most model users soon develop sets of input files to describe the range of conditions they are modeling. From these typical data sets, only minor file modifications are then necessary to describe each new application.

REFERENCES[1]

AWWA. 1988. New Dimensions in Safe Drinking Water. American Water Works Association.

Bates, A.K. 1990. Climate in Crisis: The Greenhouse Effect and What We Can Do. The Book Publishing Company, Summertown, Tennessee.

---

[1] Not all listed publications have been cited. In some places, publications appeared to present conflicting information. This is due to the difficulties of obtaining precise information on worldwide issues or due to the ideological bent of the authors. We have taken the liberty of reporting data that provide a general consensus of information reported in the literature. In future chapters, only publications which have been cited are presented in the list of references at the end of each chapter.

Beecher, J.A., and A.P. Laubach. 1989. Compendium on Water Supply, Drought, and Conservation. Report NRRI 89-15, National Regulatory Research Institute, Columbus, Ohio.

Bilger B. 1992. Global Warming: Earth at Risk. Chelsea House Publishers, New York.

Brown, L.R. (Project Director). 1992. State of the World. A Worldwatch Institute Report on Progress Toward a Sustainable Society, W.W. Norton & Company. New York.

Federal Council for Science and Technology. 1962. Report by an Ad Hoc Panel on Hydrology and Scientific Hydrology, Washington, D.C.

Gabler, R. 1987. Is your water safe to drink? Consumer Reports Books, Consumers Union, Mount Vernon, New York.

Hillel, D.J. 1992. Out of the Earth: Civilization and the Life of the Soil. The Free Press, MacMillan, Inc, New York.

Horton, T., and W.M. Eichbaum. 1991. Saving the Chesapeake Bay. The Chesapeake Bay Foundation, Island Press, Washington, D.C.

ICID. 1987. Editorial. ICID Bulletin vol. 38 (2), International Commission on Irrigation and Drainage, New Delhi, India.

Iseri, K.T. and W.B. Langbein. 1974. Large Rivers of the United States. U.S. Geological Survey Circular 686, Washington, D.C.

Jorgensen, E.P. (Editor). 1989. New Strategies for Groundwater Protection: The Poisoned Well. Sierra Club Legal Defense Fund, Island Press, Washington, D.C.

Knisel, W.G. (ed.). 1980. CREAMS: A Field-Scale Model for Chemicals, Runoff, and Erosion from Agricultural Management Systems. Conservation Research Report No. 26. Washington, D.C.: USDA-Science and Education Administration.

Kollar, K.L., and P. MacAuley. 1980. Water Requirements of Industrial Development, J. Am. Water Works Assoc., vol. 72, no. 1.

Laflen, J.M., W.J. Elliot, J.R. Simanton, C.S. Holzhey and K.D. Kohl. 1991. WEPP soil erodibility experiments for rangeland and cropland soils. J. Soil and Water Conserv. 46(1):39–44.

Leonard, R.A., W.G. Knisel, and D.A. Still. 1987. GLEAMS: Groundwater loading effects of agricultural management systems. Transactions of the ASAE 30(5):1403–1418.

Madison, R.J. and J.O. Brunett. 1985. Overview of the Occurrence of Nitrate in Ground Water in the United States. National Water Summary 1984. Water Supply Paper 2275, U.S. Geological Survey, Washington, D.C., pp 93–105.

Nearing, M.A., G.R. Foster, L.J. Lane and S.C. Finkner. 1989. A Process-based Soil Erosion Model for USDA-Water Erosion Prediction Project Technology. Transactions of the ASAE 32(5):1587–1593.

Patrick R., E. Ford, and John Quarles. 1987. Groundwater Contamination in the United States (2nd Edition). University of Pennsylvania Press, Philadelphia.

Postel, S. 1984. Water: Rethinking Management in a Age of Scarcity. Worldwatch Paper 62, Worldwatch Institute, Washington, D.C.

Renard, K.G., G.R. Foster, F.A. Weesies, and J.P. Porter. 1991. RUSLE Revised Universal Soil Loss Equation. J. of Soil & Water Conserv. 46(1):30–33.

Silver, C.S., and R.S. DeFries. 1990. One Earth, One Future: Our Changing Global Environment. National Academy of Sciences, National Academy Press, Washington, D.C.

Tarrant J. (Editor). 1991. Farming and Food. Oxford University Press, New York.

U.S. Department of Commerce. 1987. Statistical Abstract of the United States. U.S. Department of Commerce, Washington, D.C.

USGS. 1984. National Water Summary 1983-Hydrologic Events and Issues. U.S. Geological Survey Water-Supply Paper 2250, Washington, D.C.

USGS. 1986. National Water Summary. The Association of State and Interstate Water Pollution Control Administrators, in cooperation with the EPA. America's Clean Water: The States' Nonpoint Source Assessment, 1985. U.S. Geological Survey, Washington, D.C.

van der Leeden, F., F.L. Troise and D.K. Todd. 1991. The Water Encyclopedia (2nd Edition). Lewis Publishers Inc., Chelsea, Michigan.

Wischmeier, W.J. and D.D. Smith. 1978. Predicting rainfall erosion losses, a guide to conservation planning. Agriculture Handbook No. 537. Washington, D.C.: USDA-Science and Education Administration.

Woolhiser, D.A. 1982. Hydrologic system synthesis. In Haan, C.T., H.P. Johnson, and D.L. Brakensiek (eds.). Hydrologic Modeling of Small Watersheds. pp 3–16 St. Joseph, MI: ASAE.

World Resources Institute. 1986. World Resources 1986. Basic Books.

Problems

1.1 Select an environmental system of interest to you (such as a forest, crop production system, urban area, surface mine, etc.) and then list the components of the hydrologic cycle which are most likely to cause water contamination problems for this system. Briefly discuss how the land use you have selected will influence the components of the hydrologic cycle you listed. As you learn more about hydrology in the next few chapters, you might wish to refer back to your answer to this question.

1.2 In the future, it is anticipated that in the United States, the amount of water used for irrigation and livestock will be four times higher than the levels reported in Figure 1.3. Also, the amount of water used for industry and mining will double. For this future use, determine the consumptive use, return flow, and amount of water not withdrawn from streams and groundwater systems. Assume other water uses do not change.

1.3 Table 1.1 provides information on the world water balance by continent. Determine the evaporation rates in North America and Africa as percentages of the precipitation in each of these continents. Why do you think the rate is higher in Africa?

1.4 The chapter contains a quote from Hillel (1992) on water use changes. Convert the quantities expressed in this quote in cubic meters to cubic feet and also to gallons. (Note that 1 cubic meter is 35.27 ft^3 and there are 7.48 gallons in a cubic foot.)

1.5 Based on the information in Table 1.6, determine the urban water use by a family of four if a dishwasher is used twice daily, toilets use 50% less water, one car is washed monthly, and water use for lawns and swimming pools is only 30% of reported levels. What percentage reduction in the reported total daily water use has been achieved?

1.6 Select from Table 1.7 one industry which is of interest to you, then visit a library and determine the annual number of units of production for this industry in the United States. Determine the total annual water use and consumptive use for this industry. Convert this volume into cubic feet and then, acre-ft. Suggest how the needed consumed amounts of water might be reduced. (An acre is 43,560 ft^2. Acre-ft are often used to express large volumes of water in the environment. For example, 5 acre-ft can be visualized as a depth of 5 feet of water covering an area of 1 acre. The volume contained in 5 acre-ft would be 5 times 43,560 ft^2 or 217,800 ft^3.)

1.7 Conduct a literature review and identify common computer models which are used to determine: (1) soil erosion; (2) surface runoff from single storm events; (3) daily information on runoff, evaporation, and infiltration; (4) steady-state groundwater flow; (5) the hydrology of watersheds; and (6) water quality information for either agricultural or surface mining land uses.

CHAPTER 2

Precipitation

William J. Elliot

2.0. INTRODUCTION

Precipitation is any form of solid or liquid water that falls from the atmosphere to the earth's surface. Rain, drizzle, hail, and snow are examples of precipitation. In the U.S., rain is the most common form of precipitation, and snow is the next most common.

Precipitation is formed from water vapor in the atmosphere. As air in the atmosphere cools, its capacity to hold water decreases. When air's capacity to hold water is reached, it is said to be saturated. For example, at 70°F, a given volume of saturated air contains nearly four times the water vapor that it contains at 32°F. Figure 2.1 shows the relationship between air temperature and the amount of water contained at saturation. Moist, unsaturated air can become saturated as it cools. Continued cooling beyond saturation will cause the transition or condensation of some water vapor in the air into liquid or solid water. The condensation process can be seen when droplets of water form on the outside of a glass of a cold beverage because the moist air next to the glass has cooled below the temperature of saturation, or dew point.

If all the atmospheric water were precipitated at one time, it would result in an average depth of only 1 inch of water on the earth's surface. However, a one-hour storm has resulted in water depths in excess of 4 inches over land areas of a few square miles. Such local storms result from lateral flow of moist air from surrounding areas to the storm center.

A meteorologist is generally concerned with the flow of moist air masses from their sources (lakes, oceans, transpiration from land areas) and their associated effects on precipitation and temperature. A hydrologist considers such phenomena in relation to water supplies and movement to, on, and beneath the earth's surface. Hydrologists usually become involved in meteorology in the planning, study, and evaluation of the interaction between the climate, surface water, and subsurface water.

2.1. CAUSES OF PRECIPITATION

The atmosphere around the earth is a dynamic system of air masses in constant motion and collision. The weather system is driven by solar energy from the sun.

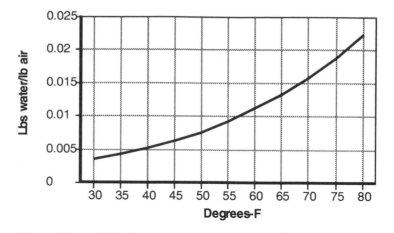

Figure 2.1. The relationship between water content of air at saturation and air temperature.

Generally, daytime radiation is absorbed nearer the earth's tropics, and radiant energy losses occur mainly during nighttime hours and from the earth's polar regions. These energy gains and losses result in keeping dynamic air masses in constant motion, continually exchanging water vapor with the land, vegetation and water on the earth's surface. The movement and collision of air masses lead to atmospheric instability, often resulting in precipitation.

There is always some water vapor in the air, and there is always some condensation occurring in the atmosphere, even on the fairest of days. Clouds may be composed of water vapor, water droplets and/or ice crystals. The precipitation process begins by the condensation of water molecules on precipitation nuclei such as smoke, dust, or sea salt particles. As clouds cool, the amount of condensed water increases. Precipitation begins when the air is cooled, increasing cloud formation, and the condensed water droplets or ice crystals reach a size that causes them to fall toward the earth's surface. Some of these droplets will collect additional condensed water vapor as they fall. Others may evaporate and return to the atmosphere.

In the atmosphere, the cooling of the air is mainly caused by lifting the air mass. At higher elevations, atmospheric pressure and temperature are lower. Both the reduced pressure and lower temperature reduce the temperature of the rising air mass. Air may be lifted by three processes (Figure 2.2).

2.1.1. Frontal Precipitation

Precipitation occurs when a warm or light air mass meets a cold or heavy air mass, and the lighter air rides up over the heavier air. The zone where the two air masses meet is commonly called a front, and the resulting precipitation is frontal. Frontal precipitation is the dominant type of precipitation in the north central U.S. and other continental areas.

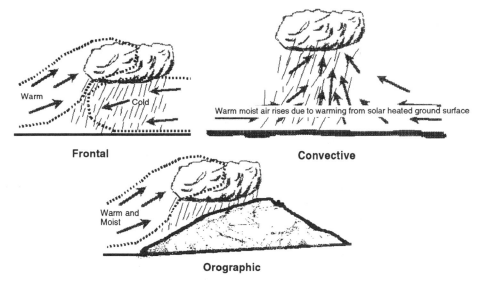

Figure 2.2. Main mechanisms causing air to rise and cool, resulting in precipitation: frontal, convective and orographic.

2.1.2. Convective Precipitation

Air expands when heated by solar energy and becomes lighter than the air around it. The lighter air rises by convection, causing convective precipitation. In humid climates, convective precipitation frequently occurs on hot mid-summer days in the form of late afternoon thunderstorms. Convective precipitation also results from the movement of air into a low-pressure atmospheric system. Air at the center of the low pressure region rises, causing surface air to flow to the center. As the air rises, it can cool until it reaches saturation. Severe storms of rain and hail may follow. Convective storms are common during the summer in the central U.S. and other continental climates with moist summers. Some thunderstorms are associated with this convergent air flow, as are hurricanes, which produce very heavy rainfall.

2.1.3. Orographic Precipitation

Air currents force air masses to rise over hills or mountains. Precipitation that results from this process is called orographic precipitation. This is the dominant source of precipitation in the mountains in the western U.S., and contributes to precipitation in most mountainous areas.

2.2. STORM DESCRIPTION

A storm is described by several key parameters. The most common parameter is the total amount of precipitation, or depth in inches, feet, millimeters, or meters. The time from the beginning of the storm until the end of the storm is the duration. The rate of precipitation, or intensity, is found by dividing the amount of precipitation that occurs during a given time period by the length of that period. Common units of intensity are inches per hour, millimeters per hour, or centimeters per hour. Less common, but of importance to some studies is the area covered by a given storm. The areas tend to be very large and are given in square miles or square kilometers.

2.3. PRECIPITATION DATA

Precipitation data are necessary for most land use plans. The supply of water for human, agricultural, or industrial use, the disposal of waste water, and the control of excess rainfall are key elements in most planning processes. In addition, recreational areas may include lakes or ponds which require precipitation data for both supply and overflow considerations.

In agriculture, precipitation data can indicate when and where a lack or a surplus of water for crops may be expected. Provision for irrigation and/or drainage systems may be necessary. In forestry, precipitation predictions assist planners in determining forest growth rates and downstream water yields.

State and federal governments have collected and published precipitation data for many years. These publications are available in most libraries (Chapter 12, Exercise 1). Data sets can also be purchased on computer tapes or CD-ROM formats from government and private sources. These data aid researchers in studying characteristics and distributions of precipitation events. Precipitation records may report amounts of precipitation by years, months, days, and sometimes shorter time periods, and indicate the form, like rain or snow. From time to time, special reports are published to provide precipitation data on specific topics like flood events of major importance, rainfall rate-duration-frequency, or droughts.

In many hydrological studies, the hydrologist will determine the precipitation patterns in the area of concern. These patterns will be a valuable guide for work on flood problems, water supply, and soil erosion. The general storm precipitation patterns in the midwestern U.S. are outlined in Table 2.1.

2.4. GEOGRAPHICAL AND SEASONAL VARIATIONS

Rainfall distribution across the U.S. is dependent on the air masses dominating a given region (Figure 2.3). Annual precipitation (average for over 30 years) in the U.S. varies from less than 5 inches in California, Nevada, and New Mexico to over 100 inches in western Washington (Figure 2.4), where moist air from the Pacific Ocean flowing over the mountains results in orographic precipitation. Much of the precipitation in the central region comes from atmospheric water from the Gulf of Mexico. In the central U.S., average annual amounts range from less than 20 inches along the eastern foothills of the Rocky Mountains to nearly 80 inches in the southeastern mountains. Precipitation east of the Appalachians is largely influenced by water from the Atlantic Ocean (Figure 2.3).

Seasonal distribution of precipitation varies widely across the nation (Figure 2.5). Along the west coast, most of the precipitation falls during the winter, with maximum monthly values of 4 to 6 inches. Minimum values of less than 1 inch occur in

Table 2.1.   General storm precipitation patterns in the midwestern U.S.

Precipitation Characteristic	October to March	April to September
Form	Snow and Rain	Rain
Rate	Low	High
Storm Duration	Long	Short
Area Covered	Large	Small
Raindrop Size	Small	Large

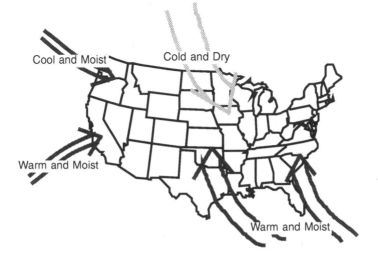

Figure 2.3.   Major wet and dry air masses influencing the precipitation patterns in the U.S.

July and August. In the central U.S. the wettest months are during the summer. In the east, precipitation is generally evenly distributed throughout the year, except for Florida where the winter precipitation is lowest and the summer rainfall averages over 6 inches per month.

Storms with longer durations will have a greater total amount of precipitation. Shorter duration storms, however, will have greater intensities. The maximum amount of rainfall for periods of 1 minute to 1 year, recorded over the world, prior to 1950, is listed in Table 2.2, along with the calculated average intensities for the given storms.

2.5. TIME TRENDS

Time trend studies are conducted to detect whether there is a tendency for precipitation to increase or decrease as a continuous trend or in cycles of years (Chapter 12, Exercise 1). Historic records of annual precipitation may be used, but as year-to-year values of precipitation usually vary widely, it is difficult to judge whether

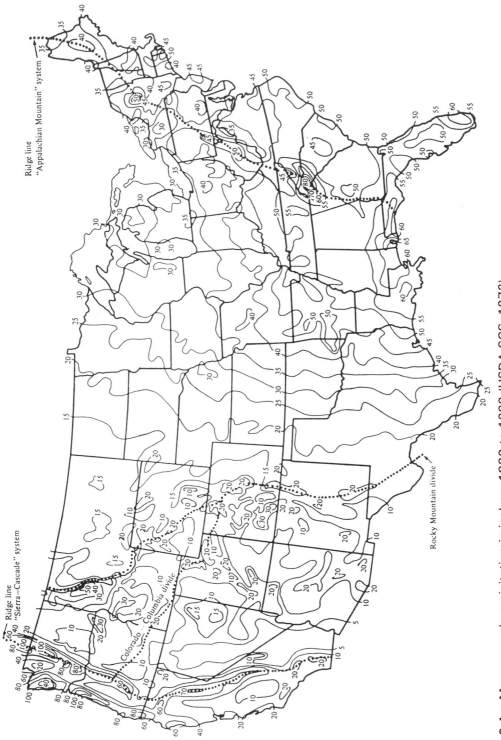

Figure 2.4. Mean annual precipitation in inches, 1889 to 1938 (USDA SCS, 1972).

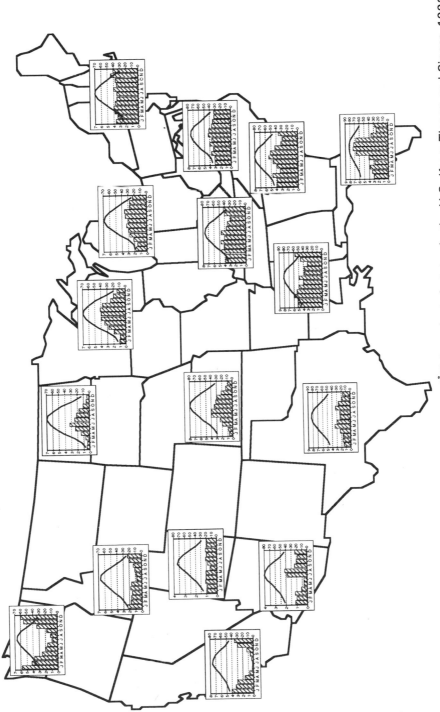

Figure 2.5.   Monthly precipitation (inches) and mean temperature (˚F) for selected states in the U.S. (from Tiegen and Singer, 1992).

Table 2.2.   World's greatest point rainfall depths and corresponding average intensities.

Time	Depth, in.	Intensity, in./hr	Location
1 min	1.23	73.8	Maryland, U.S.
15 min	7.80	31.2	Jamaica
2 hrs 10 min	19.0	8.8	W. Virginia, U.S.
15 hrs	34.5	2.3	Pennsylvania, U.S.
24 hrs	46.0	1.9	Philippine Islands
48 hrs	65.8	1.4	Taiwan
7 days	131.	0.8	India
1 year	1,042.	—	India

McCuen, 1989.

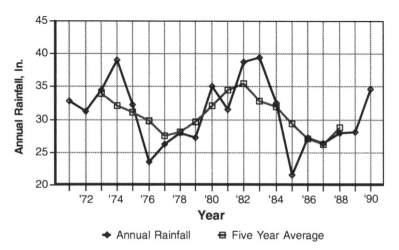

Figure 2.6.   Annual precipitation and 5-year average annual precipitation for Priest River Experimental Forest, Idaho.

these time trends are occurring. Sometimes hydrologists calculate 5-year moving averages to help reveal time trend changes. For example, the first 5-year average in the record from 1970 to 1994 is for the period 1970 to 1974; the second from 1971 to 1975; and the last from 1990 to 1994. These moving averages eliminate the sharp fluctuations in the annual values, making it easier to discern trends (Figure 2.6).

Rain gage data in the U.S. have been recorded for less than 100 years, so it is not possible to use precipitation data to study trends over a long period of time. However, records of tree-ring growth rates cover hundreds of years and have been used by hydrologists to determine long term rainfall trends. Times of reduced growth rate indicate periods of relatively low precipitation, so these records may be used to detect historic precipitation trends.

2.6. STORM AREA PATTERNS

Rainfall amounts, durations, and intensities vary spatially within the area covered by a given storm. Large area storms tend to be more uniform in distribution and to

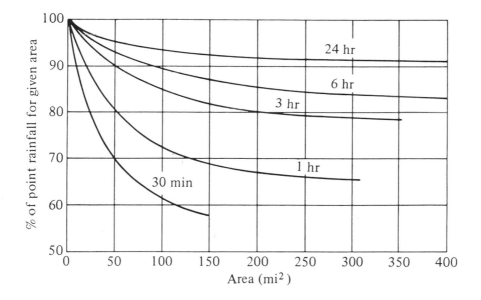

Figure 2.7. Relationship between storm area, storm duration, precipitation at the center of a storm and average precipitation (Myers and Zehr, 1980).

have longer durations. Small area storms tend to have much higher rainfall amounts near the center of the storm compared to the edges, and tend to be of shorter duration. Figure 2.7 shows the general relationship between the area covered by the storm, the duration, and the average amount of precipitation as a percent of the maximum precipitation at the center. Hydrologists generally select a maximum storm for a point, but will reduce the size of storm for larger watersheds, based on Figure 2.7. For example, for an area of 50 square miles, the average rainfall depth for a 30-minute event is about 68% of the maximum precipitation and the average depth of a 24-hour event is about 96% of that at the storm center.

Example 2.1: Using Figure 2.7, determine the average depth of storm rainfall for a 100-mi^2 area given following maximum rainfalls: 1.60 in. in 3 hours; and 2.30 in. in 24 hours.

Solution: Average depth of rainfall for a 100-mi^2 storm:

3 hrs: 1.60 in. × 0.84 =   1.35 in.
24 hrs: 2.30 in. × 0.93 =   2.16 in.

Answer: The average depth of rainfall for 3 to 24 hr storms on a 100 mi^2 area are 1.35 inches and 2.16 inches, respectively.

2.7. MEASUREMENT OF PRECIPITATION

Measurement of rainfall at a particular site, point rainfall, is sufficient for studies where spatial distribution is of no concern. Point observations may be used to define rain or snowfall for a specific area, such as a watershed, or to identify the characteristics of storms, irrespective of defined land area.

Rain gages must be established in such a way that their sample is a measure of true precipitation at a point and is essentially unaffected by the surroundings. The Environmental Science Service Administration, ESSA (formerly U.S. Weather Bureau) checks its sites for possible wind eddy currents from trees, buildings, or other objects, or from sharp topographic ridges or valleys. As a general rule, no object should be closer to the gage than twice the height of the object above the receiver surface. If the top of the nearest object is 30 ft above the gage receiver surface, the gage should be located at least 60 ft from the object. As strong wind currents affect the accuracy of most common gages, it might be desirable to locate gages in sheltered spots, like clearings of sufficient size in an orchard or grove of trees.

Hydrologic studies of precipitation involving areas or watersheds of many square miles require data from a number of gages. If there are no gages in the study area, then the hydrologist may follow the less than satisfactory approach of using precipitation data from nearby gages. If the hydrologist is initiating a watershed study, a network of gages can be established to meet the needs. However, the establishment of the ideal network is likely to be limited by economics and accessibility. Figure 2.8 shows that as the area per gage increases, so does the standard error associated with each gage reading. Smaller watersheds also require a greater density of gages than larger basins to limit statistical error.

There are numerous recommendations for gage spacings that depend on reasons for gaging, topography, and climate. Generally more rugged areas require more gages, as do climates that tend to experience a significant amount of precipitation from small-area convective thunderstorms. Flatter areas receiving the majority of significant rainfall from frontal storms generally tolerate a wider gage spacing. Brakensiek et al. (1979) developed a guide for planning precipitation gage networks for watershed studies (Table 2.3). The distribution of gages over the study area should be fairly uniform, one gage representing about the same area as any other. Local conditions, such as accessibility, may make it impractical to distribute the gages evenly.

Example 2.2: How many gages are required for a 300-acre watershed?

Solution: From Table 2.3, three gages are recommended for 100 acres and four for 600 acres, so three gages are adequate.

Answer: Three rain gages are adequate for 300 acres.

2.8. MEASUREMENT OF SNOWFALL

The measurement of snow on the ground, termed "snow pack," is important in hydrology for determining water supply and flood runoff potential. In the inter-

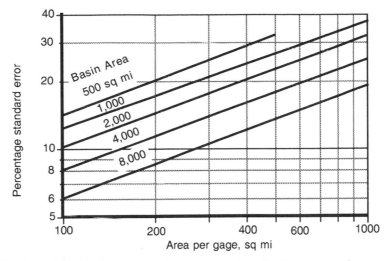

Figure 2.8. Relationship between rain gage spacing, basin area, and percentage standard error of rain gages.

mountain west, snow may be 80 to 90% of the annual precipitation. Foresters are frequently involved in snow-pack measurements because in the U.S., much of the high-elevation snow country is forested and about 75% of the West's water supply comes from forest land.

Hydrologists are concerned with snow pack depth and its water equivalent (water depth after melting). Freshly fallen snow has a water equivalent ranging from 5 to 20% or 0.5 to 2.0 inches of water per 10 inches of snow, averaging 10%. Snow still on the ground in late summer may reach a density of 60%.

Inventories of snow-pack depth and density are made at periodic intervals throughout the snow season in order to estimate total water supply in the watershed snow pack. Several snow courses, usually having five to ten points, 50 ft apart, are laid out over a watershed to represent the expected snow packs. The depth of snow is measured, a core is taken, and a known amount of snow in the sampling core is weighed to determine its water equivalent (SCS, 1984).

Example 2.3: The snow depth at one site was recorded as 5.0 ft and its measured density was 20%. What is the equivalent depth of water?

Solution: The equivalent depth of water would be:

$$\text{Depth} = 5 \text{ ft} \times 12 \text{ in./ft} \times 20\% = 12 \text{ in.}$$

Answer: The equivalent depth of water is 12 inches.

Snow pillows of white neoprene, placed on the ground surface before the snow season, may be used for snowpack inventories. Pillows vary from 6 to 12 ft in diameter and 2 to 4 in. deep. The pillows are filled with an antifreeze solution

Table 2.3.   Guide for network gage numbers (Brakensiek et al., 1979).

Size of Watershed	Number of Gage Sites
40 acres	2
100 acres	3
600 acres	4
5 mi^2	10
10 mi^2	15
100 mi^2	50
300 mi^2	100

(alcohol-water mix) and the pressure on the pillows due to the snowpack is recorded. The pressure indicates the water equivalent of the snowpack on top of the pillow. Snowpack depth and water equivalent data may be compiled into estimates of total watershed water storage.   The expected snowmelt seasonal water supply can be estimated from these data.

2.9. RAIN GAGES

The purpose of rain gages is to measure the depth of rain falling on a horizontal surface. They may also be used, with some difficulty, to measure snowfall. There are several common types of non-recording and recording rain gages (Figures 2.9, 2.10, and 2.11). Standard non-recording rain gages are of low cost and usually free of maintenance problems. The simplest are tapered plastic and test tube-type rain gages. These gages may be attached to fence posts or set vertically in the ground. They are fairly reliable and their data (to 0.1 in. accuracy) are of some value to hydrologists in areas where no full-size standard gages are located. Many of these tube gages are made of glass and should not be used in freezing weather because water freezing in the gage may break the glass.

The most common rain gage used in the U.S. is the standard 8-in. diameter, sharp-edged horizontal receiver placed about 36 in. above the ground surface. In areas where deep snow is expected, the receiver surface is raised so that it will always be exposed. During seasons of no snow, a funnel inside the receiver passes rain water down into a small measuring tube or storage bucket. The small hole in the funnel outlet reduces evaporation of rain water from the tube or bucket. The funnel is removed in the snow season. These gages are not good for measuring snowfall because wind tends to blow the snow across the receiver, allowing only part of the snow to fall into the receiver. Wind shields are advisable where measurements of snowfall are important (Figure 2.11). Measurements of snow depth on the ground and of its water equivalent are generally used for hydrologic studies in areas of high snowfall. Propane heaters may be used in the winter to prevent freezing, particularly with recording rain gages.

In standard rain gages, rainfall amounts are obtained by measuring the depth of water in the collecting tube with a scale provided with the gage, or a ruler graduated in inches, and dividing this number by 10 (since the area of the tube is 1/10th the area of the gage catch surface). Readings are accurate to 0.01 in. Some rain gages

Figure 2.9. Weather station at the Priest River Experimental Forest in northern Idaho. The instrumentation from left to right is: Tipping bucket rain gage, sunshine recorder, solar radiation pyranometer, standard 8-inch rain gage, screened box containing thermometers for maximum, minimum, and wet-bulb temperatures, and strip-chart recording rain gage. The striped post with a diamond sign in the background is one of the points on the snow survey course around this site. (Photograph by W.J. Elliot)

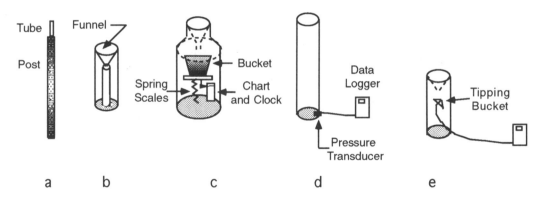

Figure 2.10. Types of rain gages: a) test tube on post, b) standard 8-inch, c) strip chart recorder, d) pipe with pressure sensor, and e) tipping bucket recording rain gage with electronic data logger.

may use a graduated cylinder calibrated to directly measure the rainfall amount. Others measure the volume in the gage, which is then converted to a depth by dividing by the collecting area of the gage. Observations are usually made at the beginning of the work day (usually 8 a.m., and sometimes 8 p.m.), and at the same time on weekends and holidays.

Recording rain gages may be of several types (Figure 2.10). Traditionally, weighing recording gages were preferred in small-area research, and many are still in use today. This gage has a spring scale beneath the collecting bucket platform that is calibrated to show the rainfall depth on a paper chart. The chart is rotated by a spring-driven clock at speeds of 1 revolution in 6, 9, 12, 24, or 192 hours. The

Figure 2.11.   Weather station at the Manitou Experimental Forest near Denver, Colorado. The instrumentation from left to right is: screened box with thermometers; strip-chart recording rain gage; tower with tipping bucket rain gage, electronic temperature sensor, solar panel to recharge system batteries, antenna to relay recordings to central site via satellite, and box with data logger instrumentation; and 8-inch standard rain gage with windshields. (Photograph by W.J. Elliot)

record on the chart shows the accumulation of rainfall depth over time. A variation on the weighing bucket is the use of a single large diameter pipe, about 5 ft high, with an electronic pressure sensor in the base regularly recording the pressure with an electronic data logger. The record of changes in pressure with time allows the hydrologist to determine the cumulative rainfall depth over extended periods of time. Careful interpretation of the results is necessary to account for the evaporation of water from the pipe that will occur between storms. A concentrated antifreeze solution can be added to these gages during the winter so any snow falling in the gage will melt, allowing a direct reading of water equivalent.

The tipping bucket gage connected to a data logger is the most common type of recording rain gage presently being installed. Two small calibrated buckets of equal size and weight are balanced on opposite sides of a fulcrum. Rain falling through a collector funnel into one bucket causes it to tip and empty. The rain water is then delivered to the other bucket until it is full, tips back and empties, and so on. At each tip, a signal is sent to an electronic data logger. The data logger calculates the amount of rainfall that occurs for the time increment desired, and provides the user with the resulting rainfall distribution. Table 2.4 is a summary from a tipping bucket rain gage for one storm from an erosion study in a forest in the southeastern U.S. The time increment was originally 1 minute, but the data were reduced to show 5-minute amounts.

2.10. POINT PRECIPITATION

Point precipitation data may be in the form of daily totals from non-recording gages or for smaller time increments from recording gages. Processing point pre-

Table 2.4.   Distribution of rainfall during a storm on the University of Georgia White-hall Forest in June, 1990 (Brown, 1993).

Time, min	Amount, mm
5	1.5
10	5.6
15	13.0
20	5.8
25	5.1
30	9.7
35	4.8
40	1.3

cipitation data from non recording gages can give summaries, maximums, and minimums for daily, monthly, seasonal, and annual time periods. Table 2.5 gives an example of a 20-year annual summary.

Processing data from a recording rain gage is more complex than from non-recording gages, but provides more detailed information to determine rainfall intensity distribution within a storm and storm duration. For some hydrologic projects, hourly rainfall data are sufficient. For others, 30-min or even 5-min rainfall amounts may be desired. Rainfall rates may be plotted with time to produce a hyetograph of rainfall. Figure 2.12 shows hyetographs for 5-min and 20-min increments from the data given in Table 2.4.

Example 2.4: For the rainfall amounts shown in Table 2.4, calculate a) the total rainfall amount; b) the duration; c) peak intensities for 5 min, 10 min, and 30 min; and d) the average intensity.

Solution:

a) The total rainfall amount is the sum of the individual amounts in Table 2.4.

$$\text{Amount} = 1.5 + 5.6 + 13.0 + 5.8 + 5.1 + 9.7 + 4.8 + 1.3 = 46.8 \text{ mm}$$

$$46.8 \text{ mm} \div 25.4 \text{ mm/in.} = 1.84 \text{ in.}$$

b) The duration is the length of time from the beginning of the storm until the end of the storm. In Table 2.4, the time is given in minutes from the beginning of the storm, which gives a total duration of 40 minutes.

c) The 5-minute peak intensity occurs between 10 and 15 minutes when 13 mm of precipitation is measured:

$$\text{Intensity} = 13 \text{ mm} \times \frac{60 \text{ min/hr}}{5 \text{ min}} = 156 \text{ mm/hr}$$

Table 2.5.   Annual precipitation for Los Angeles, California, 1934 to 1953.

Year	Inches	Year	Inches	Year	Inches	Year	Inches
1934	14.6	1939	13.1	1944	19.2	1949	8.0
1935	21.7	1940	19.2	1945	11.6	1950	10.6
1936	12.1	1941	32.8	1946	11.6	1951	8.2
1937	22.4	1942	11.2	1947	12.7	1952	26.2
1938	23.4	1943	18.2	1948	7.2	1953	9.5

or

$$\frac{156 \text{ mm/hr}}{25.4 \text{ mm/inch}} = 6.14 \text{ inches/hr}$$

The 10-min intensity can be found by comparing each 10-min period to find the period which receives the greatest amount of rainfall. The period is between 10 and 20 minutes, with a total rainfall of 13.0 + 5.8 mm = 18.8 mm. The intensity is

$$\frac{18.8 \text{ mm}}{25.4 \text{ mm/inch}} \times \frac{60 \text{ min/hr}}{10 \text{ min}} = 4.44 \text{ inches/hr}$$

By the same procedure, the maximum 30-min rainfall intensity is determined. The maximum the 30-min rainfall between 5 and 35 minutes is:

$$5.6 + 13.0 + 5.8 + 5.1 + 9.7 + 4.8 = 44.0 \text{ mm}$$

so the 30-min intensity is:

$$30\text{-min intensity} = \frac{44 \text{ mm}}{25.4 \text{ mm/inch}} \times \frac{60 \text{ min/hr}}{30 \text{ min}} = 3.46 \text{ inches/hr}$$

d) The average intensity is the total rainfall amount divided by the storm duration:

$$\text{Average Intensity} = 1.84 \text{ inch} \times \frac{60 \text{ min/hr}}{40 \text{ min}} = 2.76 \text{ inches/hr}$$

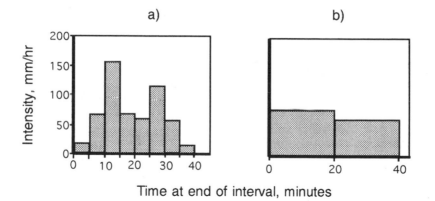

Figure 2.12.   Hyetographs of the storm presented in Table 2.4 showing a) 5-minute and b) 20-minute breakpoint data.

Answer: The total rainfall amount is 1.84 inches and the duration is 40 minutes. The peak intensities for 5, 10, and 30 minutes are 6.14, 4.44, and 3.46 inches/hr, respectively. The average intensity is 2.76 inches/hr.

2.11. RAINFALL EROSIVITY

The ability of rainfall to detach and transport soil particles is the rainfall erosivity. Soil erosion researchers have postulated that the erosivity of a given storm can be estimated from the product of the maximum 30-min intensity and the amount of energy in the storm, or the EI value. The total storm energy is found by calculating the amount of energy per inch within each time increment and multiplying that energy by the rainfall that fell during that time increment. Figure 2.13 gives the relationship between rainfall energy and the rainfall intensity. The total energy is the sum of each of the time increments. This total is multiplied by the maximum 30-min intensity and, to make the units smaller, the product is divided by 100.

Example 2.5: Calculate the EI value of the storm summarized in Table 2.4.

Solution: The following table is set up to determine the energy for each of the time increments. The table has been shortened for this example by considering two 20-minute increments rather than each five-minute increment separately. Generally, the storm should be divided into increments with approximately similar intensities.

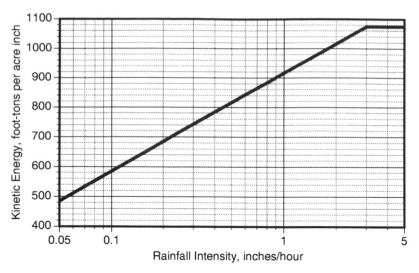

Figure 2.13.   Kinetic energy versus rainfall intensity (E = 916 + 331 log$_{10}$ (intensity), If i > 3, E = 1074 ft-tons/acre-in.) (Based on Wischmeier and Smith, 1978).

Time (min)	Depth (mm)	Amount (in.)	Duration (min)	Intensity (in./hr)	Energy per in.	Total Energy ft-tons/acre
0 to 20	25.9	1.02	20	3.06	1074	1095
20 to 40	20.9	0.82	20	2.47	1040	853
Totals	46.8	1.84	40			1948

The total energy is multiplied by the maximum 30-minute intensity of 3.46 in./hr and the result is divided by 100:

$$EI = 1948 \times 3.46/100 = 67.4 \quad 100\text{-ft-tons/acre} \times \text{in./hr}$$

Answer: The EI value is 67.4   100-ft-tons/acre-hr. × in./hr

2.12. TIME SEQUENCE PATTERNS

The time distribution of rainfall rates in a storm may appear random. Upon closer examination, thunderstorms often begin with relatively high intensities, followed by periods of decreasing intensity. Winter rainfalls often have relatively low intensity, with little variation in intensity throughout the storm. Other storms may have the maximum intensity at an intermediate time, or a delayed time. The delayed intensity distribution is more common on the west and southeast coasts of the U.S. Figure 2.14 shows four typical hyetographs of rainfall patterns. The distribution of runoff rates from a given storm can be influenced by the rainfall pattern, with delayed patterns tending to result in greater runoff rates.

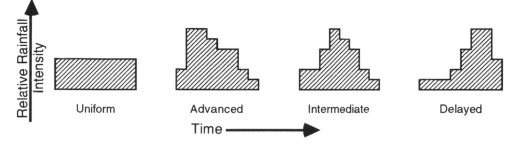

Figure 2.14.   Hyetographs of typical storm patterns.

2.13. AVERAGE PRECIPITATION OVER AN AREA

Hydrologists frequently need to determine the rainfall distribution over an area with multiple rain gages (Exercise 12.1). The distribution may be for specific storms, days, months, or even years. It may be necessary to determine an average depth, or a spatial distribution of depths.

There are three methods commonly used to determine the average rainfall amount from a series of rain gages. If the rain gages are evenly distributed, a simple arithmetic average will be adequate. If the gages are not evenly distributed, the most common method of determining an average depth is the Thiessen method. The use of the Thiessen method is illustrated in Figure 2.15: a) The locations of each rain gage in, and immediately adjacent to, the area of interest are located on a map. The rainfall depth at each gage is noted. b) Straight dashed lines are drawn between each adjacent gage site. c) Solid perpendicular bisectors to these lines are constructed so that the area around each gage is enclosed by the bisectors and/or the area boundary. The enclosed areas around each gage are known as Thiessen polygons. The entire area within each polygon is closer to the rain gage in that polygon than to any other rain gage. The rainfall depth measured by the gage within each polygon is assumed to be representative of the rainfall for that polygon. d) The areas of each polygon are calculated by geometry, with a planimeter, with graph paper, or with a computer-aided drawing package. The average rainfall for the entire area is then assumed to be a weighted average of the observed rainfalls, calculated by Equation 2.1:

$$P = \frac{\sum\limits_{i=1}^{n} A_i \times P_i}{\sum\limits_{i=1}^{n} A_i} \qquad (2.1)$$

where P represents the average depth of rainfall in the watershed with a total area of

$$\sum\limits_{i=1}^{n} A_i \,,$$

and A_i is the area of the ith polygon with precipitation of P_i in that polygon.

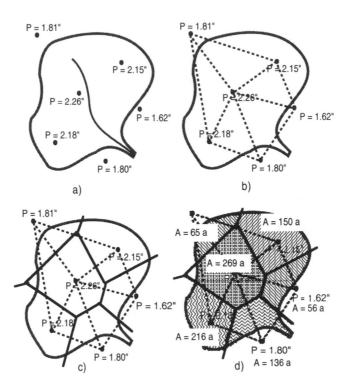

Figure 2.15.   Use of Thiessen method to find average precipitation. a) Distribution of rain gages in a watershed located on a map; b) connection lines drawn between rain gage positions; c) lines perpendicular to connection lines drawn until they intersect to form polygons; and d) areas calculated within each polygon.

Example 2.6: Use the Thiessen Method and an arithmetic average to determine the average rainfall for the watershed shown in Figure 2.15 for the depths and areas shown.

Solution: By the Thiessen Method, the areas represented by the various rain gages are determined and used in Equation 2.1 as follows:

$$P = \frac{(65 \times 1.81) + (150 \times 2.15) + (269 \times 2.26) + (216 \times 2.18) + (56 \times 1.62) + (136 \times 1.8)}{65 + 150 + 269 + 216 + 56 + 136} = 2.08 \text{ in.}$$

The arithmetic average is the mean of the observations:

$$P = \frac{1.81 + 2.15 + 2.26 + 2.18 + 1.62 + 1.8}{6} = 1.97 \text{ in.}$$

Answer: The average rainfall based on the Thiessen method and arithmetic average method are 2.08 and 1.97 inches, respectively.

Another method of determining average rainfall is the isohyetal method. With this method, lines of equal rainfall (isohyets) are drawn on the watershed similar to contour lines (See Appendix C). From the resulting map, a weighted average based on the area within each of the contour lines can be calculated. This method may have some benefit in mountainous regions where rainfall variation with elevation is fairly consistent, making isohyets a better representation of the rainfall distribution than Thiessen polygons.

The rain gage distribution in Example 2.6 is fairly uniform with relatively small areas for each rain gage. The average rainfall calculated should be relatively reliable. For watersheds where there are no rain gages within their boundaries, averages by any method is less reliable.

2.14. RAINFALL FREQUENCY DISTRIBUTIONS

Hydrologists need to estimate the probability that a given rainfall event will occur in order to assist planners in determining the likelihood of the success and/or failure of a given project (Exercise 12.1). The main parameters that are needed to describe rainfall frequency distributions are the duration and intensity, and the return period. The return period is the average period of time in years expected between either high intensity storms, or between very dry periods. Generally, plans for drains, ditches, and grassed waterways consider an event with a return period of 5 to 10 years sufficient, whereas hydrologic plans for urban areas are more commonly concerned with rainfall events that occur only once in 100 years, or even less frequently for large cities.

For projects that are concerned with storm runoff from small to medium size watersheds (under 1 to 100 mi^2), plans generally require a storm event with a duration varying from 5 minutes to 24 hours. Larger watersheds (over 100 mi^2) may require storm durations or cumulative rainfall amounts for time periods of up to one month. For river basin projects with large scale irrigation, drainage, river flood, or river drought flows, rainfall amounts for time periods of one month to one year may be needed.

If long term records are available, such as 100 years, it may be sufficient to determine the highest or lowest rainfall from that record. Because data from recording rain gages have not generally been available for over 40 years, it is frequently necessary to estimate a 50 or 100 year event from 20 or fewer years of rainfall records. There are numerous methods for estimating the return period for a given rainfall duration. The Hazen method will be demonstrated in this text. The Hazen method is commonly used by the Soil Conservation Service and is similar to other methods. The method consists of determining the statistical distribution of rainfall amounts for the duration of interest, plotting that distribution on log-probability graph paper, and interpolating or extrapolating from the graph to determine the storm associated with the return period of interest.

The first step in the Hazen method is to assemble as many years as possible of rainfall records for the duration of interest. An example set of data are given in Table 2.5 for annual rainfall records. Analyses of annual data are useful for planning water supply and waste disposal projects.

The annual values are first listed in order from the highest to the lowest, as shown in Table 2.6. A ranking number is then given each rainfall amount with 1 for the highest, 2 for the second, etc. From the ranking, a plotting position is determined from:

$$Fa = \frac{100\ (2n - 1)}{2\ y} = \frac{100}{\text{Return Period}} \tag{2.2}$$

where Fa is the plotting position or probability of occurrence (%) for each event, y is the total number of events, and n is the rank of each event. The precipitation amounts are plotted against the probability of recurrence on log-probability graph paper (Figure 2.16). A straight line is drawn through the plotted points. The line can be extended to obtain larger return periods. The size of event for a given return period can be estimated from the graph.

Example 2.7: For the annual rainfall given in Table 2.5, calculate the 2-year and 100-year annual rainfall amounts.

Solution: Rank the values as done in Table 2.6. For each ranking, calculate the probability of occurrence (Fa) using Equation 2.2. The first value for Fa would be:

$$Fa = \frac{100\ [(2 \times 1) - 1]}{2 \times 20} = 2.5$$

The return period is found from the second part of Equation 2.2:

$$\text{Return Period} = \frac{100}{Fa} = \frac{100}{2.5} = 40 \text{ years}$$

The values for Fa and return period are found for each of the other years by the same process. The precipitation is then plotted against Fa on log-normal probability paper as shown in Figure 2.16. Figure 2.16 shows that the two-year amount would be about 15 in.

Extending the prediction line in Figure 2.16 until it intersects the 100 year return period shows that the wettest year in 100 would experience about 38 in. of rainfall, or, in a given year, there is a 1% chance that there will be more than 38 in. of rainfall based on these data.

Table 2.6.   Numerical ranking of annual precipitation, probabilities of occurrence, and return periods.

Rank	Precipitation in.	Probability (Fa) percent	Return Period years
1	32.8	2.5	40.0
2	26.2	7.5	13.3
3	23.4	12.5	8.0
4	22.4	17.5	5.7
5	21.7	22.5	4.4
6	19.2	27.5	3.6
7	19.2	32.5	3.1
8	18.2	37.5	2.7
9	14.6	42.5	2.4
10	13.1	47.5	2.1
etc.			

Note: Return period = 100/probability

Answer: The 2 year and 100 year annual rainfall amounts are 15 and 38 inches, respectively.

In Example 2.7, the probabilities of the occurrence of larger events were predicted. The extreme wet year probabilities are useful in planning reservoir spillway capacities and river flood routing projects. On the lower end of the probability scale, the probabilities of dry years are used to assist in planning volumes of storage reservoirs or the capacity of irrigation systems that may be needed to meet demands during dry years. Low river flow estimates are also needed to assist in determining the quality of sewage discharge that a river can accept without adversely affecting the river ecosystem.

Sometimes, hydrologists need to estimate the probability that a given return period storm will occur at least once within a given number of years, e.g., what is the probability that a 100 year storm will occur at least once during the next ten years? The relationship to determine that probability is:

$$P(T,n) = 1 - \left[1 - \frac{1}{T}\right]^{n} \qquad (2.3)$$

where P(T, n) is the probability that a T year return period storm will occur at least once during n years (Barfield et al., 1983).

Example 2.8: What is the probability that a 50 year storm will occur during the first 5 years following the construction of a drainage ditch?

Solution: Apply equation 2.3 with T = 50 years and n = 5 years:

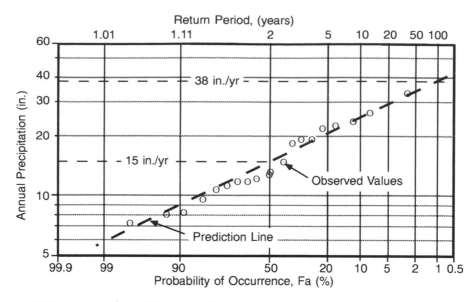

Figure 2.16.   Log-probability plot of data shown in Table 2.5 for Example 2.7, plotting annual rainfall versus probability of occurrence, *Fa*.

$$P(50,5) = 1 - \left[1 - \frac{1}{50}\right]^{5} = 0.096 \ (9.6\%)$$

Answer: The probability that a 50 year storm will occur during the 5 year period following construction of a drainage ditch is 9.6%.

2.15. RAINFALL DISTRIBUTIONS FOR SHORTER TIME PERIODS

Most hydrologic studies with smaller watersheds require rainfall-frequency data for time periods or durations shorter than one year. Rainfall amounts for these shorter periods can be found by analyzing weather records in a manner similar to that demonstrated above, but more frequently, weather data are available from local Soil Conservation Service offices, other state agencies, or state universities. For example, Appendix C shows the 24-hr rainfall amounts for return periods of 10, 25, and 100 years. Similar maps for shorter durations are also available (Hershfield, 1961). It is recommended that the reader consult local sources of information for values appropriate for any site-specific plans that are being developed. Table 2.7 is an example of storms for a range of durations and return periods for Columbus, OH. One method for determining rainfall estimates for periods varying from 5 minutes to 24 hours follows.

Figure 2.17 shows the relationship between storm duration, return period, and rainfall intensity for St. Louis, MO. Rainfall amount can then be found by multiply-

Table 2.7.   Rainfall amounts equaled or exceeded for periods of 5 minutes to 24 hours for expected return periods of 2 to 100 years for Columbus, Ohio.

Duration	Rainfall Amounts for Return Periods (in.)					
	2 yrs	5 yrs	10 yrs	25 yrs	50 yrs	100 yrs
5 min	0.35	0.45	0.51	0.59	0.61	0.71
10	0.55	0.72	0.83	0.98	1.08	1.18
15	0.65	0.88	1.05	1.22	1.38	1.50
30	0.90	1.20	1.40	1.70	1.87	2.07
1 hr	1.10	1.50	1.75	2.10	2.32	2.60
2	1.26	1.72	2.00	2.40	2.65	3.00
4	1.42	1.93	2.26	2.68	2.96	3.30
8	1.61	2.16	2.52	2.96	3.30	3.64
12	1.80	2.36	2.74	3.20	3.53	3.88
24	2.14	2.76	3.18	3.75	4.08	4.50

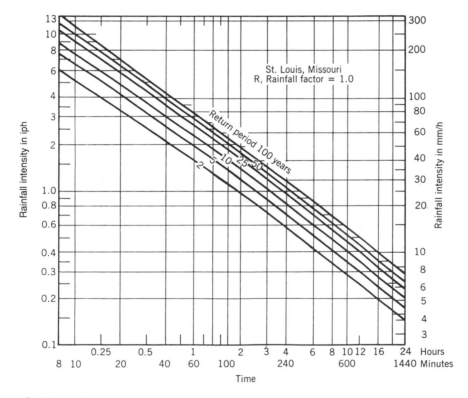

Figure 2.17.   Rainfall rate-duration-frequency distribution for St. Louis, Missouri (Hershfield, 1961 and Weiss, 1962).

ing the duration and intensity. The value obtained from Figure 2.17 for St. Louis is then multiplied by a factor from Figure 2.18 to estimate a storm of the same duration and return period for a different locality in the continental U.S. For example, Columbus, OH, would have a storm intensity approximately 0.75 the size of that predicted for St. Louis, MO for the same duration and return period.

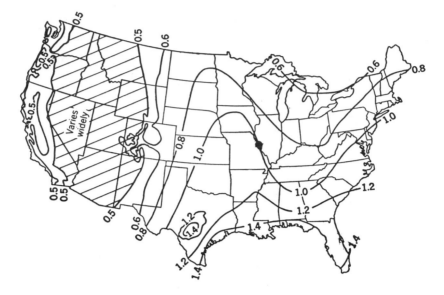

Figure 2.18.   Factor to adjust rainfall amount determined for St. Louis (from Figure 2.17) for other sites in the U.S. (Hamilton and Jepson, 1940)

Example 2.9: Estimate the amount of rainfall expected once in 25 years for a watershed near Columbus, OH for a storm with a duration of 30 minutes.

Solution: From Figure 2.17, determine that the 25-year, 30-minute duration storm has an intensity of 4.1 in. per hour, or a total rainfall of:

$$P = 4.1 \text{ inches/hr} \times \frac{30 \text{ min}}{60 \text{ min/hr}} = 2.05 \text{ inches}$$

For Columbus, Figure 2.18 shows that both the intensity and the amount will be 0.75 of that for St. Louis: $i = 4.1$ in./hr $\times 0.75 = 3.1$ in./hr and $P = 2.05$ in. x 0.75 = 1.54 in.

Answer: The amount of rainfall in the 25-year, 30-minute duration storm is 2.05 inches.

Rainfall rate-duration-frequency data for Columbus, OH, from the U.S. Weather Bureau are presented in Table 2.7. From the table, it can be seen that the 25-year, 30-minute precipitation amount is given as 1.7 in., which agrees reasonably well with the simplified prediction based on Figures 2.17 and 2.18.
    There are seasonal differences in monthly rainfall amounts, as shown in Figure 2.5. These differences are also found in shorter duration storm occurrences. Most large storms occur in the summer in the Central U.S., and in the winter or early

spring in the Western U.S., following the seasonal distributions of total rainfall. The east coast will frequently experience its most severe storms during hurricane season in the late summer.

2.16. PROBABLE MAXIMUM PRECIPITATION

In some situations, the cost of failure of a structure due to precipitation is so great that planners desire to ensure that the structure can store and/or pass the runoff from the probable maximum precipitation (PMP) that is likely to occur. PMP amounts are determined from long term records using principles described in Section 2.14, as well as considering the extreme fundamental meteorological events that are likely to lead to such an event. Figure 2.19 is an example of a map that is available to assist in determining the PMP for a given duration and area of coverage. Values from Figure 2.19 are generally adjusted to include areas of watershed and other site-specific conditions. Readers should consult specialized references for a more detailed discussion on this topic (e.g., Viessman et al., 1977).

REFERENCES

Barfield, B.J., R.C. Warner, and C.T. Haan. 1983. Applied Hydrology and Sedimentology for Disturbed Areas. Oklahoma Technical Press, Stillwater. 603 pp.

Brakensiek, D.L., H.B. Osborn, and W.J. Rawls, coordinators. 1979. Field Manual for Research in Agricultural Hydrology. USDA, Agriculture Handbook, 224, 550 pp, illustrated.

Brown, R.E., 1993. Runoff and sediment production from forested hillslope segments in the Georgia Piedmont. MS Thesis, University of Georgia, Athens.

Hamilton, C.L. and H.G. Jepson. 1940. "Stock-Water Developments; Wells, Springs, and Ponds," USDA Farmers' Bu. 1859.

Hershfield, D.N. 1961. Rainfall Frequency Atlas of the United States. U.S. Weather Bureau Tech. Paper 40, May. Washington, D.C.

McCuen, R.H. 1989. Hydrologic Analysis and Design. Prentice-Hall, NJ. 867 pp.

Myers, V.A. and R.M. Zehr. 1980. A Methodology for Point-to-Area Rainfall Frequency Ratios. NOAA Technical Report NWS 24. National Oceanic and Atmospheric Admin., U.S. Dept. of Commerce, Washington, D.C.

Schwab, G.O., D.D. Fangmeier, W.J. Elliot, and R.K. Frevert. 1993. Soil and Water Conservation Engineering, Fourth Edition. John Wiley & Sons, New York. 507 pp.

Teigen, L.D. and F. Singer. 1992. Weather in U.S. Agriculture: Monthly Temperature and Precipitation by State and Farm Region, 1950–1990. USDA Statistical Bulletin No. 834. 129 pp.

U.S.D.A. S.C.S. 1972. SCS National Engineering Handbook on Hydrology, Washington, D.C.

U.S.D.A. S.C.S. 1984. Snow Survey Sampling Guide. Agriculture Handbook Number 169. Washington, D.C.

U.S. Weather Bureau. 1947. Generalized Estimates of Probable Maximum Precipitation for the United States East of the 105th Meridian for Areas to 400 Square Miles and Durations to 24 Hours. Hydrometeorological Rept. No. 23.

U.S. Weather Bureau. 1960. Generalized Estimates of Probable Maximum Precipitation for the United States West of the 105th Meridian for Areas to 400 Square Miles and Durations to 24 Hours. Tech. Paper 38.

Viessman, Jr., W, J.W. Knapp, G.L. Lewis, and T.E. Harbaugh. 1977. Introduction to Hydrology, Second Edition. Harper & Row, New York. 704 pp.

Weiss, L.L. 1962. A general relation between frequency and duration of precipitation, Monthly Weather Rev., pp 87–88, March.

Wischmeier, W.H. 1959. A rainfall erosion index for a Universal Soil Loss Equation. Soil Science Society of America Proceedings 23:246–249.

Wischmeier, W.H. and D.D. Smith. 1978. Predicting Rainfall Erosion Losses—A Guide to Conservation Planning. *USDA Handbook* No. 537. 58 pp.

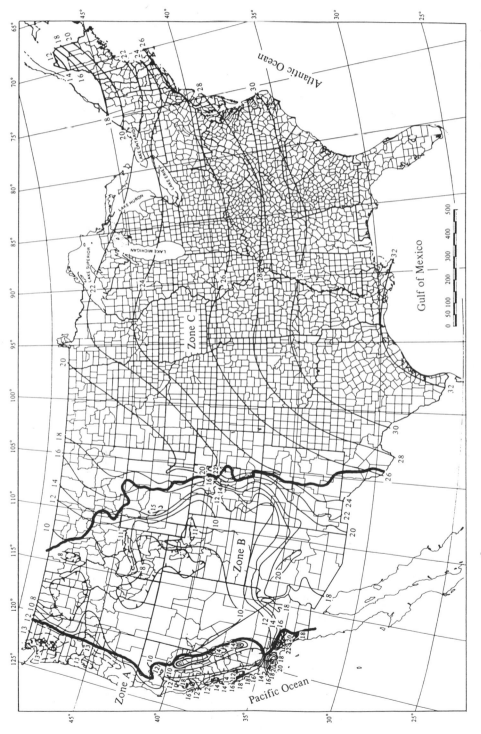

Figure 2.19.   The 10-mi² or less PMP for 6-hr duration (in.). (U.S. Weather Bureau, NOAA.)

Problems

2.1 The maximum rainfall for a 4-hr storm is calculated to be 0.85 in. What would be the average rainfall amount and average intensity if the watershed area is 80 mi^2?

2.2 A forest researcher wishes to locate a series of rain gages around the edge of a newly harvested area. The height of the surrounding unharvested trees is 100 ft. What is the minimum distance that should be allowed between the unharvested trees and the rain gages?

2.3 A hydrologic study is planned for a watershed of 2.5 mi^2. How many rain gages are recommended for this study (1 mi^2 = 640 acres)?

2.4 On the snow course in the Priest River Experimental Forest in Idaho, on Feb. 24, 1994, one of the snow depths was measured and found to be 7.5 ft. The density was measured as 35%. What is the water equivalent?

2.5 From the data provided in the following table, calculate the storm duration, total rainfall amount, maximum intensities for 10 and 30 minutes, average intensity, and erosivity.

Rainfall distribution for a storm on the University of Georgia Whitehall Forest, in August, 1990 (Brown, 1993).

Time, min	Amount, mm	Time, min	Amount, mm	Time, min	Amount, mm
10	3.0	110	2.8	210	2.0
20	0.8	120	10.9	220	0.3
30	0.5	130	3.0	230	0.8
40	3.3	140	2.0	240	0.3
50	0	150	1.8	250	0.5
60	0	160	0.3	260	0.3
70	0	170	0.8	270	0.3
80	0.3	180	0.8	280	0.3
90	0.3	190	9.9	290	0.5
100	0.5	200	12.2	300	0.8

2.6 The author designed an irrigation reservoir to withstand a 50-year storm event while living in Scotland. He lived there for three years after the reservoir was constructed, and the reservoir withstood the design storm runoff once. What was the probability that this reservoir would experience this 50-year storm at least once in those three years?

2.7 Compute the average rainfall for a 1,000-acre watershed by the arithmetic average method and by the Thiessen method with data from the following table.

Rain Gage	Acres per Gage	Rainfall, in.
A	400	2.1
B	300	2.5
C	250	2.3
D	50	2.7

2.8 Determine the expected annual rainfall for a return periods of 2, 20, and 100 years for the Priest River Experimental Forest in northern Idaho from the following annual rainfall records.

Year	Rainfall, in.	Year	Rainfall, in.
1971	32.73	1981	31.53
1972	31.16	1982	38.72
1973	34.43	1983	39.45
1974	39.04	1984	32.54
1975	32.26	1985	21.55
1976	23.61	1986	27.21
1977	26.25	1987	26.33
1978	28.01	1988	27.97
1979	27.31	1989	28.09
1980	35.08	1990	34.63

Infiltration and Soil Water Processes

Andy D. Ward and Jay Dorsey

3.1. INTRODUCTION

Infiltration plays an important role in nature and human activities. Water infiltrating into the soil profile provides water for vegetative food and fiber production, contributes to underground water supplies which sustain dry-weather streamflow, and decreases surface runoff, soil erosion, and the movement of sediment and pollutants into surface water systems. Infiltration is defined as the passage of water through the surface of the soil, via pores or small openings, into the soil profile. Infiltration directly affects deep percolation, groundwater flow, and surface runoff contributions to the hydrologic balance on a watershed. Accounting for infiltration is fundamental to understanding and evaluating the hydrologic cycle.

Infiltration processes are complex and difficult to quantify. Infiltration methods may be classified as theoretical equations based on the physics of soil water movement; empirical equations which are based on parameters with physical significance; empirical equations which contain parameters with no physical significance; and *in situ* measurement methods. Much research has been devoted to measuring infiltration and developing and verifying empirical methods. In recent years, advances in computer technologies have made feasible the solution and application of complex equations which describe infiltration processes.

This chapter presents a description of soil water relationships, infiltration processes, factors affecting water movement through soils, methods for estimating infiltration rates, and methods for measuring infiltration rates and soil physical properties which influence soil water movement.

3.2. SOIL WATER RELATIONSHIPS

A soil profile consists of a mixture of solid, liquid, and gaseous materials. Solid materials include soil particles of different sizes, shapes, and mineral composition. Other solid materials in the soil profile are organic matter from plants, animals, and microorganisms. Soils are commonly classified based on the size of the soil parti-

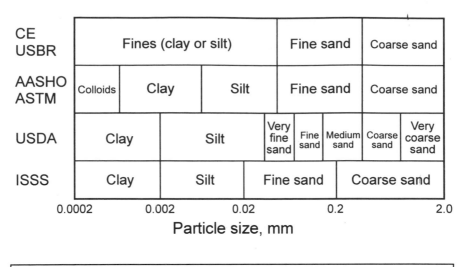

Figure 3.1. Classification of soil based on particle size.

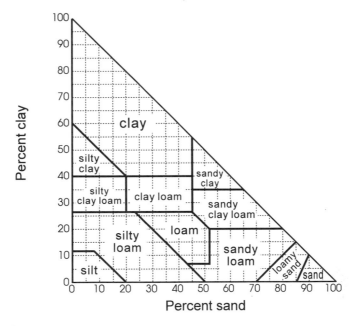

Figure 3.2. USDA triangle for determining textural classes.

cles. For example, sand has large particles while clay has very small particles.
Commonly used classification systems are presented in Figure 3.1. Soils usually
contain a mixture of clay, silt, and sand particles. The USDA has established a
classification scheme based on different particle size mixtures (Figure 3.2). For
example, a soil which is 30% clay, 10% silt, and 60% sand is classified as a sandy
clay loam.

Between the soil particles and organic matter are open spaces called voids or pores. Water which fills part or all of the pores is commonly called soil water. When the pores are completely filled, the soil is described as being saturated. Usually a small part of the water, called hygroscopic soil water, is held so tightly by molecular attraction to the surface of soil particles that it is not removed under normal climatic conditions. Pores that are empty or only partially filled with water contain oxygen, water vapor, and other gases. Chemicals and solid materials associated with climatic processes and surface activities might also be found in a soil profile. These materials and chemicals might be attached to soil particles, trapped in pores, or in the soil water.

If we approximate this complex system as three phases consisting of air, water, and solids, as illustrated in Figure 3.3, then several useful relationships can be established. The total volume V_t will be the sum of the volumes for the three phases:

$$V_t = V_a + V_w + V_s \qquad (3.1)$$

where the subscripts, a, w, and s, refer to the air, water, and solid phases. The volumes might be expressed in ft^3 or cm^3 (cc) or as a depth (inches, feet, or cm) per unit cross-sectional area. For example it is not uncommon in agriculture to talk about the change in the soil water content of a soil profile as being some measured number of inches. To convert this amount into a volume it is necessary to multiply it by the surface area of the land being considered.

The volume of the voids or pores will be:

$$V_v = V_a + V_w \qquad (3.2)$$

If we substitute Equation 3.2 into Equation 3.1, rearrange the equation, and divide the whole equation by V_t a relationship is established for the porosity:

$$n = \frac{V_v}{V_t} = 1 - \frac{V_s}{V_t} \qquad (3.3)$$

Porosity is an important property in problems involving water volumes or water movement. It is commonly used in calculations made by hydrologists and agricultural drainage engineers.

It is often necessary to know the actual soil water content in a soil profile. Soil water contents can be expressed either by volume or by mass. The soil water content by volume, θ_v, is the volume of water in a soil sample divided by the total volume of the sample:

$$\theta_v = \frac{V_w}{V_t} \qquad (3.4)$$

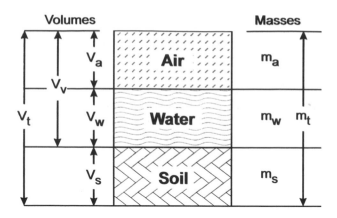

Figure 3.3. Soil matrix constituents.

Normally the volumetric soil water content is expressed as a percentage and the answer from Equation 3.4 will need to be multiplied by 100.

The fraction or percentage of the pores which are filled with water is called the degree of saturation. The volumetric soil water content, degree of saturation, and porosity are related as follows:

$$\theta_v = S\,n \qquad\qquad (3.5)$$

where S is the degree of saturation, and n is the porosity. Use of the relationships presented in Equations 3.1–3.5 are illustrated in Example 3.1.

Example 3.1: Determine the maximum depth of water that can be stored in the top three feet of a soil profile with a porosity of 0.5. Also, determine the volumetric soil water content if the degree of saturation is 60%.

Solution: From Equation 3.3, the volume of voids $V_v = n\,V_t$. Therefore:

$$V_v = 0.5 \times 3 \text{ ft} \times 12 \text{ inches} = 18 \text{ inches/unit area}$$

The volumetric soil water content when the degree of saturation is 60% can be determined from Equation 3.5:

$$\theta_v = S\,n = 0.6 \times 0.5 = 0.3 \ (\text{in/in or in}^3/\text{in}^3)$$

Answer: The top three feet of the soil profile can store 18 inches of water. The volumetric soil water content is 0.30 (in^3/in^3) when the degree of saturation is 60%.

The total mass, m_t, of a soil profile is:

$$m_t = m_a + m_w + m_s \qquad (3.6)$$

where m_a is the mass of air, m_w is the mass of water, and m_s is the mass of solids. The mass of the air is negligible and the total mass is approximated as the sum of the water and soil masses.

In Equation 3.4 the soil water content was expressed by volume. However, the most common way of determining the soil water content is to obtain a sample of the soil, weigh the sample, dry the sample, and then reweigh the dry sample. This procedure is discussed in more detail at the end of the chapter. When the soil water content is determined as a function of mass or weight, it is called the gravimetric soil water content, θ_g, and is equal to:

$$\theta_g = \frac{m_w}{m_s} \qquad (3.7)$$

Use of Equation 3.7 is illustrated in Example 3.2.

Example 3.2: A sample of soil was placed in a soil tin with a mass of 25 g. The combined mass of the soil and tin was 140 g. After oven drying, the soil and tin had a mass of 120 g. Calculate the soil water content by mass.

Solution: By subtracting the mass of the tin, the mass of wet sample is calculated to be 115 g (140–25 g), and the mass of the dry sample is 95 g (120–25 g). Therefore, the mass of water in the sample is 20 g (115 g–95 g). Using Equation 3.7, the soil water content by mass is:

$$100 \times \frac{20}{95} = 21\%$$

Answer: The soil water content by mass (or weight) is 21%.

The behavior of a soil is very dependent on the bulk density of the soil profile. The bulk density, ρ_b, is the mass per unit volume and is related to other soil physical properties as follows:

$$\rho_b = \frac{m_t}{V_t} \qquad (3.8)$$

The dry bulk density is the mass of dry soil divided by the total volume:

$$\rho_{dry} = \frac{m_s}{V_t} \qquad (3.9)$$

The dry bulk density of the profile can be related to the density of soil particles and porosity as follows:

$$\rho_{dry} = \rho_p (1 - n) \qquad (3.10)$$

The density of the soil particles, ρ_p, is defined as the mass of dry soil divided by the volume of the soil (m_s/V_s). The density of most soil particles is 160–170 lb/ft^3 (2.6–2.75 g/cm^3). If the soil particle density is not measured, a value of 165 lb/ft^3 (2.65 g/cm^3) is often assumed.

The ratio of the density of the soil particles to the density of water is called the specific gravity:

$$G_s = \frac{\rho_p}{\rho_w} \qquad (3.11)$$

where the density of water is usually assumed to be 1.0 g/cm^3 or 62.4 lb/ft^3. Use of several of these equations is illustrated in Example 3.3.

Example 3.3: The dry bulk density of a soil sample is 84.24 lb/ft^3 and the density of the soil particles is 168.48 lb/ft^3. Calculate the porosity of the soil profile and specific gravity of the soil particles. Also, determine the bulk density of the sample if it was saturated.

Solution: In Equation 3.10, substitute 84.24 lb/ft^3 for the dry bulk density, ρ_{dry}, and 168.48 lb/ft^3 for the density of the soil particles, ρ_p, then:

$$\rho_{dry} = \rho_p (1 - n)$$

and,

$$84.24 = 168.48 (1 - n)$$

dividing both sides by 168.48 lb/ft^3 and rearranging gives $n = 0.5$.

The specific gravity of the soil particles is found by substituting into Equation 3.11 the particle density of 168.48 lb/ft^3 and the density of water, 62.4 lb/ft^3:

$$G_s = \frac{\rho_p}{\rho_w} = \frac{168.48}{62.4} \quad \left(\frac{lb/ft^3}{lb/ft^3}\right)$$

Therefore, the specific gravity of the soil particles is 2.7.

If the sample is saturated, all the pores are filled with water and the volume of water will be equal to the volume of pores. From Equations 3.3, the volume of water will be equal to the porosity times the total volume:

$$n = \frac{V_v}{V_t} \text{ and } V_w = V_v$$

Therefore,

$$V_w = 0.5 \, V_t$$

The mass of water is equal to the density of water ($62.4 \, lb/ft^3$) times the volume of water. The mass of soil is the density of the soil particles ($168.489 \, lb/ft^3$) times the volume of soil or the dry bulk density ($84.24 \, lb/ft^3$) times the total volume (Equation 3.9).

Therefore, using Equation 3.8 and 3.6, the saturated bulk density is:

$$\rho_b = \frac{m_t}{V_t} = \frac{m_w + m_s + m_a}{V_t}$$

and,

$$\rho_b = \frac{0.5 \times 62.4 \times V_t + 84.24 \times V_t + 0}{V_t} = 115.44 \quad (lb/ft^3)$$

Answer: The porosity, n, is equal to 0.5 and the specific gravity, G_s, is 2.7. V_t in the numerator and denominator cancels and $\rho_b = 115.44 \, lb/ft^3$.

3.3. INFILTRATION PROCESSES

The downward movement of water through a soil profile occurs due to tension and gravitational forces in the soil matrix. The soil matrix is generally heterogeneous and consists of a labyrinth of pores of varying shape and size connected by porous fissures and channels. Water is held in the soil matrix by tension forces at the air–water interfaces in pores. These tension forces are also frequently called suction or capillary forces. To illustrate water movement due to tension forces, dip a dry blotter or paper towel into water and note how quickly water is sucked into the dry material against the pull of gravity. Also note that the rise of water is slower as the height of the wetting front above the water surface increases.

A common example of capillary flow (flow due to tension forces) occurs when one's finger is pricked and a blood sample is taken. The blood sample is drawn into the thin sample tube due to capillary forces.

Soil water tension varies from less than 1 inch of water head for a soil near saturation to as much as 10,000,000 inches of head for a very dry condition. The effect of surface tension in a soil matrix during drainage can be described by considering water held in a single pore within the soil profile connected to the groundwater table (Figure 3.4). The figure describes an equilibrium state at the meniscus with radius r_1. If it is assumed that the meniscus with radius r_1 supports a column of water, h, then the gravitational forces will equal the surface tension forces and:

$$\pi r_1^2 \, h\rho_w \, g = 2\pi r_1 \tau \, \cos\alpha \qquad (3.12)$$

where ρ_w is the density of water, g is the gravitational constant, τ is the surface-tension force, and α is the angle of contact of the meniscus with the soil.

The surface tension force is a function of the area of the meniscus. The smaller the radius, the longer the column of water that can be supported. The smallest pores will fill first as they exert the largest surface tension forces. If the water table falls, h will increase and Equation 3.12 will no longer hold true. The largest pores containing water will begin to drain until a smaller controlling meniscus is reached and equilibrium between the suction forces and gravitational forces is attained. For the soil profile illustrated in Figure 3.4, a slight drop in the water table might cause no drainage from the pore with radius r_2, because the water in this pore is controlled by the two smaller openings with radius r_1 and r_3. Drainage would only occur when the water table fell far enough that the small opening r_3 began to drain because it could not maintain the water column to the water table. It is evident, therefore, that the soil water content under drainage conditions is different from that under adsorption (wetting) for a given soil water tension.

During wetting, the small pores fill first, while during drainage and drying, the large pores empty first. This causes hysteresis in the system and is illustrated in Figure 3.5. Hysteresis will vary, depending on the wetting and drying history of the soil.

Retention and movement of water during wetting or drainage is a function of the shape and size of the pores. When water is applied, soil air is displaced from the pores, the soil water content increases and the soil tension decreases. This results in decreased infiltration rates. Provided the amount of water applied is high enough, this process will continue until the soil is saturated (all the pores are filled). At saturation, the soil suction will be zero. However, in most natural environments a small amount of air will be trapped in the pores and prevent complete saturation. The final degree of saturation will probably only be 80–90%.

Following wetting there will be a redistribution of soil water and pores will drain due to capillary and gravity flow. When gravity flow becomes negligible, the soil water content of the profile will be at field capacity. Field capacity typically occurs at soil suctions of 0.1–0.33 bars (Figure 3.6). However, it will vary depending on the wetting and drying history of the soil, soil texture, porosity, and the subsurface

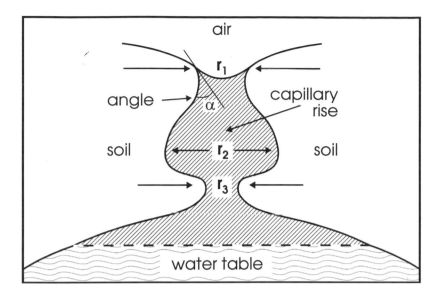

Figure 3.4. Water retention in a heterogenous porous media.

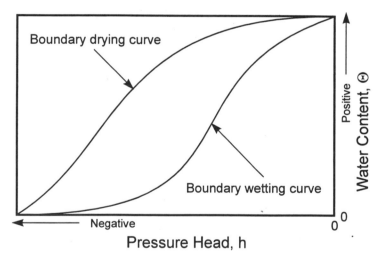

Figure 3.5. Hysteresis during drying and wetting.

characteristics of the soil profile. The elusive nature of field capacity is illustrated in Figure 3.7.

Without realizing it, you have probably experienced a situation where gravity drainage occurred following wetting of a soil. Consider watering a plant in a pot and adding the water until you see water ponded at the top of the container. If you come back a short time later, you discover that the water has drained through the potting soil, filled the saucer below the pot, and perhaps overflowed onto your table or desk.

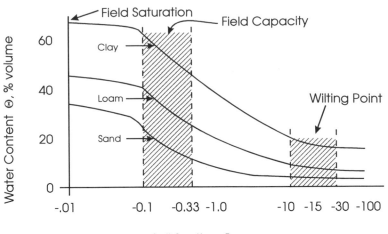

Figure 3.6. Soil water and soil suction relationships to field capacity wilting point and plant available water (Courtesy of S.W. Trimble, UCLA).

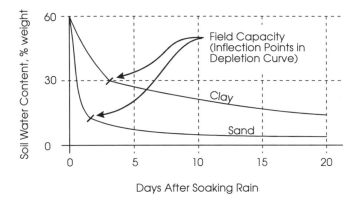

Figure 3.7. The elusive nature of field capacity (Courtesy of S.W. Trimble, UCLA).

Redistribution of water due to capillary flow will continue after gravity flow ceases. Usually, upward movement of soil water will occur due to evapotranspiration. The depth that plants can remove soil water will depend on the vegetation, profile characteristics, and climatic conditions. The soil water content in the root zone will eventually reach the wilting point unless there is further wetting, or the soil water content within 1–3 feet of the root zone is near saturation. The soil suction at wilting point is typically 10 to 15 bars and is the point at which plants begin to wilt and die. For sand, the soil water content at wilting point (WP) is very low. For clay it is much higher (Figure 3.6 and 3.8). Soil water in excess of the amount at the wilting point is available for evapotranspiration (ET) and the maximum amount is termed plant available soil water (PAW) (Figure 3.8) which is related to field capacity and wilting point as follows:

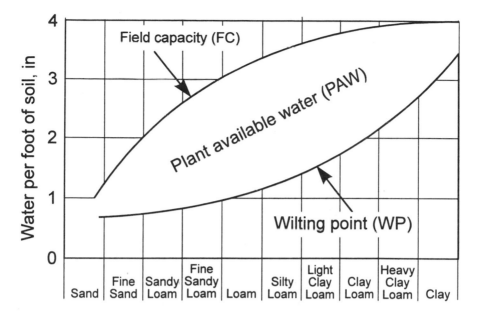

Figure 3.8. Water holding characteristics of soils with different texture.

$$\theta_{paw} = \theta_{fc} - \theta_{wp} \qquad\qquad (3.13)$$

where θ_{paw}, θ_{fc}, and θ_{wp} are the volumetric or gravimetric plant available soil water content, and the soil water contents at field capacity and wilting point, respectively.

Knowledge of the wilting point, field capacity, and plant available soil water content is particularly important in agriculture. In arid areas, this type of information is needed to design irrigation systems and determine irrigation schedules. For most plants there is a reduction in yield when the soil water content falls below a critical value, which is 40–70% of the plant available soil water content. A typical relationship is shown in Figure 3.9 and use of this type of information is illustrated in Example 3.4.

Example 3.4: Determine the amount of water needed to increase the soil water content in the root zone of a clay loam soil from the critical value for plant growth to field capacity. The critical value occurs at 50% of the plant available water, the root zone depth is 2 ft., and the soil has the soil water properties presented in Figure 3.8 at the interface between a clay loam and heavy clay loam.

Solution: From Figure 3.8 the soil water content at wilting point is 2 inches of water in each foot of the soil profile. The soil water content at field capacity is 3.8 inches of water in each foot of the soil profile. Therefore, by using Equation 3.12, the plant available water content is:

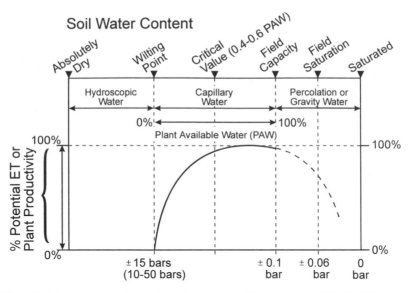

Figure 3.9. Yield responses to soil water content (courtesy of S.W. Trimble, UCLA).

$$\theta_{paw} = (3.8 - 2.0) \text{ inches of water per foot of soil profile}$$

The root zone is 2 ft deep, and PAW $= 1.8 \times 2 = 3.6$ inches.

Answer: If the critical value is 50% of the PAW, then 1.8 inches of water (0.5 x 3.6 inches) are needed to increase the soil water content in the root zone from the critical value to field capacity.

Bodman and Coleman (1943) showed that soil water movement into a uniform dry soil under conditions of surface ponding could be divided into four zones as shown in Figure 3.10. The saturated zone usually only extends from the surface to a depth of less than one inch. Below the saturated zone is the transition zone, which represents a zone of rapid decrease in soil water content. Below the transition zone is a zone of nearly constant soil water called the transmission zone. This zone increases in length as the infiltration process continues. Below the transmission zone is the wetting zone. The wetting zone maintains a nearly constant shape and moves downward as the infiltration process proceeds. The wetting zone culminates at the wetting front, which is the boundary between the advancing water and the relatively dry soil. The wetting front represents a plane of discontinuity across which a high suction gradient occurs.

3.4. FACTORS AFFECTING WATER MOVEMENT THROUGH SOILS

Infiltration may involve soil water movement in one, two, or three dimensions, although it is often approximated as one-dimensional vertical flow. Water movement

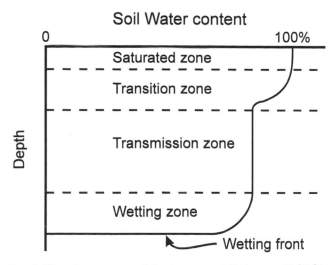

Figure 3.10.   The infiltration zones of Bodman and Coleman (1943).

into and through a soil profile is dependent on many interrelated factors. Factors of importance include surface conditions, subsurface conditions, factors which influence surface and subsurface conditions, flow characteristics of the water or fluid, and hydrophobicity (Figure 3.11).

3.4.1. Surface Conditions

Vegetation and land use practices have a marked effect on infiltration. Surfaces such as paved roads, sidewalks, parking lots, and buildings allow negligible infiltration. In more natural environments, the type of vegetation, season of the year, and land management practices (such as tillage) that temporarily modify the near-surface soil conditions will greatly influence infiltration processes. Some practices, such as no-till (conservation tillage), are used to increase water movement into the soil. In other situations, such as sites used for land disposal of toxic wastes, efforts are made to compact underlying soil in order to minimize water movement into the soil profile.

If the ability of the soil profile to transmit infiltrating water is not limiting, soil surface conditions usually govern infiltration. This is illustrated by curves of cumulative infiltration for hay cover and bare ground during the first 60 minutes of a rainstorm (Figure 3.12). Hay provides a ground cover which absorbs the energy of falling raindrops and prevents soil puddling or packing.

It is important to note that surface runoff occurs as soon as the infiltration rate is less than the water application rate and surface depressions have been filled with water. Once water is lost due to runoff, it is not available for infiltration. Tillage practices that leave the soil surface rough with many pockets for water storage are likely to have more infiltration than where the surface has been worked down and smoothed by tillage. In the latter case, infiltration occurs primarily while it is raining. In the first case with the rough soil surface, infiltration continues after the rain

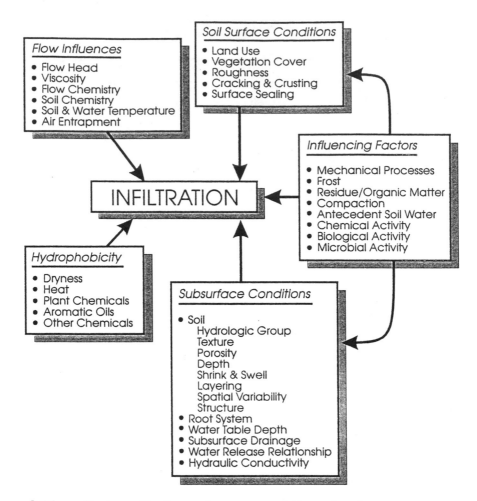

Figure 3.11.   Factors affecting water movement through soils.

has stopped until all stored water is absorbed. This prolonged time for infiltration opportunity may be many minutes and, in some cases, significant amounts of increased infiltration result.

A crust will form at the surface of some soils as they dry. Often cracks will form between crusted areas. The crusts will inhibit water movement into the soil matrix. Gravity flow will occur down the cracks provided a positive head of water is established at the top of the crack. This might be due to water ponding at the surface or surface runoff which flows across a crusted area and into a crack. Cracks might also form in the absence of crusting and gravity flow might occur in the absence of cracks. If flow through a pore occurs primarily due to gravity flow, the pore is commonly called a macropore. Worm holes, coarse sands and gravels, and dry organic matter (residue) extending to the soil surface might all result in gravity flow. Water movement through macropores will result in more rapid wetting at deeper depths. In some cases wetting of the soil matrix might be due primarily to lateral movement of water which has ponded in cracks or upward movement of water from an impeding layer due to water which reached the impeding layer by macropore flow.

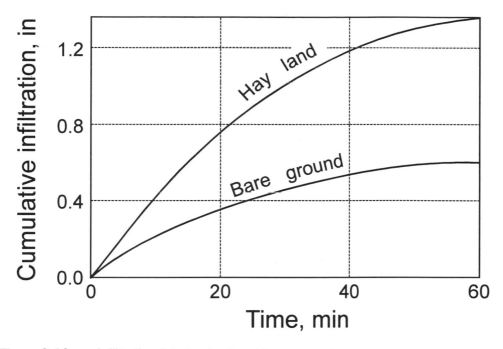

Figure 3.12.   Infiltration into hayland and bare ground.

Infiltration can also be impeded by surface sealing due to physical and chemical processes during a hydrologic event. As the soil surface becomes wet, a chemical reaction can cause a temporary bond between soil particles. For example, the author has conducted research on sandstone spoil materials from Kentucky surface mines. Infiltration through loosely packed air dried spoils was very rapid. However, after wetting and drying, it was noted that infiltration rates were extremely slow during any future applications of water. A very fine seal formed at the soil surface but water would flow readily through any breaks in the seal.

During high intensity rainfall or surface applications of water, soil might be detached and then moved across the soil surface. Some of the detached soil might then be redeposited in cracks and large pores. This deposition of soil in cracks and pores will tend to seal the surface, and infiltration or percolation will be reduced.

3.4.2. Subsurface Conditions

As mentioned earlier, soil texture, bulk density, heterogeneity, cracks, and surface conditions will all influence water movement. Also, hydraulic conductivity is an important property of soils and is defined as the ability of a soil to transmit water under a unit hydraulic gradient. Hydraulic conductivity is often called permeability and is a function of soil suction and soil water content. Fine grained soils tend to have lower hydraulic conductivity values than coarse grained soils. However, fine grained soils such as clays have smaller pores than coarse grained soils such as sands. The smaller the pores, the greater the suction forces. Therefore water might move readily into a dry, fine-grained soil even though the hydraulic conductivity is low.

If there is no spatial variability in soil properties, the soil is described as homogeneous. Most soils are heterogeneous because they exhibit a considerable amount of variability in properties both laterally and vertically. Soils are also described as isotropic or anisotropic. Isotropic soils exhibit the same hydraulic conductivity in all directions, while anisotropic soils have different values in the vertical and lateral directions.

Land surfaces can be likened to a blotter (Figure 3.13) or a sponge (Figure 3.14). Infiltration rates of the blotter are low. In order to visualize this situation, apply water onto a blotter which is tilted. A plastic squeeze bottle can be used to apply water to the blotter. Note how little water goes into the blotter and how much runs off. The pores of the blotter are very fine and transmit water slowly. Now, apply water from the squeeze bottle onto a sponge. Note how little water runs off, even though the water application is high. Most of the sprinkled water infiltrates quickly into the sponge. Infiltration rates are high and might exceed 2 in./hr, as found in sandy areas or in woodlands on well-drained soil. The blotter might represent a slowly permeable clay with very few large pores or a well-drained silt loam where the surface had been compacted by excessive tillage and puddled and compacted by rainfall. Its infiltration capacity might be less than 0.2 in./hr. Now place the blotter on top of the sponge (Figure 3.15) and apply water. Note the effect of the simulated thin, compacted layer of low infiltration capacity overlying a soil with high infiltration rates. Water will not enter the coarse textured subsurface layer until the suction forces at the interface between the two layers is equal.

The sponge on top of the blotter (Figure 3.16) exemplifies a common field situation where a grass-covered, well drained soil overlays a layer of soil of low permeability. This situation affects infiltration more than the sponge alone or the blotter on top of the sponge. In this case, at the start of the storm, infiltration rates into the sponge are high. However, the blotter restricts the percolation rate, causing infiltrated water to accumulate in the pores of the sponge; depleting its capacity for storing more water; and slowing down the infiltration rate until it corresponds with that of the blotter beneath.

Initial infiltration rates are higher for dry soil than for wet soil. This is illustrated in Figure 3.17. Infiltration rates decrease as the soil water content increases. This is because soil suction reduces with increases in soil water content. Some clay soils will swell during wetting and will then shrink during drying. Swelling will inhibit infiltration while shrinking will create cracks and increase macropore flow.

3.4.3. Hydrophobicity

Water falling on the surface of a hydrophobic soil will bead up rather than enter the soil due to suction forces. This phenomenon occurs due to waxy organic materials on the soil surface which create a negative contact angle between the waxes and any applied water. This effect normally occurs following brushland, range, and forest fires. Organic matter at and above the soil surface vaporizes during the fire and then condenses on the bare burned soil as waxy materials.

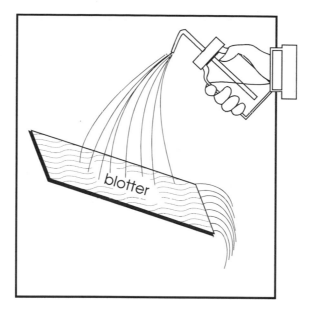

Figure 3.13.   Infiltration capacity of blotter is low and there is much runoff.

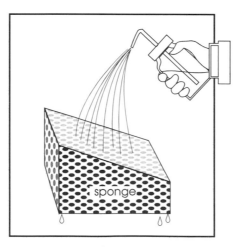

Figure 3.14.   Infiltration capacity of sponge is high and there is little runoff.

3.4.4. Flow Characteristics

Viscosity of water can affect infiltration. The colder the temperature, the higher the viscosity and the slower the infiltration. In watershed studies the viscosity of water flow is often neglected by the practicing hydrologist. However in most laboratory research, viscosity of water is a significant factor. The effect of frost on infiltration will depend on the soil water content at the time of freezing. A very wet, frozen soil can be practically impervious, causing a condition commonly called concrete frost. A dry forest or hayland frozen soil is likely to be porous, providing nearly normal infiltration rates and a condition called lattice frost.

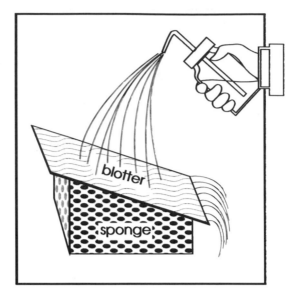

Figure 3.15.   Infiltration capacity of the sponge is limited by the overlying low permeable layer.

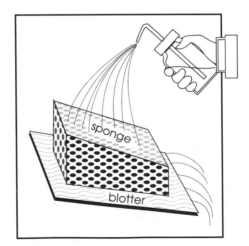

Figure 3.16.   Infiltration capacity of the sponge is limited by the underlying layer.

As water infiltrates through the soil, not all air in the pores is freely displaced by the flow and air bubbles are trapped in the pores. Trapped air will block some pores and will retard infiltration as air bubbles try to move upwards against the direction of water flow. If you observe a puddle on bare soil immediately following a storm event, it is possible that you will see bubbles of air coming to the surface of the puddle.

The soil water content following a rainfall event will depend on the rainfall depth, duration, and intensities. Intermittent light rain will produce higher soil water contents than if the same amount of rain occurred in a short period. This is due, in

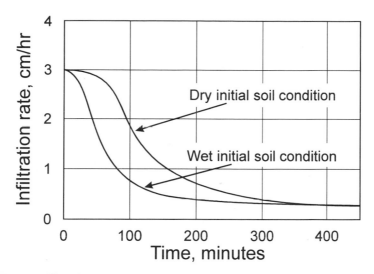

Figure 3.17.   Infiltration in top soil and spoil materials for (A) dry initial soil water conditions and (B) wet initial soil water conditions.

part, to the fact that during light intermittent rain, it is easier for air bubbles to escape upwards throughout the soil profile.

Entrapment of air is also dependent on the physical properties of the soil and in particular soil pore sizes. Very small pores will have a larger amount of air entrapment. This is because the suction force pulling water into the pores is larger than it would be for larger pores. These high suction forces oppose upward movement of air and, when combined with the small pore sizes, result in air being trapped or pushed downward. The likelihood that air gets pushed downwards into the soil profile ahead of the wetting front increases and this causes a build up of air pressure which opposes the downward movement of water.

3.5. SOIL WATER BALANCE

Changes in soil water storage occur throughout the year. Increases occur due to precipitation, irrigation, and subsurface inflows. Depletion is caused by percolation, gravitational drainage, and evapotranspiration. A typical cycle for Ohio is illustrated in Figure 3.18.

A water balance equation describing these changes for any period of time is expressed as:

$$\Delta SM = P + IR - Q - G - ET \tag{3.14}$$

where: ΔSM is the change in soil water storage in the soil profile, P is precipitation, IR is irrigation, G is percolation water, ET is evapotranspiration, and Q is surface runoff. All quantities are expressed as a depth (inches or mm) of water over a study area for a specific period of time. Monthly water budget data in Table 3.1 illustrates

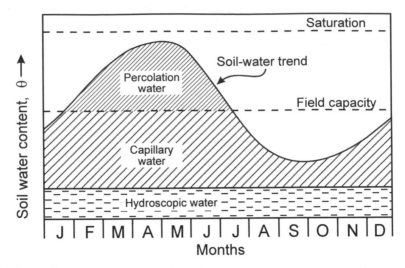

Figure 3.18.   Generalized seasonal soil-water content trends and occurrence of percolation water.

changes in soil water as related to factors given in Equation 3.14.   An increase in soil water of 2.78 inches in January resulted from precipitation (*P*) of 3.48 inches, runoff (*Q*) of 0.31 inches, evapotranspiration (*ET*) of 0.31 in, and percolation to groundwater of 0.08 inches. These data were obtained on an 8-ft deep natural soil profile lysimeter where the entire soil block was weighed automatically at 10-minute intervals.

For Ohio, soil water storage (Figure 3.18 and Table 3.1) typically increases through January and February, stays at high levels in March, depletes throughout the warm growing season, and then increases in the fall (September to December). Excessive rainfall in August started the trend of increasing soil water for the remainder of the year. The soil water content curve in Figure 3.18 is only a trend, and actual changes in soil water fluctuate markedly from day to day above and below the trend.   Three rain periods in June caused soil-water increases (accretion) which interrupted the general depletion trend for the month. In January there was a change in soil water content of 2.78 inches. This represented the net amount of water that was added to the soil profile during the month of January.

3.6. ESTIMATING INFILTRATION RATES

3.6.1. Horton Equation

One of the most widely used infiltration models is the three parameter equation developed by Horton (1939):

$$f = f_c + (f_o - f_c) \, e^{-\beta t} \tag{3.15}$$

where f is the infiltration rate at time t, f_o is the infiltration rate at time zero, f_c is the final constant infiltration capacity and β is a best fit empirical parameter.

Table 3.1.   Monthly accretion, depletion, and storage of soil water (inches) as determined by monolith lysimeters, Coshocton, Ohio, 1962.

Month	Precipitation (P)	Runoff (Q)	Evapotran-spiration (ET)	Percolation (G)	Total depletion	Profile Storage change (ΔS)
	inches	inches	inches	inches	inches	inches
Jan	3.48	0.31	0.31	0.08	0.70	2.78
Feb	4.10	0.57	0.71	0.83	2.17	1.93
Mar	3.52	0.02	1.12	2.41	3.55	-0.03
Apr	1.73	0.00	2.44	1.49	3.93	-2.20
May	2.86	0.01	5.71	0.83	6.55	-3.69
Jun	1.84	0.00	4.25	0.30	4.55	-2.71
Jul	2.72	0.00	4.56	0.02	4.58	-1.86
Aug	2.12	0.00	3.59	0.01	3.60	-1.48
Sep	5.36	0.00	2.88	0	2.88	2.48
Oct	2.39	0.00	2.25	0	2.25	0.14
Nov	3.19	0.00	0.74	0	0.74	2.45
Dec	3.42	0.00	*-0.34	0	-.34	3.76
Year	36.73	0.91	28.28	5.97	35.16	1.57

* Negative ET resulted from blowing snow.

Advantages of the method are that the equation is simple and usually gives a good fit to measured data because it is dependent on three parameters. Main disadvantages are that the method has no physical significance and field data are required to calibrate the equation. The equation does not describe infiltration prior to ponding.

Horton's equation has seen widespread application in storm watershed models. The most commonly used model which uses Horton's method is the Environmental Protection Agency Storm Water Management Model (Huber, 1981).

*3.6.2. Green-Ampt Equation

In 1911, Green and Ampt developed an analytical solution of the flow equation for infiltration under a constant rainfall. The method is developed directly from Darcy's law and assumes a capillary-tube analogy for flow in a porous soil. The equation can be written as:

$$f = K(H_o + S_w + L)/L \tag{3.16}$$

where K is the hydraulic conductivity of the transmission zone, H_o is the depth of flow ponded at the surface, S_w is the effective suction at the wetting front, and L is the depth from the surface to the wetting front. The method assumes piston or plug flow and a distinct wetting front between the infiltration zone and soil at the initial water content (refer back to Figure 3.10).

The Green-Ampt method is often approximated by the equation:

$$f = \frac{A}{F} + B \qquad\qquad (3.17)$$

where f is the infiltration rate, F is the accumulative infiltration, and A and B are fitted parameters which depend on the initial soil water content, surface conditions, and soil properties. This form of the method is used in the DRAINMOD water table management simulation model (NCCI, 1986).

Example 3.5: The results of an infiltration test are presented in Table 3.2. Determine the empirical parameters A and B in the Green-Ampt equation. Determine the infiltration rate and accumulated infiltration after applying water for 6 hours.

Solution: The empirical parameters A and B can be estimated by substituting values of the infiltration rate, f, and the accumulative infiltration, F, into Equation 3.17. As there are two unknown parameters, values need to be substituted into the equation for two different times to form a pair of equations which are then solved simultaneously.

Substitute values of f and F at times 0.5 hr and 4.0 hr.

at 0.5 hr:

$$1.41 = \frac{A}{0.85} + B \qquad \text{(inches/hr)}$$

at 4.0 hr:

$$0.53 = \frac{A}{3.58} + B \qquad \text{(inches/hr)}$$

Subtracting the two equations from each other will eliminate B and gives:

$$(1.41 - 0.53) = \frac{A}{0.85} - \frac{A}{3.58}$$

Multiplying both sides by (0.85×3.58) gives:

$$0.88 \times 0.85 \times 3.58 = 3.58\ A - 0.85\ A$$

Therefore, $2.73\ A = 2.68$ and $A = 0.98$ in^2/hr. Substituting this value in Equation 3.17 gives:

$$0.53 = \frac{0.98}{3.58} + B \qquad \text{(inches/hr)}$$

Table 3.2.   Infiltration test data for Example 3.5.

Time (hrs)	Infiltration Rate (in./hr)	Accumulative Infiltration (in.)
0.0	2.0	0.0
0.5	1.41	0.85
1.0	1.09	1.47
2.0	0.69	2.36
4.0	0.53	3.58

and B is equal to 0.26 in./hr.

Therefore the Green-Ampt equation can be written:

$$f = \frac{0.98}{F} + 0.26 \quad \text{(inches/hr)}$$

After 6 hours the infiltration rate and accumulative infiltration are unknown. Therefore, a trial and error solution needs to be developed. We know that the infiltration rate will be less than the value of 0.53 in./hr at 4 hours. Assume a value of 0.4 in./hr.

From 4 to 6 hours the average infiltration rate will be (0.53 in/hr + 0.40 in./hr)/2 which is 0.465 in./hr. Therefore the infiltration volume during these 2 hours will be 2 hours × 0.465 = 0.93 in. The accumulated infiltration at 6 hours will be the accumulated infiltration at 4 hours (3.58 in.) plus 0.93 in., which gives 4.51 inches.

Substitute $F = 4.51$ inches in the Green-Ampt equation:

$$f = \frac{0.98}{4.51} + 0.26 \quad \text{(inches/hr)}$$

and $f = 0.48$ in./hr. This value is larger than the assumed value of 0.4 in./hr. Repeat the calculation with an infiltration rate at 6 hrs which is between 0.4 and 0.48 in./hr. An estimate of 0.47 in./hr gives an accumulative infiltration of 4.58 inches which, when substituted into the Green-Ampt equation, gives a similar infiltration rate of 0.474 in./hr.

Answer: The Green-Ampt parameter A = 0.98 in.^{2}/hr and B = 0.267 in./hr. After six hours the infiltration rate is 0.47 in./hr and the accumulative infiltration is 4.58 inches.

The Green-Ampt equation was derived for application when ponded infiltration occurred through a homogeneous soil with a uniform initial soil water content. If the equation parameters are estimated based on fitting infiltration results from field infiltrometer tests, good estimates of infiltration can also be obtained in heterogeneous soils. Applications of this nature result in good estimates because the equation is treated as a fitted parameter procedure rather than because it accounts for physical processes.

The preceding methods are only applicable to infiltration during ponded conditions. In many situations, application of these models is limited. Mein and Larson also (1973) developed a two stage form of the Green-Ampt equation to account for conditions prior to ponding.

*3.6.3. Physically Based Methods

In 1856, Henry Darcy developed the basic relationship for describing the flow of water through a homogeneous soil. Darcy used sand as the porous media and concluded that the rate of flow was proportional to the hydraulic gradient. Darcy's law may be written as:

$$q = -K \frac{\partial \phi}{\partial z} \tag{3.18}$$

where q is the flux, or volume of water moving through the soil in the z-direction (vertically or laterally) per unit area per unit time, and $\partial \phi / \partial z$ is the hydraulic gradient in the z-direction (vertically or laterally). Darcy's experiments were limited to one-dimensional flow, but his results have been found to be applicable to three-dimensional flow. Further discussion is presented in Chapter 9.

The hydraulic conductivity is also a function of the soil water content, θ, and the pressure head, ϕ. Generally, the hydraulic conductivity is dependent on the direction of flow and the soil is defined as being anisotropic. For anisotropic soils, a subscript is used to denote the direction of flow. The hydraulic conductivity is usually expressed as $K(\phi)$, a function of the pressure head, or as $K(\theta)$, a function of the soil water content.

A theoretical differential equation for unsaturated flow is obtained by combining Darcy's equation with the continuity equation. This equation is referred to as the diffusion equation or Richards equation (Swartzendruber, 1969). The equation is expressed as:

$$\frac{\partial \theta}{\partial t} = \frac{\partial}{\partial} \left[(D(\theta) \frac{\partial \theta}{\partial z} \right] - \frac{\partial K(\theta)}{\partial z} \tag{3.19}$$

where $D(\theta)$, the soil water diffusivity, is defined as $K(\phi) \, \partial z / \partial \theta$. The difficulty with solving the equation is the nonlinear relationship between hydraulic conductivity, pressure head, and soil water content.

Several solutions of the equation can be developed based on the boundary conditions for the system being evaluated. For a saturated soil, $\partial \theta / \partial t$ is zero, $K(\theta)$ reaches a constant and the pressure head becomes positive. Equation 3.19 reduces to the Laplace equation for one-dimensional saturated flow (Jury et al., 1991).

In recent years many researchers have examined solutions of Richard's equation. Most of these solutions are only defined for a particular aspect of the infiltration process or are approximate solutions obtained by using finite difference or finite element methods.

3.7. A PERSPECTIVE ON INFILTRATION METHODS

Based on the authors' research, review of the literature, and consulting experiences, it appears that while numerous equations have been developed to model the infiltration process, no single equation works well for all situations. Most methods are inadequate because they require knowledge of equation parameters which are difficult to determine on a site specific basis.

The Richards equation and Green-Ampt methods best describe infiltration processes in profiles where a piston type of infiltration process occurs and where air effects may be modeled by determining the model parameters at field saturation. Thomas and Phillips (1979) and Quisenberry and Phillips (1976) have shown that in many agricultural soils, a piston type of flow is not the primary mechanism of infiltration. Their results, which are based on field measurements, indicate that initially, macropore flow through fissures in the profile dominate. Surface sealing effects have also been found to be very important, especially in clayey soils.

In the last decade the scientific community has placed considerable emphasis on better understanding flow processes through soil profiles. The results of much of this research are presented in scientific journals.

3.8. MEASUREMENT OF SOIL PROPERTIES

Understanding the hydrology of environmental or agricultural systems (e.g., subdivision layout, landfill siting, drainage system design) often requires the measurement of one or more soil properties. Though you may not be doing the measurement work yourself, it is important to understand measurement procedures, their application and limitations. Many agencies and industries have standard methods of measurement which are outlined in individualized procedures (or available in a book of standard methods; for example, see ASTM, 1993 or Klute, 1986). This overview should help you understand why a test is being used for a certain situation, or why some other test may be more appropriate. Such tests should be judged on (in order of importance): 1) relevance to the system being studied (i.e., purpose for which the measurements are being made and the nature of the physical system); 2) reliability/repeatability/accuracy; 3) skills and knowledge of the person/people conducting the test; and 4) time and equipment considerations (availability/cost).

3.8.1. Particle Size Analysis

The size distribution of individual soil particles can tell much about a soil. Many of the soil characteristics (hydrologic and otherwise) reported in soil surveys and other reports/studies have been attributed to a soil based on its particle-size distribution. This characteristic of a soil can give a general idea about the suitability of a given soil for agricultural drainage, or reservoir or landfill siting, other purposes, or the susceptibility to chemical leaching or soil erosion.

The measurement of particle size is based on the dispersion/destruction of soil aggregates into discrete units and the separation (and measurement) of these particles into size groups by sieving and sedimentation. The size distribution of larger particles (> 0.05 mm diameter) is determined by passing the sample through a series of nested sieves of decreasing opening size.

The size distribution of smaller particles is found from the relationship between settling velocity and particle diameter known as Stokes' Law:

$$\nu = \frac{g \, (\rho_s - \rho_l) \, X^2}{18 \, \eta} \qquad (3.20)$$

where ν, is velocity of the fall, η is fluid shear viscosity, ρ_s and ρ_l are particle and liquid densities, respectively, and X is the "equivalent" particle diameter. Gee and Bauder (1986) describe the theory and process in some detail. The two primary methods of particle-size analysis are the hydrometer method and the pipet method.

In the *hydrometer method,* the soil sample (50 g $\pm$ in 250 mL of water) is dispersed by using a combination of chemical (Na-HMP) and mechanical (mixer or shaker) or ultrasonic means. The sample is then poured into a 1-liter (graduated or Buoyoucos) cylinder which is topped off with water to the 1-liter mark. The sample is shaken to completely disperse the sample throughout the cylinder. A hydrometer (a device which measures specific gravity of a liquid) is introduced to the cylinder and readings are taken at specific times. Gee and Bauder (1986) suggest 30 s, 1, 3, 10, 30 min, and 1, 1.5, 2, and 24 h. Measuring the specific gravity (or density of the soil solution) allows calculation of the percentage of the soil sample that has settled out at a given time.

In the *pipet method,* the soil sample is dispersed in a similar fashion to the hydrometer method and again introduced into a 1-liter cylinder. A small subsample is taken at time t and depth h using a pipet. At that point, all particles coarser than X have settled below the sampling depth, and the percentage of the original soil left in suspension can be determined.

3.8.2. Particle Density

The particle density is measured by either the pycnometer method or the submersion method (Blake and Hartge, 1986b). Both methods are based on the difference in weight between a volume of water and that same volume with some of the water displaced by a known weight of soil.

The particle density can be calculated from the relationship:

$$\rho_p = \frac{\rho_w \, (W_s - W_a)}{(W_s - W_a) - (W_{sw} - W_w)} \qquad (3.21)$$

where: ρ_p = particle density (lb/ft^3)
          ρ_w = density of water (lb/ft^3)
          W_s = weight of pycnometer (or weighing dish) plus oven-dry soil
          W_a = weight of pycnometer (or weighing dish)
          W_{sw} = weight of pycnometer (or weighing dish) filled with soil and water
          W_w = weight of pycnometer (or weighing dish) filled with water

3.8.3. Bulk Density

Four general methods are available for measuring bulk density (Blake and Hartge, 1986a): 1) core methods; 2) excavation methods; 3) the clod method; and 4) radiation method. The first three methods are based on direct measures of weight and volume, whereas the last is based on the empirical relationship between density and the amount of gamma radiation that is transmitted through or reflected (backscattered) by the soil.

In *core methods,* a volume of soil is collected by using a specially designed cylinder (called a coring tube) and drop hammer or a hydraulic probe (Figure 3.19). Knowing the diameter of the coring tube and the length of the sample allows us to calculate sample volume, V. The sample is weighed and the bulk density is calculated by using Equation 3.8. The dry bulk density can be determined from the same sample by drying it at 105°C for 24 hours and reweighing. The oven-dry weight is taken to be the weight of the soil solids only and the dry bulk density is calculated by using Equation 3.9. Extreme care must be taken such that the sample volume is not compacted by the sampling process.

In *excavation methods,* a hole is "excavated" with all of the removed soil being retained. The volume of the hole is then determined by placing a plastic bag or balloon in the hole and filling it with water (or air) until the fluid reaches the surface of the surrounding soil. The displaced soil is either weighed on a portable scale in the field or taken (in a sealed container to prevent water loss) to the laboratory to be weighed. Again, dry bulk density can be determined by drying the soil in the laboratory to determine oven-dry weight. The advantage of this method is that sampling errors are minimal if the test is performed properly. A major disadvantage is the amount of work involved, especially if bulk density of layers below the surface are desired.

In the *clod method,* naturally forming soil units (peds or clods) are coated with a substance, such paraffin, that prevents water from entering the clod. The clods are coated by tying a string to the clod and dipping it into a heated vat of the liquified waterproofing agent. Upon cooling, the waterproofing agent will solidify and form a waterproof seal. Each clod is weighed and is then lowered (submersed) into a beaker or graduated cylinder filled with water. The amount of water displaced (as measured by the rise in water level) is equal to the volume of the clod. Equation 3.8 is then used to determine the bulk density of the individual clod. The dry bulk density can also be found by this method by oven-drying the clods before coating and using Equation 3.9. The advantage of this method is that it gives very specific density information about soil structural units (clods) of various sizes. However, the method has several drawbacks. This method will not work for soils, such as organic

Figure 3.19.   Uhland drop hammer bulk density sampler.

soils, with bulk density less than the density of water because the clods will float instead of submerging. Many soils, and particularly cultivated or sandy soils, do not have stable structural units to perform the test on. The layer of waterproofing agent introduces a small error which will vary, depending on its thickness and the size of the clod. This method is also very time consuming.

In the *radiation method,* an instrument known as a gamma density gauge uses a radioactive source (e.g., Cesium[137]) of gamma radiation to indirectly estimate soil density. The radioactive source emits gamma radiation at a known rate. The rate that this radiation is transmitted through or reflected (backscattered) by the soil is related to the density of the surrounding soil, as well as the texture, chemical composition, and water content of the soil. The gamma density gauge uses either two probes (one containing an emitter, the other a receiver) which are placed at a fixed spacing to measure transmitted radiation; or a single probe (containing both emitter and receiver) which measures reflected radiation. Though calibration curves are typically provided with commercial gamma density gauges, the actual relationship will depend somewhat on local conditions. The gamma density gauge may be used in conjunction with the neutron probe (see section on measurement of soil water content above) to determine soil density and soil water properties *in situ*.

3.8.4. Soil Hydraulic Properties

The properties that describe the movement of water into and through the soil are among the most important in designing agricultural and environmental management systems. They are also among the most difficult to quantify. Measurement and prediction techniques for these hydraulic properties are discussed individually below. The property or combination of properties measured should depend on the system to be analyzed and/or designed. Table 3.3 offers a quick reference on which properties should be used for a particular application.

3.8.5. Soil Water Content

Methods for measuring soil water content can be divided into direct measures and indirect measures (Gardner, 1986; Jury et al., 1991). Direct methods are based on determination of weight before and after oven-drying. To directly measure soil water content, the wet sample is weighed, dried in an oven, and reweighed. The sample can be dried in a conventional oven (at 105°C for 24 hr) or in a microwave oven (see Gardner, 1986). The gravimetric water content, θ_g, is then calculated using Equation 3.7 (Jury et al., 1991).

Indirect measures depend on the change in physical and chemical properties of the soil with the change in soil water content. Indirect methods include: 1) measures of electrical conductivity and capacitance; 2) radiation methods; and 3) reflectometry. Gardner (1986) described the process of determining soil water content by means of *electrical conductivity methods*. Electrodes are imbedded in blocks of some porous material (typically gypsum). The blocks are then placed in the soil at the desired depth. The porous blocks reach a degree of wetness which is in equilibrium with the surrounding soil. The conductance of the block is then determined by connecting a specially designed electrical equipment (commercially available) which contains a power source and a circuit (called a Wheatstone bridge) which measures resistance. The resistance (the inverse of conductivity) is empirically related to the water content of the block, and thus to the soil. The resistance of the blocks is calibrated against a soil matric potential. To find soil water content of a given soil, the relationship between soil water content and matric potential must be established. The method is useful for distinguishing between wet and dry conditions, but is not well suited for measuring small changes in soil water content.

In *radiation methods,* soil water is determined by the attenuation of either gamma or neutron radiation. In gamma ray attenuation, a radioactive source emits gamma radiation at a known rate. The rate that this radiation is transmitted through the soil is related to the water content and density of the soil. By measuring the soil density at a known soil water content, that component can be removed from the calculations and a relationship between radiation transmitted and water content is established (for further details see Jury et al., 1991). Gamma radiation methods are normally used for measuring changes in soil water content with time.

In neutron attenuation (Gardner, 1986; Jury et al., 1991), a probe (containing both emitter and detector) which measures "thermalized" radiation is lowered into an access tube in the ground. The radiation source emits high-energy neutrons which collide with nuclei of atoms in the surrounding soil. The energy of the neutrons is

Table 3.3.   Application of soil hydraulic properties.

Application	Measurement Techniques
Above Water Table (Unsaturated Hydraulic Conductivity)	
Vadose Zone Models Evaporation Rate	For soil-water/soil suction-type techniques. See Green et al. (1986); Klute and Dirksen (1985); or Bouwer and Jackson (1974).
Below Water Table (Saturated Hydraulic Conductivity)	
Agricultural Drainage System Design	Auger Hole (Amoozegar and Warrick, 1986)
Groundwater Exploration/Well Investigation	Pumping Test/Slug Test (Freeze and Cherry, 1979)
Surface Applications (Infiltration)	

Irrigation System Design
    Sprinkler
    Surface
	Sprinkler Infiltrometer (Peterson and Bubenzer, 1986)
	Furrow Infiltration Test (Kincaid, 1986)

Urban/Suburban Drainage	Sprinkler or Ring Infiltrometer (Bouwer, 1986)
Infiltration Models	Sprinkler, Ring, Furrow Infiltrometer
Erosion Models	Sprinkler Infiltrometer
Combination (Saturated Hydraulic Conductivity and Infiltration)	
Water Table Management Models	Sprinkler or Ring Infiltrometers plus Auger Hole
Pesticide/Nutrient Movement to Ground and Surface Water	Sprinkler or Ring Infiltrometers plus shallow well pump-in method

substantially changed only by hydrogen nuclei (present mostly in the form of water). The percentage of neutrons that have had their energy changed (thermalized) can be recorded by the detector. Calibration curves are then developed to relate the output to water content (removing background sources of hydrogen nuclei such as organic matter and kaolinite).

In *time domain reflectometry* (Jury et al., 1991), the dielectric constant of the soil is related to the soil water content. A special apparatus is used which sends a step pulse of electromagnetic radiation along dual probes inserted into the soil. The pulse is "reflected" and returned to the source with a velocity characteristic to a specific dielectric constant. This instrument can be calibrated to give soil water content based on the measured dielectric constant.

3.8.6. Soil Suction and Soil Water Release/Retention Characteristics

The movement of water into and out of the soil matrix is controlled by suction (tension) forces. Unfortunately, accurate measurement of soil suction is difficult and time consuming. Soil suction relationships are a function of the pore size distribution, the soil structure, soil water content, organic matter content, and soil solution properties. The soil texture and the porosity of the soil profile will have the most influence on soil suction properties of the profile.

The most common approach to obtain a relationship between soil suction and soil water content is to obtain equilibrium conditions for different soil water content conditions whereby both soil suction and soil water content can be measured. This is most easily accomplished by drying a wet soil because the changes in soil water content and soil suction are slow. However, it should be recalled that there is hysteresis during wetting and drying (see Figure 3.5). Therefore, information obtained during drying needs to be adjusted if used to describe wetting. Details in field and laboratory procedures to determine relationships between soil water content and soil suction are presented by Klute (1986).

The general approach that is used in many laboratory procedures is to: (a) obtain disturbed or undisturbed soil samples from the field; (b) place the soil samples on a porous plate in a pressure cell or chamber; (c) wet the samples to saturation; (d) raise the pressure in the cell to a selected value above atmospheric pressure; (e) leave the sample until equilibrium has been reached and water flow from the sample has ceased (normally at least 24 hours); (f) weigh the sample and obtain the gravimetric soil water content; and (g) repeat this procedure at several different pressures (Figure 3.20). If knowledge of wilting point and field capacity are required, the test will be conducted at pressures of 0.1–0.33 bars and 15 bars, respectively.

Field measurements of soil suction are made with tensiometers or pressure transducers. Common types of tensiometers are shown in Figure 3.21. They consist of a porous cup which is connected to a top and a device to measure pressure changes as water moves into or out of the porous cup. As water moves out of the cup into the soil profile a partial vacuum is created in the tube. The movement of water into the soil will depend on the size of soil pores in the vicinity of the porous cup and the water content of the pores. As the soil dries, more water will be "sucked" from the porous cup and the partial vacuum will increase. If the soil water content increases, perhaps due to a rainfall event, water might flow back into the porous cup. Tensiometers will normally function only if the soil suction is less than about 0.8 bars.

3.8.7. Infiltration

Infiltration into soil is commonly measured by the cylinder (single-ring or double-ring) infiltrometer or by a sprinkler infiltrometer. The cylinder infiltrometer is the most common method used (and misused) because it is relatively inexpensive and simple. This method, used properly, can provide useful information for situations where the infiltrating water will be ponded on the surface. In situations where the system to be studied involves erosion, runoff, and infiltration of rainfall, the sprinkler infiltrometer should be used. To determine infiltration information for surface irrigation systems, the border or furrow irrigation methods described by Kincaid (1986) should be employed.

When using the cylinder infiltrometer method (Bouwer, 1986), a large ring (> 1m diameter) is installed in the soil surface. The infiltration rate is measured using a Mariotte device that maintains a shallow head on the exposed surface (Figure 3.22). The final infiltration rate is taken to be the rate of infiltration when soil suction at the bottom of the ring equals zero (saturated condition). Water flow is assumed to be one-dimensional (in the vertical direction), though flow is in reality three-dimensional. Users often attempt to "buffer" the effect of horizontal infiltration by using a

Figure 3.20.   Pressure chamber for making laboratory measurements of soil water release relationships.

Figure 3.21.   Tensiometer methods for measuring *in-situ* soil suction.

"double-ring" infiltrometer. A larger ring is placed around the inner ring (Figure 3.23), and the same water level is maintained in the inside and outside rings. However, many users of this methodology assume that by using the second ring, they can use smaller diameter rings with the same accuracy. Bouwer (1986) stated that " ... *when lateral capillary gradients below the cylinder infiltrometer cause the flow to diverge, it does not help to put a smaller cylinder concentrically inside the big*

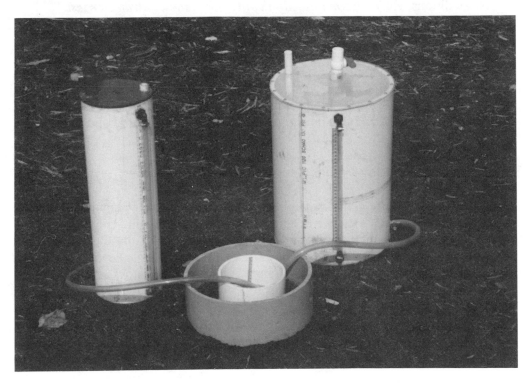

Figure 3.22.   Ring infiltrometer with a Mariotte hydraulic head device.

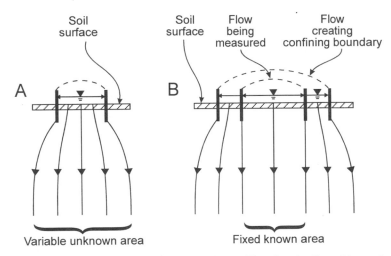

Figure 3.23.   Divergence of streamlines during infiltration buffered by using a double ring infiltrometer.

cylinder and measure the infiltration rate in it, in hopes of getting a measure of the true vertical infiltration rate." Bouwer (1986) also cautions that *"while cylinder infiltrometers in principle are simple devices ... the results can be grossly in error if the conditions of the system....* " (i.e., water quality, temperature, soil conditions, surface conditions, etc.) *"are not exactly duplicated in the infiltrometer test."*

The sprinkler infiltrometer as described by Peterson and Bubenzer (1986) provides the most detailed information on the infiltration process. In this method a rainfall simulator is used to simulate natural rainfall characteristics (rate of application, drop size). The amount and rate of infiltration is determined by monitoring the rate of runoff from a plot of known size and subtracting runoff from rainfall. This method also allows the user to study soil erodibility characteristics. However, this method requires specialized equipment which can be costly to build and operate.

3.8.8. Unsaturated Hydraulic Conductivity

Unsaturated hydraulic conductivity is used to model water movement and water content distributions above the water table. Unsaturated hydraulic conductivity is a function of soil-water content and soil suction and measures of these properties are used to estimate the conductivity. Estimation of unsaturated conductivity requires soil-water and soil-suction values over a range of soil conditions, which may require monitoring soil-water conditions over several weeks. Techniques for estimating unsaturated hydraulic conductivity are described by Klute and Dirksen (1986) for laboratory techniques; Green, et al. (1986) for field techniques; or Mualem (1986) for predictive techniques.

3.8.9. Saturated Hydraulic Conductivity

Saturated hydraulic conductivity can be found directly by measuring water movement through a soil sample, or indirectly, by estimating from associated soil properties. Methods that measure saturated hydraulic conductivity directly can be divided into two categories, (1) laboratory techniques, and (2) field techniques. Field techniques can be further divided into methods that require a shallow water table and those that do not. Klute and Dirksen (1986) stated that choosing a method for determining hydraulic conductivity depends on such factors as available equipment, nature of the soil, kind of samples available, the skills and knowledge of the experimenter; the soil-water suction range to be covered; and the purpose for which the measurements are being made.

The method for measuring hydraulic conductivity should be selected so that the soil region and flow direction used in the hydraulic conductivity measurement adequately represents the soil and flow direction in the actual system to be characterized (Bouwer and Jackson, 1974). Because hydraulic conductivity is difficult to measure, and information on other soil properties is often readily available or simpler to determine, methods have been developed to estimate hydraulic conductivity from related soil properties, such as soil texture, porosity, and bulk density. Studies that address estimation of hydraulic conductivity from related properties include Rawls et al. (1982), Puckett et al. (1985), Wang et al. (1985), Bouma (1986), and Baumer et al. (1987).

The *core sample method* determines saturated hydraulic conductivity on samples by either a constant head or falling head test. Core samples are taken in the field and kept from drying. In the laboratory, the samples are saturated from the bottom to prevent air entrapment.

For the constant head test, water is supplied to the bottom of the core samples at constant hydraulic head. The volume of outflow is measured with time. The hydraulic conductivity is found by rewriting Darcy's law as:

$$K = \frac{q \, L}{A \, h} \qquad (3.22)$$

where q is the rate of outflow, L is the length of the core sample, A is the core cross-sectional area, and h is the depth of the constant head applied.

In the falling head test, used with soils of low permeability, the hydraulic conductivity is determined from the rate of change of velocity of a falling water column. A manometer (cross-sectional area a) is placed on top of the sample core (cross-sectional area A) and filled with water. The hydraulic conductivity is determined as:

$$K = \frac{a \, L}{A(t_2 - t_1)} \ln \frac{h_1}{h_2} \qquad (3.23)$$

where L is the length of the soil core, h_1 is the height of the water column at time t_1, and h_2 is the height of the water column at time t_2.

The laboratory core technique is relatively cheap, does not require a water table, and enables sampling for layers or anisotropy. However, this technique is limited by soil disturbances during handling, small sample size, and loss of head at the water/soil interface. Hydraulic conductivity values resulting from this method are typically lower than those calculated from field measurements and there is often a large variability between samples.

When using field techniques for determining saturated hydraulic conductivity above the water table, the hydraulic conductivity calculations are based on infiltration into unsaturated soil. Even under the best of circumstances, some air is entrapped during infiltration and the soil does not become fully saturated. These methods give a value for the field saturated hydraulic conductivity which may be significantly different between different methods, sites, and initial water conditions.

The percolation test is the simplest test to conduct. A hole is dug to the desired depth, water is ponded in the bottom of the hole to saturate the soil, and the vertical velocity of water entering the soil is measured. This technique relates well to subsurface seepage.

The shallow well pump-in method otherwise known as the dry Auger hole method or well permeameter method, measures the rate of flow of water from a (cased or uncased) Auger hole when a constant height of water is maintained in the hole. A float valve is usually used to maintain the water level with a large water tank providing the water supply. Hydraulic conductivity values are calculated using the steady state outflow rate and a shape factor determined from nomographs or equations. The position of the water table or impermeable layer below the bottom of the well must be known. This technique is easy to use, but is limited by the time requirements needed to reach steady state and to replicate measurements. A self-contained version of the shallow well pump-in method has been developed by researchers at Guelph University (Reynolds and Elrick, 1985). The Guelph permeameter, which is now commercially available, uses a built-in Mariotte bottle to control the depth of water in the Augured hole. The saturated hydraulic conductivity is determined from the difference in outflow rates for two different water depths.

In the air-entry permeameter (Bouwer, 1986), a small covered cylinder is driven into the ground. Water is applied to the cylinder until all air is driven out. A large constant head is kept in a reservoir at the top of the cylinder until saturation is reached at the bottom of the cylinder. The water level is then allowed to fall and the conductivity is calculated from falling head equations. The air-entry permeameter measures field saturated hydraulic conductivity in the vertical direction. An improved (for quickness and ease of use) version of the air-entry permeameter, called the velocity head permeameter, has been developed to measure hydraulic conductivity in either the vertical or horizontal direction. A description of this method, now commercially available, can be found in Merva (1979).

In the presence of a shallow water table, the Auger hole method is the most widely used method for determining hydraulic conductivity of soils in the field. A hole is augured to the desired depth below the water table and water is allowed to rise until equilibrium is reached. The hole is then pumped or bailed and the rate of rise of the water level in the hole is measured. The saturated hydraulic conductivity is calculated as:

$$K = \frac{\pi R^2}{S h} \frac{dh}{dt} \qquad (3.24)$$

where R is the radius of the Auger hole, S is a function of hole geometry found from nomographs, h is the depth of water in the Auger hole, and dh/dt is the rise in the water level over time increment dt.

The Auger hole method is simple in conception and practice, is quick and cheap, and measures a large sample area compared to other techniques. It is also one of the most reliable methods. However, extensive variability in repeated measurements can result from soil heterogeneity, depth of Auger hole, water table depth, and depth of water removed from the Auger hole. This technique is not suitable for layered soils and is unreliable in some cases due to macropores or side-wall failure.

Other methods can be found in Bouwer and Jackson (1974) and Klute and Dirksen (1986). Several tests are commonly used for exploring the saturated hydraulic conductivity of *groundwater* systems. These are the *slug test* and the *pumping test*. *Laboratory core methods* (see above) may also be used to describe individual layers either in or above the aquifer. For more information on measuring saturated hydraulic conductivity for groundwater systems, see Freeze and Cherry (1979).

REFERENCES

ASTM. 1993. Annual Book of ASTM Standards, Section 4: Construction - Soil and Rock, Building Stones; Geotextiles. Vol. 04.08. American Society for Testing and Materials, Philadelphia.

Baumer, O.W., R.D. Wenberg, and J.F. Rice. 1987. The use of Soil water retention curves in DRAINMOD. Proc. 3rd Intl. Wkshp. on Land Drainage. Dept. of Agr. Eng., Ohio State University, Columbus, OH. pA1–A10.

Blake, G.R. and K.H. Hartge. 1986a. Bulk density. Ch. 13 in A. Klute (ed.), Methods of Soil Analysis, Part 1: Physical and Mineralogical Methods, 2nd Edition. American Society of Agronomy, Madison, WI.

Blake, G.R. and K.H. Hartge. 1986b. Particle density. Ch. 14 in A. Klute (ed.), Methods of Soil Analysis, Part 1: Physical and Mineralogical Methods, 2nd Edition. American Society of Agronomy, Madison, WI.

Bodman, G.B. and E.A. Coleman. 1943. Moisture and energy conditions during downward entry of water into soils. Soil Sci. Soc. Am. Proc. 7:116–122.

Bouma, J. 1986. Using soil survey information to characterize the soil water state. J. Soil. Sci. 37:1–7.

Bouwer, H. 1986. Intake rate: cylinder infiltrometer. Ch. 32 in A. Klute (ed.), Methods of Soil Analysis, Part 1: Physical and Mineralogical Methods, 2nd Edition. American Society of Agronomy, Madison, WI.

Bouwer, H. and R.D. Jackson. 1974. Determining Soil Properties, in Drainage for Agriculture. ASA Monograph No. 17. J.van Schilfgaarde (ed.), American Society of Agronomy, Madison, WI.

Cassel, D.K. and D.R. Nielsen. 1986. Field capacity and available water content. Ch.36 in A. Klute (ed.), Methods of Soil Analysis, Part 1: Physical and Mineralogical Methods, 2nd Edition. American Society of Agronomy, Madison, WI.

Elrick, D.E. and W.D. Reynolds. 1986. An analysis of the percolation test based on three-dimensional, saturated-unsaturated flow from a cylindrical test hole. Soil Science 142(5): 308–321.

Freeze, R.A. and J.A. Cherry. 1979. Groundwater. Prentice-Hall, Englewood Cliffs, NJ.

Gardner, W.H. 1986. Water content. Ch. 21 in A. Klute (ed.), Methods of Soil Analysis, Part 1: Physical and Mineralogical Methods, 2nd Edition. American Society of Agronomy, Madison, WI.

Gee, G.W. and J.W. Bauder. 1986. Particle-size analysis. Ch. 15 in A. Klute (ed.), Methods of Soil Analysis, Part 1: Physical and Mineralogical Methods, 2nd Edition. American Society of Agronomy, Madison, WI.

Green, R.E., L.R. Ahuja, and S.K. Chong. 1986. Hydraulic conductivity, diffusivity, and sorptivity of unsaturated soils: field methods. Ch. 30 in A. Klute (ed.), Methods of Soil Analysis, Part 1: Physical and Mineralogical Methods, 2nd Edition. American Society of Agronomy, Madison, WI.

Hillel, D. 1982. Introduction to Soil Physics. Academic Press, Orlando, FL.

Horton, R.E. 1939. Analysis of runoff plot experiments with varying infiltration-capacity. Trans. Am. Geophysical Union, Hydrology Papers:693–711.

Jury, W.A., W.R. Gardner, and W.H. Gardner. 1991. Soil Physics (5th Edition). John Wiley & Sons, Inc., New York.

Kincaid, D.C. 1986. Intake rate: border and furrow. Ch. 34 in A. Klute (ed.), Methods of Soil Analysis, Part 1: Physical and Mineralogical Methods, 2nd Edition. American Society of Agronomy, Madison, WI.

Klute, A. and C. Dirksen. 1986. Hydraulic conductivity and diffusivity: laboratory methods. Ch. 28 in A. Klute (ed.), Methods of Soil Analysis, Part 1: Physical and Mineralogical Methods, 2nd Edition. American Society of Agronomy, Madison, WI.

Mein, R.G. and C.L. Larson. 1973. Modeling Infiltration During a Steady Rain. Water Resour. Res. 9(2):384–394.

Merva, G.E. 1979. Falling head permeameter for field investigation of hydraulic conductivity. ASAE paper No. 79-2515, American Society of Agricultural Engineers, St. Joseph, MI.

Mualem, Y. 1986. Hydraulic conductivity of unsaturated soils: prediction and formulas. Ch. 31 in A. Klute (ed.), Methods of Soil Analysis, Part 1: Physical and Mineralogical Methods, 2nd Edition. American Society of Agronomy, Madison, WI.

Peterson, A.E. and G.D. Bubenzer. 1986. Intake rate: sprinkler infiltrometer. Ch. 33 in A. Klute (ed.), Methods of Soil Analysis, Part 1: Physical and Mineralogical Methods, 2nd Edition. American Society of Agronomy, Madison, WI.

Puckett, W.E., J.H. Dane, and B.F. Hajek. 1985. Physical and mineralogical data to determine soil hydraulic properties. Soil Sci. Soc. Am. J. 49: 831–836.

Quisenberry, V.L. and R.E. Phillips. 1976. Percolation of Surface-Applied Water in the Field. Soil Sci. Soc. Am. J., 40:484–489.

Rawls, W.J., D.L. Brakensiek, and K.E. Saxton. 1982. Estimation of soil water properties. Trans. ASAE 25(5): 1316–1320, 1328.

Reynolds, W.D. and D.E. Elrick. 1985. In-situ measurement of field-saturated hydraulic conductivity, sorptivity, and the α-parameter using the Guelph parmeameter. Soil Science 140(4):292–302.

Swartzendruber, D. 1969. The Flow of Water in Unsaturated Soils. pp. 215–292. In R.M. DeWiest (ed.) Flow Through Porous Media. Academic Press, New York.

Thomas, G.W. and R.E. Phillips. 1979. Consequences of Water Movement in Macropores. J. Environ. Qual., 8(2):149–152.

Wang, C., J.A. McKeague, and G.C. Topp. 1985. Comparison of estimated and measured horizontal Ksat values. Can. J. Soil Sci. 65:707–715.

Ward, A.D., L.G. Wells and R.E. Phillips. 1983. Infiltration through reconstructed surface mine spoils and soils. Transactions of the ASAE, Vol 26, No 3: 821–832, St Joseph, MI.

Wells, L.G., A.D. Ward, I.D. Moore, and R.E. Phillips. 1986. Comparison of Four Infiltration Models in Characterizing Infiltration through Surface Mine Profiles. Transactions of the ASAE, Volume 29, Number 3. ASAE, St. Joseph, MI. pp. 785–793.

Problems

3.1 Based on the USDA classification scheme, determine the textural class of a soil which is 20% clay and 40% sand.

3.2 How much clay would need to be added to the soil in Problem 3.1 in order to make it a clay loam?

3.3 Determine the soil water content in percent by weight for a 120 g sample in a moist condition which, when oven-dried, weighed 90 g.

3.4 The combined weight of a soil sample and a tin is 50.0 grams. The tin weighs 20.0 grams. After oven drying, the soil sample and the tin have a weight of 45 grams. Determine the soil water content (by weight) of the soil.

3.5 Determine the porosity of a soil profile if it has a dry bulk density of 1.8 g/cc and the density of the soil particles is 2.7 g/cc. What would the bulk density of the soil be if it was saturated?

3.6 Based on field tests the bulk density of a soil profile was found to be 99.8 lb/ft^3. The specific gravity of the soil particles is 2.8 and the dry bulk density of the profile is 87.4 lb/ft^3. Determine the porosity and soil water content (by weight) of the profile.

3.7 The porosity of a soil is 0.4 (cc/cc). Determine the volumetric soil water content as a fraction if the degree of saturation is 60%. Also, determine the depth of water contained in the top 50 cm of a soil with a degree of saturation of 60% and porosity of 0.4.

3.8 Determine the soil water content at field capacity and plant available soil water content in inches for a 30 inch soil column. The field capacity by volume is 25% and the wilting point is 10%.

3.9 The field capacity of a soil is 35% and the wilting point is 15% by weight. The dry bulk density of the soil is 78 lbs per cu ft. Determine the available water content in percent by weight and by volume. Also determine the plant available soil water content in the top 2 ft of soil. Express the answer as a depth of water in inches.

3.10 Determine the particle density of a soil based on the following information from a laboratory pycnometer test: $W_s = 32$ g; $W_a = 20$ g; $W_{sw} = 37.5$ g; and $W_w = 30$ g (use Equation 3.21).

3.11 Horton's infiltration is fitted to the infiltration curve, for the dry initial conditions, which was presented in Figure 3.17. Determine the initial infiltration rate, final infiltration rate, and the best fit empirical parameter β.

3.12 Determine the daily changes in soil water content for the hydrologic conditions presented in Table 3.4. What depth of irrigation water should be added after 10 days to raise the soil water content of the root zone back to the initial value of day 1? The initial soil water content in the root zone was 12 inches of water. Assume that all percolation passes through the root zone and is not stored.

Table 3.4. Daily accretion and depletion of soil water for use in Problem 3.12.

Day	Accretion	Depletion		
	Precipitation (P)	Runoff (Q)	Evapotranspiration (ET)	Percolation (G)
	inches	inches	inches	inches
1	0.0	0.0	0.15	0.0
2	1.5	1.0	0.05	0.2
3	0.0	0.0	0.10	0.0
4	0.0	0.0	0.15	0.0
5	1.0	0.3	0.05	0.1
6	0.0	0.0	0.10	0.0
7	0.0	0.0	0.15	0.0
8	0.0	0.0	0.10	0.0
9	0.0	0.0	0.15	0.0
10	0.5	0.2	0.05	0.1
11	0.0	0.0	0.10	0.0
12	0.0	0.0	0.15	0.0

3.13 The root zone (1m deep) of a silt clay loam soil has an average dry bulk density of 1.35 g/cc and the specific gravity of the soil particles is 2.7. Based on laboratory measurements it has been established that the degree of saturation at field capacity and wilting point are 60 % and 30%, respectively. Irrigation is initiated by an automatic irrometer system when 50% of the plant available water has been depleted. The profile is irrigated at a constant rate of 10 mm/hr for 5 hours. Using the Green-Ampt equation, determine: a) the infiltration rate at the end of the irrigation application period; b) the volume of runoff; and c) the soil water content of the root zone when gravity flow has ceased. Based on a double ring infiltration test, the Green-Ampt parameters have been determined to be A = 30 mm/hr and B = 3 mm/hr. (Hint: The problem is best solved on a spread sheet and will require several iterations. Divide the irrigation application time into 30 minute time blocks. Assume the initial infiltration rate is the irrigation application rate. Obtain a first estimate of the accumulative infiltration in the first 30 minutes by assuming a constant infiltration rate. You will then need to estimate infiltration rates at the end of each time block and compare them with values determined from Equation 3.15.)

CHAPTER **4**

Evapotranspiration

Sue E. Nokes

4.0. INTRODUCTION

Evapotranspiration (ET) is the process which returns water to the atmosphere and therefore completes the hydrologic cycle. Students seem to have an intuitive feel for the hydrologic cycle and the idea that evapotranspiration completes the cycle is easily believed. This was not always the case, however. As late as the mid-1500s there was still a dispute over the source of water in rivers and springs. The age-old theory was that streams originated directly from sea-water or from air converted into water. In the mid-1500s, Bernard Palissy, a French potter, stated that rivers and springs could not have any source other than rainfall. Edmond Halley (1656–1742) proved the other half of the concept of the hydrologic cycle by showing that enough water evaporated from the earth to produce sufficient rainfall to replenish the rivers (Biswas, 1969).

Because evapotranspiration is a major component of the hydrologic cycle, this chapter will explore the processes by which ET occurs. In addition, we will discuss techniques for estimating the amount (depth) and rate (depth/time) of evapotranspiration. You may be asking yourself why you would ever need to calculate evapotranspiration. If you work with water or plants, chances are you will need to be able to estimate ET. The rate and amount of ET is the core information needed to design irrigation projects, and it is also essential for managing water quality and other environmental concerns. Policy makers need to know how estimates of ET are determined because these methods are used in litigation and in negotiations of compacts and treaties involving water. In order to predict meltwater yields from mountain watersheds or to plan for forest fire prevention, estimates of ET are needed. In urban development, ET calculations are used to determine safe yields from aquifers and to plan for flood control. Anyone involved with resource management will likely need to understand the methods available for estimating evapotranspiration.

Evapotranspiration can be divided into two sub-processes: evaporation and transpiration (Figure 4.1). Evaporation essentially occurs on the surfaces of open water such as lakes, reservoirs, or puddles, or from vegetation and ground surfaces. Transpiration involves the removal of water from the soil by plant roots, transport of the water through the plant into the leaf, and evaporation of the water from the leaf's interior into the atmosphere.

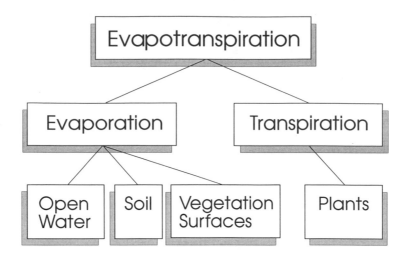

Figure 4.1. Evapotranspiration divided into sub-processes.

4.1. EVAPORATION PROCESS

Evaporation occurs when water is changed in state from a liquid to a gas. Water molecules, both in liquid and in gas, are in constant motion. Some water molecules possess sufficient energy to break through the water surface into the atmosphere and some water vapor molecules may cross back into the liquid. When more molecules are leaving the liquid than returning, net evaporation is occurring (Figure 4.2). The energy a molecule needs to penetrate the water surface is called the *latent heat of vaporization* (approximately 540 cal/g of water evaporated at 100°C). Evaporation therefore requires a supply of energy to provide the latent heat of vaporization, typically provided by solar radiation.

Consider a newly-closed container half full of water. With a sufficient supply of energy, water will evaporate; initially at a high rate, but eventually the container will equilibrate and no net rate of evaporation will occur. When the container was first closed, the concentration of water molecules in the air space above the water was very small. The newly-evaporated water molecules could move (diffuse) easily into the air space. The rate of diffusion of water molecules is proportional to the difference in concentration between water molecules at the water surface and in the air space. As more molecules evaporate and move into the air space, the concentration of water molecules in the air space increases. Since the concentration of water molecules at the water surface is essentially constant, and the concentration in the air space is increasing, the difference between the two concentrations decreases, and the rate of evaporation decreases. The net evaporation will eventually cease when the container equilibrates.

When net evaporation ceases, the air is said to have reached the *saturation vapor pressure* (e_s). If evaporation is to continue after the saturation vapor pressure is reached, some mechanism to remove water vapor from the air above the evaporating surface is needed. In an open system, such as a lake or crop field, a layer of water vapor can build up adjacent to the water or leaf surface and reduce evaporation. For

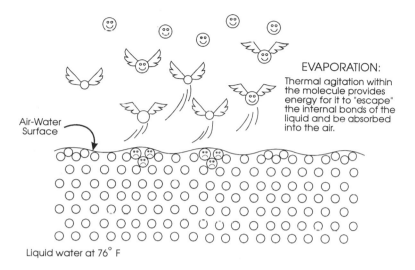

Figure 4.2. Schematic of the evaporation process.

evaporation to continue, air movement (typically provided by wind) is needed to remove the vapor.

Therefore, two important factors in the evaporation process are:

1. A source of energy to supply the latent heat of vaporization.
2. A concentration gradient in the water vapor, typically provided by air movement, which removes the water vapor adjacent to the evaporating surface.

*Fick's First Law of Diffusion

Fick was one of the first people to quantify the movement of molecules from a region of higher concentration to a region of lower concentration (Nobel, 1983), such as water molecules moving from a water surface into air. He developed Fick's first law of diffusion:

$$J_j = -D_j \frac{\partial c_j}{\partial x} \qquad (4.1)$$

where J_j (the flux density) is the amount of species j crossing a certain area per unit time and is typically expressed in units such as moles of particles per m^2 per second. D_j is the diffusion coefficient of species j (analogous to resistance in electrical circuits). The term $\partial c_j / \partial x$ represents the concentration gradient of species j and is the driving force that leads to molecular movement (Nobel, 1983).

Evaporation from Open Water

Evaporation from open water has been well studied. Open water in this context refers to lakes, ponds, reservoirs or evaporation pans (see Sec. 4.3). While it is true that evaporation from lakes and reservoirs may be of great local interest (for instance to predict summer water levels) the justification for studying open water in such great detail is that open water provides a reproducible surface. The state of crops and soils is dynamic, which complicates evaporation from these surfaces. The surface of open water is relatively unchanging, and the dependence of evaporation on weather conditions alone is easier to study. Crops and soil evaporation losses can be estimated from open water evaporation predictions (Penman, 1948).

*Evaporation from Bare Soil

Evaporation from bare soil is similar to evaporation from open water if the soil is saturated. If the bare soil is not saturated, the process is more complex because water evaporates deeper in the soil and the vapor must diffuse out into the atmosphere. The rate of evaporation from bare soil is typically divided into two distinct stages. During the first stage, the soil surface is at or near saturation. The rate of evaporation is controlled by heat energy input and is approximately 90% of the maximum possible evaporation based on weather conditions. The duration of the first stage is influenced by the rate of evaporation, the soil depth, and hydraulic properties of the soil. This stage typically lasts one to three days in midsummer (Jensen et al., 1990).

The second stage, sometimes called the falling stage, begins once the soil surface has started to dry. At this stage evaporation occurs below the soil surface. The water vapor formed in the soil reaches the soil surface by diffusion or mass flow caused by fluctuating air pressures. The evaporation rate during this stage is no longer controlled by climatic conditions, but rather soil conditions such as hydraulic conductivity (Jensen et al., 1990).

4.2. EVAPOTRANSPIRATION FROM SOIL AND PLANTS

When plants are introduced into the system, the complexity of measuring or predicting ET increases because plants are transpiring, in addition to water evaporating from the soil surface or from the canopy surfaces.

Transpiration

Transpiration is defined by Kramer (1983) as the loss of water from plants in the form of vapor. You will recognize from this definition that transpiration is basically an evaporation process. The evaporation of water in the leaves is responsible for the ascent of plant water and the rate at which water is taken in through the roots. The rate at which water moves through a plant is critical to the plant's functioning because water is the vehicle which carries nutrients and minerals into the plant. So how does evaporation, which occurs in the leaf, influence what happens in the roots of the plant?

Figure 4.3 shows the structure of a typical leaf. An epidermis is present on both the upper and lower sides of a leaf. The epidermal cells usually have a waterproof

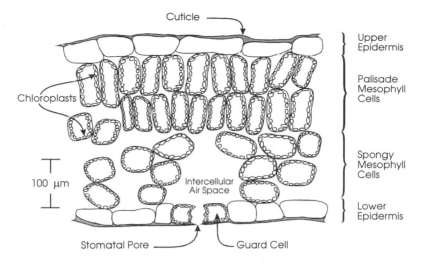

Figure 4.3. Schematic transverse section through a leaf.

cuticle on the atmospheric side. Between the two epidermal layers is the leaf meso-phyll, usually differentiated into the elongated, orderly palisade cells, and the more loosely packed, spongy mesophyll cells. The unstructured arrangement of the spongy mesophyll cells allows for the intercellular formation of air spaces that are the primary sites for transpiration. The pore which connects the intercellular spaces with the atmosphere is called a stoma or stomata. The plural forms are stomata or stomates respectively. Each stoma can open or close by changing the size of the guard cells at the edges of the pore.

The leaf mesophyll cells are saturated and are typically in contact with the inter-cellular spaces. Water from a saturated surface in contact with the atmosphere will evaporate if there is a source of energy and a difference in concentration of vapor in the air (or "room" for new water molecules to enter the air). The needed energy comes primarily from direct solar radiation. Evaporation will increase the vapor pressure in the intercellular space, resulting in a vapor pressure gradient between the leaf and the atmosphere under most atmospheric conditions. This gradient will cause vapor to diffuse out of the stomates into the atmosphere. The larger the vapor pressure gradient, the higher the rate of diffusion and the faster evaporation will occur (Fick's first law of diffusion). As a practical example of this, consider a hot summer day when the temperature is around 90°F. Would you be more comfortable when the relative humidity is 60% or 95%? Why? Most people are more comfort-able at 60% because there is more room in the air for water vapor. The sweat on your skin evaporates faster, which, in turn, cools you. The latent heat of vaporiza-tion is absorbed during the change of state of water, and this removes heat from the air/skin system.

Once some water is evaporated from the interior cells in a leaf, more liquid water moves in to replace that droplet (Figure 4.4). The process is similar to using a straw to drink. The plant has many cellular pathways that are connected and effectively act as the straws. The evaporation at the intercellular space initiates the pull. The divisions and subdivisions of the xylem (the main upwardly-conducting tissue) in the leaves allow for mesophyll cells to be no more than 3 or 4 cells away from the xylem. Transpiration effectively pulls water into the plant through the roots, up the xylem in the stem, and out the leaves.

Transpiration is considered beneficial to the plant because the process of transpiring helps the plant to absorb minerals, and helps to cool the leaves. Kramer (1983) counters these arguments however, because leaves in full sun are rarely injured by high temperatures, and transpiration merely increases the amount of water moved. Kramer also noted that many plants thrive in shaded humid areas where the rate of transpiration is very low. Numerous harmful effects of excessive transpiration resulting in water stress have been recorded. So why do plants transpire? Kramer hypothesizes that transpiration is an unavoidable evil. Unavoidable because evolution favored high rates of photosynthesis over low rates of transpiration, and the leaf structure favorable for the entrance of carbon dioxide (essential for photosynthesis) also allows for water vapor loss.

Many plants do have mechanisms to reduce transpiration if necessary. Plants can reduce the leaf area (rolling leaves or wilting) or change leaf orientation to reduce the amount of tissue exposed to sunlight (reduce the energy received). Plants can close the stomates to block the diffusion of water vapor out of the leaf (but this also blocks the entrance of CO_2). The ability of plants to constantly adjust their rate of transpiration is one of the major difficulties encountered when trying to estimate the actual transpiration rate.

Transpiration Ratio and Consumptive Use

Two terms frequently used relating to transpiration are transpiration ratio and consumptive use. *Transpiration ratio* is the ratio of the weight of water transpired to the weight of dry matter produced by the plant. This ratio is a measure of how efficiently crops use water. For example, the transpiration ratios are approximately 900 for alfalfa, 640 for potatoes, 500 for wheat, 450 for red clover, 350 for corn, and 250 for sorghum. The least efficient crop, in terms of water use, would be alfalfa because it uses 900 kilograms of water for every kilogram of dry alfalfa it produces. Sorghum is the most efficient crop listed because it uses only 250 kg of water for every kg of dry matter produced.

Consumptive use is the total amount of water needed to grow a crop (the sum of the water used in evapotranspiration plus the water stored in the plant's tissues). The term consumptive use is generally used interchangeably with evapotranspiration because the amount of water retained in plant tissue is negligible compared to the amount of evapotranspiration.

Potential Evaporation and Potential Evapotranspiration

Evaporation and especially evapotranspiration are complex processes because the rate of water vapor loss depends on the amount of solar radiation reaching the surface, the amount of wind, the aperture of the stomates, the soil water content, the soil type, and the type of plant. In order to simplify the situation, researchers have attempted to remove all the unknowns such as aperture of the stomates and soil water content, and focus on climatic conditions. The simplified calculations are termed

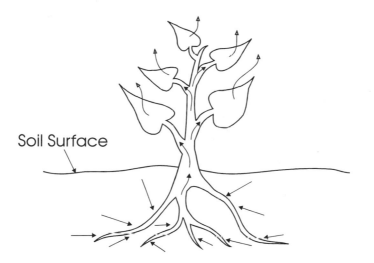

Figure 4.4. Schematic of the path of water through the plant.

potential evaporation and *potential evapotranspiration*. The definition given by Jensen et al. (1990) for potential evaporation (E_p) is the "...evaporation from a surface when all surface-atmosphere interfaces are wet so there is no restriction on the rate of evaporation from the surface. The magnitude of E_p depends primarily on atmospheric conditions and surface albedo but will vary with the surface geometry characteristics, such as aerodynamic roughness." *Surface albedo* is the proportion of solar radiation which is reflected from a soil and crop surface. Conventional abbreviations for the terms used in this section and the subsequent two sections are summarized in Table 4.1.

Penman (1956) originally defined potential evapotranspiration (E_{tp}) as "...the amount of water transpired in unit time by a short green crop, completely shading the ground, of uniform height and never short of water." Looking at Penman's definition, you can see the attempt to simplify the situation. The crop is assumed to be short and uniform, and completely shading the ground so that no soil is exposed. The crop is never short of water, so soil water content is no longer a variable and presumably the stomates would always be fully open. These conditions theoretically provide the maximum evapotranspiration rate based on the given climatic conditions.

Many authors treat potential evapotranspiration and potential evaporation as synonymous but the original intent was that potential evapotranspiration involved an actively growing crop and potential evaporation did not.

Actual Evaporation and Actual Evapotranspiration

Potential evapotranspiration and evaporation may be easier to estimate, but do not represent reality. In general, watersheds are not entirely covered by well-watered short green crops. Actual evapotranspiration (E_t) or actual evaporation (E) is the amount or rate of ET occurring from a place of interest and it is the value we want to estimate. In practice, actual ET is obtained by first calculating potential evapotranspiration and then multiplying by suitable crop coefficients to estimate the actual

Table 4.1.   Conventional abbreviations for terms relating to actual and potential evaporation and evapotranspiration.

Abbreviations for conventional evaporation and ET terms	
E_p	Potential evaporation
E_{tp}	Potential evapotranspiration
E	Actual evaporation
E_t	Actual evapotranspiration
E_{tr}	Reference crop evapotranspiration

crop evapotranspiration. Crop coefficients are usually residual terms from a statistical analysis of field data, so it is essential that the methods for estimating potential evapotranspiration be consistent with the crop coefficients. Because potential evapotranspiration has been defined vaguely (short green crop) some studies have used alfalfa and some have used grasses to measure potential evapotranspiration. Other methods correlate evaporation from free water surfaces to actual ET. The result is that the methods for determining actual ET are variable and confusing (Jensen et al., 1990). Scientists have attempted to remedy this problem by introducing reference crop evapotranspiration.

Reference Crop Evapotranspiration

Reference crop evapotranspiration (E_{tr}) is defined as "...the rate at which water, if available, would be removed from the soil and plant surface of a specific crop, arbitrarily called a reference crop" (Jensen et al., 1990). Typical reference crops are grasses and alfalfa. The crop is assumed to be well-watered with a full canopy cover (foliage completely shading the ground). The major advantage of relating E_{tr} to a specific crop is that it is easier to select consistent crop coefficients and to calibrate reference equations in new areas.

Many methods are available for quantifying evaporation and evapotranspiration. Methods are selected based on the accuracy needed, the time scale required, and the resources available. The first decision that needs to be made is whether ET will be measured or estimated from weather data. If you want to determine ET for a time period that is already elapsed, obviously your only choice is to estimate the ET from weather data records. If you want to predict ET for future time periods, and money is available to purchase instrumentation, then measuring ET would be an option.

4.3. MEASURING EVAPORATION OR EVAPOTRANSPIRATION

There are several methods available for measuring evaporation or evapotranspiration. Since vapor flux is difficult to measure directly, most methods measure the change of water in the system. Figure 4.5 shows schematically the options available for measuring potential evaporation, potential evapotranspiration, or actual evapotranspiration. An evaporation pan or ET gage can be used to measure potential evapotranspiration. Actual evapotranspiration can be measured in several ways. If

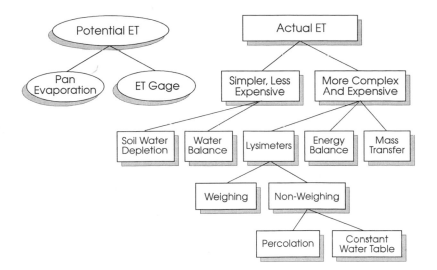

Figure 4.5. Options available for measuring potential or actual evapotranspiration.

you need a simpler, less expensive technique, measuring soil water depletion or using a water balance would be possibilities. More precise, but also more complex methods include lysimeters, either weighing or non-weighing, or using an energy balance or mass transfer technique. These methods are discussed in more detail below.

Evaporation Pan

One of the oldest and simplest ways to measure evaporation is with a pan. This involves placing a pan of water outside and recording how much water evaporates during a specific time. The type of pan (steel, plastic, glass) and the size of the pan affect how fast water will evaporate. In order to standardize pans, the U.S. Weather Bureau has chosen a standard pan, called a class A pan. This is a cylindrical container made of galvanized steel or Monel™ metal, 10 inches deep and 48 inches in diameter. The pan is placed on an open wooden platform with the top of the pan 16 inches above the soil surface. The platform site should be nearly flat, well sodded, and free from obstructions. The pan should be leveled and filled with water to a depth of 8 inches. Periodic measurements are made of the changes in water level with a hook gage set in a still well. When the water level drops 7 inches, the pan should be refilled. Lake evaporation can be estimated by multiplying pan evaporation by 0.7.

Pan evaporation data can be used to estimate actual evapotranspiration of a reference crop using (Jensen et al., 1990):

$$E_{tr} = k_p E_{pan} \qquad\qquad (4.2)$$

where k_p is a crop or pan coefficient. Many k_p values have been determined in previous studies but it is important that the study have a similar climate (humid vs. arid) and use the same pan (i.e., class A pan) with similar nearby surfaces and placement in relation to wind barriers at your site of interest. The coefficients in Table 4.2 were developed for an alfalfa reference crop. The coefficients can be used to convert pan measurements to an alfalfa reference crop ET estimate, or to convert estimates made with the other methods discussed in Section 4.5. Table 4.2 should be used with caution because the coefficients represent monthly averages and the interrelationships of methods may not be the same at all locations. To use Table 4.2, select the appropriate month from the top of the table, and the appropriate method from the left hand column, then determine the adjustment coefficient.

> **Example 4.1:** If the measured pan evaporation was 4.7 inches for June, what would the alfalfa reference crop equivalent be if computed by the combination equation?
>
> **Solution:** From Table 4.2, choose the method (E_{pan}), move to June (column 4), and select the E_{tr}/E_{pan} coefficient ($=0.92$). Then $E_{pan}(E_{tr}/E_{pan}) = 4.7 \times 0.92 = 4.3$ inches of ET from alfalfa.
>
> **Answer:** The equivalent alfalfa reference crop ET is 4.3" of water in June.

*E_{tp} Gages

Gages that estimate localized, real-time (as it is actually occurring) potential evapotranspiration are commercially available. A typical gage has a surface that simulates a well-watered leaf. The surface is exposed to the atmosphere where it will experience the same energy source and wind as the crop. Commercial literature claims excellent agreement between the potential evapotranspiration measured by the gage and potential evapotranspiration calculated by the Penman-Monteith equation for daily ET.

*Soil Water Depletion

Actual evapotranspiration from a crop can be determined by observing the change in soil water over a period of time. The average rate of E_t in mm d^{-1} between sampling dates (denoted Δt) can be calculated using the following equation (Jensen et al., 1990):

$$E_t = \frac{\Delta SM}{\Delta t} = \frac{\sum\limits_{i=1}^{n_{rz}} (\theta_1 - \theta_2)_i \Delta S_i + I - D}{\Delta t} \tag{4.3}$$

Table 4.2.   Ratios of E_{tr}/E_{pan} and $E_{tr}/E_{t(method)}$ Developed from Kimberly, Idaho, Data. (Source: *Evaporation and Irrigation Water Requirements*, Eds. M.E. Jensen, R.D. Burman, and R.G. Allen. Copyright 1990 by the American Society of Civil Engineers. Reproduced with permission of the American Society of Civil Engineers.)

Method	E_{tr}/E_{pan} and E_{tr}/E_t (method)						
	April	May	June	July	Aug	Sept	Oct
E_{pan}	0.75	0.86	0.92	0.94	0.92	0.92	0.91
E_{tr}	1.00	1.00	1.00	1.00	1.00	1.00	1.00
$E_{tp}(P)$	0.98	1.14	1.20	1.20	1.15	1.12	1.12
$E_{tr}(JH)$	1.33	1.25	1.18	1.08	1.08	1.25	1.37

E_{pan} - measured Class A pan evaporation; E_{tr} - reference ET, alfalfa, by combination equation using the 1982 Kimberly wind function (Wright, 1982); $E_{tp}(P)$ - potential ET, grass, modified Penman method using $W_f = (1.0+0.0062 u_2)$ with u_2 in km d^{-1} and vapor pressure deficit method 3 (Jensen et al., 1971); $E_{tr}(JH)$ - reference ET, alfalfa, Jensen-Haise method.

where

E_t	=	actual evapotranspiration in mm d^{-1}	

E_t = actual evapotranspiration in mm d^{-1}
ΔSM = change in soil water content
Δt = time between sampling dates
n_{rz} = number of soil layers in the effective root zone
ΔS_i = the thickness of each soil layer in mm
θ_1 = volumetric water content of soil layer i on the first sampling date (m^3 m^{-3})
θ_2 = volumetric water content of soil layer i on the second sampling date (m^3 m^{-3})
I = infiltration (rainfall - runoff) during Δt (mm)
D = drainage below the root zone during Δt (mm)

Determining accurate actual evapotranspiration by soil sampling is possible if you observe the following precautions: 1) take multiple soil moisture measurements in the field to obtain an average soil water content representative of the entire field; 2) only use this technique for sites where the depth to the water table is much greater than the depth of the root zone; and 3) use this technique for sampling periods where runoff and drainage out of the root zone are zero, because runoff and drainage values are not typically measured but are factors in the equation (Jensen et al., 1990).

*Water Balance

The water balance approach is generally used on large areas such as watersheds. The inflows and outflows are determined from streamflow and precipitation measurements and the difference between inflow and outflow over a relatively long period of time, such as a season, is a measure of evapotranspiration. The area in question must be confined so that other significant sources of inflow or outflow do not exist. The results are applicable only to the climate, cropping, and irrigation conditions similar to those in the study area.

*Lysimeters

Lysimeters are devices that allow an area of a field to be isolated from the rest of the field, yet experience similar conditions to that of the growing crop. Typically, the lysimeter is a cylinder inserted into the soil or a tank filled with soil and placed in a field. Crops are grown on the surface of the lysimeter in order to approximate the conditions of the field. Since the lysimeter is isolated from the remainder of the field, measurement of the individual components of the water balance is possible so an estimate of actual evapotranspiration can be calculated. Reference crop ET is typically determined using measurements made with lysimeters planted with alfalfa or grass. Similarly, crop coefficients can be determined by planting the crop of interest in the lysimeter and measuring ET, then relating this back to the reference crop ET.

Lysimeters can be grouped into weighing and non-weighing types. The weighing lysimeters allow changes in soil water to be determined either by weighing the entire unit with a mechanical scale, counterbalanced scale and load cell, or by supporting the lysimeter hydraulically (Jensen et al., 1990). A world-renowned example of weighting lysimeters are the large lysimeters located in Coshocton, OH.

Non-weighing lysimeters are either of the percolation type or of the constant water table type. In the percolation type, the changes in soil water are determined either by sampling or using an indirect method such as a neutron probe. The drainage or "percolation" out the bottom of the root zone is also measured. With knowledge of the rainfall, the actual transpiration for the lysimeter can be determined. The constant water table lysimeters are useful in locations where a high water table exists. With this type of lysimeter, the water table is maintained at a constant level inside the lysimeter and the water added to maintain water level is a measure of actual ET.

*Energy Balance and Mass Transfer

The energy balance and mass transfer methods of determining actual ET require measuring the average gradient of water vapor above the canopy. The instrumentation needed and the technical procedures involved typically limit these methods to research applications. Additional details can be found in Jensen et al. (1990).

4.4. WEATHER DATA SOURCES AND PREPARATION

Many empirical and physically-based methods have been developed for estimating evaporation and evapotranspiration from measured climatic data. For example, evaporation from a lake or reservoir can be estimated if wind speed, relative humidity, and temperature are known. Table 4.3 summarizes the weather data requirements for the ET estimation methods which will be discussed in the next section.

It is often difficult to find weather data for the area of interest. Many irrigation projects have weather stations from which data may be obtained, and some states have weather station networks which may be a source of data. If you are interested in data for an area in the United States, monthly summaries of each state's weather

Table 4.3.   Minimum Climatic Information Needs of ET Estimation Methods.

Method	T[1]	RH[2] or e_d[3]	Lat[4]	Elev[5]	R_s[6]	u[7]
Penman	x	x		x	x	x
Jensen-Haise	x			x	x	
SCS Blaney-Criddle	x		x			
Thornthwaite	x					

[1]Air temperature; [2]Relative humidity; [3]Actual vapor pressure of the air; [4]Latitude; [5]Elevation; [6]Solar radiation; [7]Wind speed.

are available from the National Climatic Data Center in Ashville, North Carolina, which is associated with the National Oceanic and Atmospheric Administration (NOAA). The summaries include daily precipitation and daily maximum and minimum temperatures from many observation points in each state. Daily pan evaporation and wind speed are recorded for some stations. The summaries also provide latitude, longitude, and elevation for each recording station. Many libraries subscribe to these climatological data summaries.

Once the data are located, they need to be prepared for calculations using methods identical to those used by the developers of the ET estimation methods. The most obvious preparatory task would be to ensure data are in the proper units for the equations. In addition, there are multiple ways some parameters could be calculated; the method used must be matched to the ET estimation method. Some of these methods are discussed below.

Saturation Vapor Pressure

Saturation vapor pressure, e_s, is a property of air and is a function only of temperature. We know from everyday experience that warmer air can hold more water vapor than cooler air. That is why water condenses out of air as it cools, for example on the side of your iced-tea glass in the summer. The glass is cooler than the surrounding air. When the warm, humid air comes in contact with your glass the glass cools the air. Cooler air holds less water than warm air, so water condenses out of the air onto the side of your glass.

The relationship between saturation vapor pressure and temperature is shown in Figure 4.6. Temperature is shown along the horizontal axis, and saturation vapor pressure in mm of Hg, inches of Hg, and kPa (units of pressure) is on the vertical axis. Find the temperature of interest, move vertically up to the curved line, then move horizontally until you arrive at the axis showing the saturation vapor pressure.

Alternatively, Equation 4.4 can be used to compute saturation vapor pressure in kPa if temperature is in degrees Celsius. This equation is useful if you would like to write a computer program to compute some of these values. The equation is valid for temperatures ranging from 0 to 50°C.

$$e_s = \exp\left[\frac{16.78\ T-116.9}{T+237.3}\right] \tag{4.4}$$

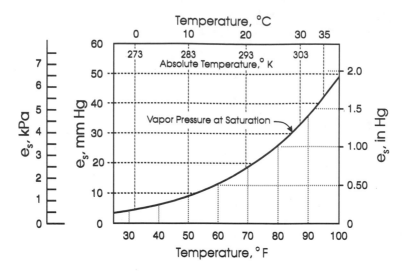

Figure 4.6.   Saturated vapor pressure as a function of temperature (Source: Schwab et al., 1992).

Example 4.2: Find e_s in kPa (using both Figure 4.6 and Equation 4.4) if the air temperature is 70°F.

Solution: From Figure 4.6, at 70°F, the e_s = 2.5 kPa. Using Equation 4.4 and the temperature in Celsius (70°F = 21.11°C), e_s = 2.5 kPa.

Answer: The saturated vapor pressure is 2.5 kPa.

Actual Vapor Pressure

Actual vapor pressure, e_d, is the vapor pressure of the air. Unlike saturation vapor pressure, actual vapor pressure cannot be determined simply by knowing the temperature of the air. To determine e_d we need to know the air temperature and either the relative humidity or the dewpoint temperature of the air. The actual vapor pressure is dependent on the amount of water currently in the air, whereas the saturation vapor pressure is based on the maximum amount of water that could be absorbed by the air. We need to know relative humidity or dewpoint temperature because they are measures of the amount of water actually in the air.

$$e_d = e_s \times \frac{RH}{100} \qquad (4.5)$$

where: e_d = actual vapor pressure
e_s = saturation vapor pressure
RH   = relative humidity in percent

Vapor Pressure Deficit

Many methods exist for calculating the vapor pressure deficit (e_s-e_d) (see Jensen et al., 1990). Three methods for calculating vapor pressure deficit are shown here.

Method 1.   Saturation vapor pressure at mean temperature minus saturation vapor pressure at dewpoint temperature, which can be written as:

$$(e_s - e_d) = e_{s_{(T_{avg})}} - e_{s_{(T_d)}} \qquad (4.6)$$

where e_d = actual vapor pressure
　　　　e_s = saturation vapor pressure
　　　　T_{avg} = mean temperature for time period of interest
　　　　T_d = mean dewpoint temperature for time period of interest

Example 4.3: Compute the vapor pressure deficit for June 17, 1993 at Piketon, Ohio using Method 1. The data are shown in Table 4.4.

Solution: From Table 4.4, T_{avg}=22.05°C, and e_s at T_{avg} is computed as follows (Eq. 4.4):

$$e_s = \exp[\frac{16.78\ (22.05) - 116.9}{22.05 + 237.3}] = \exp\left[\frac{253.1}{259.35}\right] = 2.65\ kPa$$

Next, we need the saturation vapor pressure at the dew-point temperature. If we knew the dewpoint temperature we would substitute it into Equation 4.4 and obtain the saturated vapor pressure at T_d. However, we do not know T_d, but $e_{s_{(T_d)}}$ was recorded because $e_{s_{(T_d)}} = e_d$. From Table 4.4, if we average e_d we obtain:

$$e_{d_{avg}} = \frac{\sum_{i=1}^{i=24} e_d}{24} = 1.98\ kPa$$

then,

$$e_s - e_d = 2.65 - 1.98 = 0.67\ kPa$$

Answer: The vapor pressure deficit is 0.67 kPa.

Table 4.4.   Piketon, Ohio weather data for June 17, 1993.

Hour	R_s (kJ/m^{2}h)	T (°C)	e_s (kPa)	e_d (kPa)	$(e_s - e_d)$ (kPa)	RH (%)
1:00 a.m.	0.087	16.08	1.827	1.765	0.062	96.6
2:00	0.234	15.28	1.736	1.708	0.028	98.4
3:00	0.362	14.68	1.670	1.652	0.018	99.0
4:00	0.255	14.25	1.624	1.613	0.011	99.3
5:00	0.281	14.01	1.598	1.593	0.005	99.7
6:00	1.213	13.13	1.510	1.510	0.000	100
7:00	64.25	13.16	1.513	1.513	0.000	100
8:00	368.7	14.38	1.638	1.638	0.000	100
9:00	720.0	16.81	1.916	1.810	0.106	94.7
10:00	1154.0	20.68	2.442	1.988	0.454	81.6
11:00	1665.0	23.68	2.931	2.099	0.832	71.8
12:00 noon	2164.0	26.70	3.505	2.199	1.306	62.9
1:00 p.m.	2463.0	28.26	3.837	1.925	1.912	50.2
2:00	2568.0	29.19	4.051	1.996	2.055	49.3
3:00	2439.0	30.26	4.306	1.995	2.311	46.3
4:00	2231.0	30.78	4.437	2.026	2.411	45.7
5:00	1776.0	30.96	4.483	2.103	2.380	46.9
6:00	1269.0	30.92	4.472	2.139	2.333	47.8
7:00	869.0	30.43	4.349	2.193	2.156	50.4
8:00	408.3	29.56	4.139	2.424	1.715	58.7
9:00	92.70	27.08	3.588	2.496	1.092	69.8
10:00	2.035	23.60	2.917	2.42	0.497	83.2
11:00	0.020	21.76	2.604	2.327	0.277	89.4
12:00 midnight	0.013	20.78	2.453	2.292	0.161	93.5

T_{max} = 31.4°C T_{min} = 12.7°C T_{avg} = 22.05°C

Method 2.   The vapor pressure deficit can be estimated from the saturation vapor pressure at the mean temperature times the quantity one, minus the relative humidity expressed as a proportion or:

$$(e_s - e_d) = e_{s(T_{avg})} \left(1 - \frac{RH}{100} \right) \tag{4.7}$$

Relative humidity as a percentage is by definition:

$$RH = \frac{e_d}{e_s} \times 100 \tag{4.8}$$

which, when solved for e_d, yields:

$$e_d = e_s \times \frac{RH}{100} \tag{4.9}$$

Vapor pressure deficit is:

$$(e_s - e_d) \quad \text{but} \quad e_{s_{(T_d)}} = e_d$$

Therefore, vapor pressure deficit equals:

$$e_{s_{(T_{avg})}} - e_{s_{(T_{avg})}} \frac{RH}{100} \tag{4.10}$$

So substituting for $e_{s_{(T_d)}}$:

$$= e_{s_{(T_{avg})}} (1 - \frac{RH}{100}) \tag{4.11}$$

Example 4.4: Compute the vapor pressure deficit for June 17, 1993 at Pike-ton, Ohio using Method 2. The data are shown in Table 4.4.

Solution:

$$RH_{avg} = \frac{\sum_{i=1}^{24} RH_i}{24} = =76.5\%$$

$$T_{avg} = \frac{T_{min} + T_{max}}{2} = 22.05°C$$

e_s at T_{avg} (from Figure 4.6) = 2.65 kPa

Using Equation 4.7, $e_s - e_d = 2.65 \left[1 - \frac{76.5}{100} \right] = 0.62$ kPa.

Answer: The vapor pressure deficit equals 0.62 kPa.

Method 3.   The mean of saturation vapor pressure at the maximum and minimum temperatures minus the saturation vapor pressure at the dewpoint temperature determined early in the day, typically at 8 a.m.

$$(e_s - e_d) = \frac{e_{s(T_{max})} + e_{s(T_{min})}}{2} - e_{s(T_{d\ 8am})} \tag{4.12}$$

Example 4.5: Compute the vapor pressure deficit for June 17, 1993 at Piketon, Ohio using Method 3. The data are shown in Table 4.4.

Solution: From Table 4.4, $T_{max} = 31.4°C$ and $T_{min} = 12.7°C$. Calculating $e_{s(T\ max)}$ using Equation 4.4:

$$e_{s(T_{max})} = \exp\left[\frac{16.78(31.4) - 116.9}{31.4 + 237.3}\right] = \exp\left[\frac{409.82}{268.69}\right] = 4.6\ kPa$$

Calculating e_s at T_{min} using the same equation:

$$e_{s(T_{min})} = \exp\left[\frac{16.78(12.7) - 116.9}{12.7 + 237.3}\right] = \exp\left[\frac{96.21}{250.0}\right] = 1.47\ kPa$$

Then averaging these vapor pressures:

$$\frac{e_{(T_{max})} + e_{(T_{min})}}{2} = \frac{4.60 + 1.47}{2} = 3.04\ kPa$$

From Table 4.4 we can find e_s at dewpoint temperature at 8 a.m., because it is the same quantity as e_d at 8 a.m. (1.638 kPa). Subtracting the two vapor pressures (3.03 and 1.638 kPa) we find the vapor pressure deficit calculated by Method 3 to be 1.39 kPa.

Answer: The vapor pressure deficit is 1.39 kPa.

The average of the 24 hourly vapor pressure deficits is 0.922 kPa and is the most correct estimate of daily vapor pressure deficit. You will note that Method 1 underpredicted the correct value by 32.4%, Method 2 underpredicted by 42.6%, and Method 3 overpredicted by 50.8%. The best method would be to average hourly values of vapor pressure deficit to obtain an average for the day. However, it is unlikely you will have the necessary weather data available to do this. The ET estimation methods generally indicate which method they prefer to use to calculate $(e_s - e_d)$.

Mean Temperature

Several methods are available for computing mean or average temperature for a given time period. Two of these are:

Method 1.   Average the individual mean temperatures from the next smallest time scale. For example, if you wanted a daily average of T you could average the hourly temperatures as follows:

From Table 4.4: Sum Column 3 (T) = 536.42°C. Then divide by 24 hours to obtain an average; the average daily temperature = 22.35°C.

Method 2.   Average the maximum and minimum temperatures for the period of interest; which in this example is 1 day. The daily maximum temperature was 31.4°C and the daily minimum temperature was 12.7°C. Averaging these values = 22.05°C.

Similarly, if you were interested in a monthly mean temperature, you could average the mean daily temperatures or average the maximum and minimum temperatures for the month. You will notice in our example that the two methods did not result in the same number; they differed by 1.4%. The mean temperatures needed for Method 1 are generally not available, whereas the maximum and minimum temperatures needed for Method 2 are typically available. Method 2 is the method generally used to calculate mean temperature.

*Solar Radiation

Some ET estimation methods require either R_s (solar radiation received on a horizontal plane at the earth's surface) or R_n (net radiation). R_s would be measured at a weather station, and is more likely to be recorded than R_n, which requires measuring both incoming and outgoing solar radiation.

If R_s is measured, such as in the data shown in Table 4.4, you will need to determine whether the data are being presented as instantaneous readings or as cumulative solar radiation. The units associated with the values will help you to determine this. If the units are per hour, per second, etc., then they are the average solar radiation for that time period, and are not cumulative.

If neither R_s nor R_n have been measured for a site, and these data are unavailable for a location near your area of interest, you can estimate R_s (and subsequently R_n) if you know the ratio of actual to possible sunshine and the latitude of the site. The reader is referred to Jensen et al. (1990) for a detailed description of this procedure.

The most likely scenario is that R_s has been measured and you need R_n to use Penman's method or a similar method. It is possible to estimate R_n from R_s since R_n is the net short-wave minus the long-wave components of the radiation.

$$R_n = (1 - \alpha)R_s\downarrow - R_b\uparrow \qquad (4.13)$$

where R_b is the net outgoing thermal radiation in MJ/m²d and α is the albedo or short-wave reflectance, which is dimensionless. The arrows in Equation 4.13 serve as reminders that R_s is incoming and R_b is outgoing. The short-wave reflectance or albedo, $\propto$, is typically set equal to 0.23 for most green field crops with a full cover (Jensen et al., 1990). Since we know R_s and $\propto$, R_b is all that is needed. Equation 4.14 can be used to calculate this value, in units of MJ/(m²d).

$$R_b = [a \frac{R_s}{R_{so}} + b] \ R_{bo} \tag{4.14}$$

The coefficients a and b are determined for the climate of the area of interest. For humid areas, a=1.0 and b=0; for arid areas, a=1.2 and b= -0.2; and for semi-humid areas a=1.1 and b= -0.1. R_{so} is the solar radiation on a cloudless day and can be obtained from Table 4.5 (in units of MJ m⁻² d⁻¹) based on the site's latitude. R_{bo} can be computed from Equation 4.15.

$$R_{bo} = \epsilon \sigma T^4 \tag{4.15}$$

where the Stefan-Boltzmann constant, σ, is 4.903 x 10⁻⁹ MJ m⁻² d⁻¹ K⁻⁴, and T is the mean temperature for the period of interest in degrees Kelvin (273 + T in °C). The term ϵ is the net emissivity and is calculated using the Idso-Jackson equation (Eq. 4.16) with T in degrees Kelvin.

$$\epsilon = -0.02 + 0.261 \exp[-7.77 \times 10^{-4} (273 - T)^2] \tag{4.16}$$

Example 4.6: Calculate R_n from the R_s data given in Table 4.4.

Solution:

Step 1.   The first step is to calculate ϵ, using Equation 4.16.

$\epsilon = -0.02 + 0.261 \exp[-7.77 \times 10^{-4} (273 - 295.05)^2] = 0.159$

Step 2.   Calculate R_{bo} using Equation 4.15.

$R_{bo} = \epsilon \sigma T^4 = 0.159 (4.903 \times 10^{-9}) (295.0)^4 = 5.90$ MJ/(m²d)

Step 3.   Calculate R_b from Equation 4.14 with a=1.0 and b=0 (Piketon, Ohio is a humid area). For use in Equation 4.14, R_s (Table 4.4) should be in units of MJ m⁻² d⁻¹. Totaling the second column gives R_s = 20,256.45 kJ m⁻² d⁻¹ = 20.26 MJ m⁻² d⁻¹. R_{so} is obtained from Table 4.5, knowing the latitude of the place of interest (Piketon, Ohio = 38°N), and the month of interest (in our example, June); R_{so} = 33.49 MJ m⁻² d⁻¹.

$$R_b = [1.0 \frac{20.26}{33.49} + 0] 5.91 = 3.58 \, MJ/(m^2 d)$$

Table 4.5.   Mean solar radiation (R_{so}) for cloudless skies. (Source: *Evaporation and Irrigation Water Requirements*, Eds. M.E. Jensen, R.D. Burman, and R.G. Allen. Copyright 1990 by the American Society of Civil Engineers. Reproduced with permission of the American Society of Civil Engineers.)

Mean Solar Radiation per Month for Cloudless Skies
$(MJ\ m^{-2}\ d^{-1})$

Latitude	Jan	Feb	Mar	April	May	June	July	Aug	Sept	Oct	Nov	Dec
60N	2.51	5.99	13.82	22.32	29.01	31.95	29.85	23.32	15.78	8.50	3.64	1.55
55N	4.31	8.67	16.33	24.16	29.85	32.66	30.56	25.00	18.00	10.89	5.57	3.22
50N	6.70	11.43	18.55	25.83	30.98	33.08	31.53	26.67	20.10	13.52	8.08	5.44
45N	9.34	14.36	20.64	27.21	31.53	33.37	32.36	28.05	22.06	16.04	10.89	8.25
40N	12.27	17.04	22.90	28.34	32.11	33.49	32.66	29.18	23.74	18.42	13.52	10.76
35N	14.95	19.55	24.58	29.31	32.11	33.49	32.95	30.14	25.25	20.52	15.91	13.52
30N	17.46	21.65	25.96	29.85	32.11	33.20	32.66	30.44	26.67	22.48	18.30	16.04
25N	19.68	23.45	27.21	30.14	32.11	32.66	32.24	30.44	27.63	24.28	20.39	18.30
20N	21.65	25.00	28.18	30.14	31.40	31.82	31.53	30.14	28.47	25.83	22.48	20.52
15N	23.57	26.50	29.01	29.85	30.56	30.69	30.56	29.60	29.18	26.92	24.28	22.48
10N	25.25	27.63	29.43	29.60	29.60	29.31	29.43	28.76	29.60	28.05	25.83	24.41
5N	26.92	28.47	29.85	29.31	28.18	27.76	27.93	27.93	29.73	28.76	27.21	26.25
0E	28.18	29.18	30.02	28.47	26.92	26.25	26.67	27.76	29.60	29.60	28.47	26.80
5S	28.05	29.85	29.85	27.76	25.54	24.58	25.00	26.67	28.09	29.85	30.44	29.31
10S	30.69	30.44	29.43	26.80	24.70	22.73	22.73	25.00	27.55	29.85	30.44	30.69
15S	31.53	30.69	28.76	25.54	22.32	20.81	21.48	23.86	26.59	29.73	31.28	31.95
20S	32.36	30.69	27.93	23.99	20.52	18.71	19.26	22.06	25.54	29.31	31.53	32.95
25S	32.95	30.69	27.09	22.32	18.13	16.75	17.58	20.64	24.33	28.76	32.11	33.62
30S	33.37	30.44	25.96	20.81	16.62	14.78	15.49	18.84	22.94	28.05	32.11	34.33
35S	33.49	29.73	24.58	18.97	14.53	12.56	13.40	16.87	21.48	27.21	32.11	34.88
40S	33.49	28.76	22.90	17.04	12.14	10.17	11.30	14.78	19.30	26.08	31.82	35.17
45S	33.49	27.76	21.23	14.95	9.92	7.66	8.79	12.69	18.09	24.70	31.28	35.17
50S	32.95	26.38	19.26	12.85	7.54	5.32	6.41	10.59	16.20	23.15	30.44	34.88
55S	32.36	24.83	17.17	10.47	5.32	3.22	4.19	8.25	13.90	21.48	29.60	34.33
60S	31.53	23.15	15.07	7.83	3.35	1.38	2.22	6.15	11.47	19.68	29.31	34.04

Note: August values in the southern hemisphere were corrected to obtain a smooth transition in monthly values. Also, 30-day months were assumed because when actual days per month were used, a smooth transition between January, February, and March did not occur.

Step 4.   Then R_n is calculated using Equation 4.13, assuming the generally used value for albedo of 0.23.

$$R_n = (1 - 0.23)20.26 - 3.58 = 12.0 \ MJ/(m^2d)$$

Answer: The net radiation on June 17, 1993 at Piketon, Ohio was 12.0 MJ/m^2d.

Extrapolating Wind Speed

Wind is typically slower at the ground surface and the speed increases with height. Most methods for estimating ET which require wind speed specify at what height the wind speed should be recorded. However, in practice the data have sometimes been recorded at other heights. To estimate the wind speed, u_2, at height z_2, knowing the wind speed u_1 at height z_1, Equation 4.17 can be used (Allen et al., 1989).

$$\frac{u_1}{u_2} = \frac{\ln[z_1 - 0.67h_c] - \ln[0.123h_c]}{\ln[z_2 - 0.67h_c] - \ln[0.123h_c]} \tag{4.17}$$

where h_c is the height of the vegetation, $0.67h_c$ is the height where the wind velocity approaches zero (known as the roughness height), and $0.123h_c$ is the surface roughness. The variables h_c, z_1, and z_2 are expected to have the same units, then u_2 will have identical units to u_1.

Example 4.7: Find the wind speed at 25 feet if the wind speed at 6 feet is 3 mi/hr. The vegetation is corn which is 6 feet tall.

Solution: Using Equation 4.17, with $h_c=6$ feet,

$$u_2 = 3\frac{\ln[25 - (0.67 \times 6)] - \ln[0.123 \times 6]}{\ln[6 \ (-0.67 \times 6)] - \ln[0.123 \times 6]} = 3 \times \frac{3.35}{0.987} = 10.2 \ \frac{mi}{hr}$$

Answer: The windspeed at 25 feet was estimated to be 10.2 miles/hour.

4.5. ESTIMATING EVAPORATION AND EVAPOTRANSPIRATION

Evaporation from Open Water

Monthly evaporation from lakes or reservoirs can be computed using the empirical formula developed by Meyer (Harrold et al., 1986).

$$E = C(e_s - e_d)(1 + \frac{u_{25}}{10}) \qquad (4.18)$$

where E   = evaporation in inches/month
       e_s = saturation vapor pressure (inches of Hg) of air at the water temperature 1 foot deep
       e_d = actual vapor pressure (inches of Hg) of air = $e_{s\ (air\ T)}$ x RH
       u_{25} = average wind velocity (mi/hr) at a height of 25 feet above the lake or surrounding land areas
       C   = coefficient which equals 11 for small lakes and reservoirs and 15 for shallow ponds

Example 4.8: Compute the monthly evaporation from a reservoir which had a monthly average water temperature of 60°F, measured 1 foot below the surface of the water, and a monthly average wind speed of 3 mi/hr, measured at 25 feet above the reservoir. The monthly average air temperature was 70°F and the monthly average relative humidity was 40%.

Solution: Determine the saturated vapor pressure of air at the water temperature 1 foot deep (60°F) = 0.51 inches Hg (Figure 4.6).

Next determine e_d for the air temperature (70°F) by first finding the saturated vapor pressure of air at 70°F (Figure 4.6 shows e_s @ 70°F =0.73 inches Hg). Then e_d =0.4(0.73)=0.292 inches Hg from Equation 4.5.

Since we are interested in a reservoir, C=11. Using Equation 4.18,

$$E = 11(0.51 - 0.292)(1 + \frac{3}{10}) = 3.1 \frac{inches}{month}$$

Answer: The evaporation was 3.1 inches/month.

Estimating Evapotranspiration

Figure 4.7 summarizes the decision process for selecting a method for estimating ET. It is important to know what weather data are available for the site of interest before selecting an estimation method, since each method had different climatic information requirements (Table 4.3).

The majority of the ET estimating methods were developed to predict evapotranspiration from a well-watered short green crop (typically alfalfa or grass). Of the methods described in Figure 4.7, only the SCS Blaney-Criddle method was specifically developed for estimating seasonal actual ET. The SCS Blaney-Criddle method can also be used to estimate actual monthly ET, provided the method has been locally calibrated.

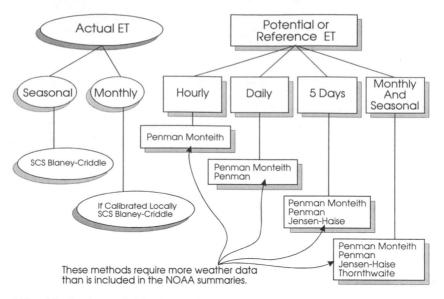

Figure 4.7. Methods available for estimating ET using climatic data.

An alternative approach to estimating actual evapotranspiration is to measure or estimate reference evapotranspiration, then adjust that value based on empirical coefficients for converting to actual ET. The reference ET values represent climatic demand, but environmental conditions such as soil water conditions and the crop canopy status need to be considered to obtain actual ET. Crop coefficients that relate actual ET to potential or reference crop ET have been derived from experimental data for particular crops, growth stages, and soil water conditions (Jensen et al., 1990). First we will explore available methods for estimating reference crop ET, then we will discuss modifying these estimates with appropriate crop coefficients to obtain estimates of actual evapotranspiration.

Many methods are available to estimate reference crop evapotranspiration. Two critical questions you need to answer are: what time scale do you need to estimate (hourly, daily, 5 day, monthly, or seasonally); and how much weather data are available? If you need to estimate hourly ET, then the Penman-Monteith method would be most applicable; however, this method requires a large amount of weather data. Daily estimations of ET also require large amounts of weather data. Two methods are shown in Figure 4.6 for daily estimates; the Penman-Monteith (estimating hourly values, then summing over 24 hours) or the Penman method. Penman's formula is one of the classics in micrometeorology and hydrology because it incorporated for the first time, a number of fundamental principles relating the evaporation from a wet surface to its equilibrium temperature (Monteith, 1981).

The next time scale shown in Figure 4.7 is 5 days, which is convenient for irrigation scheduling. Once again, if weather data are available you could compute daily ET with the Penman method or hourly ET with the Penman-Monteith method, and sum the values over a 5-day period. Typically, a 5 day time period is chosen because hourly or daily data are not available. Five-day averages can be used in the Penman and Penman-Monteith methods; however you will lose accuracy when compared to using these methods with daily or hourly weather data. The Jensen-

Haise method was specifically developed for a 5-day time period, and requires elevation and average solar radiation (Table 4.3).

Monthly estimates of ET could be obtained by all the previous methods mentioned, either by using shorter time periods and summing over a month or using monthly averages in the individual methods. The SCS Blaney-Criddle method (if calibrated locally) and the Thornthwaite method will predict evapotranspiration on a monthly basis. Less weather data are required for the Thornthwaite method, which makes the method popular, but the estimate is less accurate because of the simplified inputs (Amatya et al., 1992). Seasonal ET can be computed by all the methods mentioned previously, either summing the estimates over the season or using seasonal values for the variables in the equations.

SCS Blaney-Criddle Method

The SCS Blaney-Criddle method estimates seasonal (either growing season or irrigation season) actual evapotranspiration. This method is the standard method recommended by the USDA-Soil Conservation Service (SCS), is well known in the Western U.S., and is used extensively throughout the world (Jensen et al., 1990). The original relationships were developed around 1945 and were intended for seasonal estimates. This method may be used to obtain monthly estimates if monthly crop coefficients are locally available.

We know that the amount of evapotranspiration is related to how much energy is available for vaporizing water. The energy is provided by solar radiation, but measuring solar radiation requires instrumentation not available at most field sites. Blaney and Criddle assumed that mean monthly air temperature and monthly percentage of annual daytime hours could be used instead of solar radiation to provide an estimate of the energy received by the crop. They defined a monthly consumptive use factor, f, as:

$$f = \frac{tp}{100} \tag{4.19}$$

where t is the mean monthly air temperature in °F (avg. of daily maximum and minimum) and p is the mean monthly percentage of annual daytime hours. The 100 in the divisor converts p from a percentage to a fraction. Table 4.6 summarizes p for latitudes 0–64°N of the equator. Once f is computed for each month, then the actual ET for the season is computed by Equation 4.20:

$$U = K \sum_{i=1}^{n} f_i \tag{4.20}$$

where K is the seasonal consumptive use coefficient for a crop with a normal growing season (Table 4.7), n is the number of months in the season, and U is the seasonal consumptive use in inches/season.

Example 4.9: Estimate the evapotranspiration from alfalfa grown in Hoytville, Ohio for the 1992 growing period (June, July, August, and September) using the SCS Blaney-Criddle method.

Solution: Step 1: Obtain **p** (in %) for Hoytville, Ohio from Table 4.6. To use this table we need to know the latitude for Hoytville (41°: 13', obtained from NOAA Climatological Data, Ohio, 1992). From Table 4.6:

Latitude°	June%	July%	August%	September%
42°	10.24	10.35	9.62	8.40
40°	10.09	10.22	9.55	8.50

We need to interpolate to obtain the required values of **p**. Interpolating for June:

$$\frac{42-40}{42-41\frac{13}{60}} = \frac{10.24-10.09}{10.24-x} \text{ solving for } x = 10.18$$

The other three values were calculated by similar interpolations.

Latitude°	June%	July%	August%	September%
42°	10.24	10.35	9.62	8.40
41°:13'	10.18	10.30	9.59	8.396
40°	10.09	10.22	9.55	8.39

Step 2. Next, we need average monthly temperatures for the 4 months. These were also obtained from the NOAA summaries.

June	July	August	September
64.2°F	70.3°F	65.7°F	61.6°F

Step 3. Compute f from Equation 4.19:

$$\sum_{i=1}^{4} \frac{tp}{100} = \frac{64.2(10.18)}{100} + \frac{70.3(10.30)}{100} + \frac{65.7(9.59)}{100} + \frac{61.6(8.396)}{100} = 25.25$$

Step 4. Find K from Table 4.7 to adjust for crop. K=0.80 (choose the lower value since Hoytville is a humid area–see footnote of table).

Step 5. The estimated ET, U=KF = 0.8 (25.25")= 20.20 inches/season.

Answer: The predicted ET equals 20.2 inches/season.

Table 4.6. Monthly percentage of daytime hours, p, of the year for latitudes 0 to 64° North of the equator. (Source: *Evaporation and Irrigation Water Requirements*, Eds. M.E. Jensen, R.D. Burman, and R.G. Allen. Copyright 1990 by the American Society of Civil Engineers. Reproduced with permission of the American Society of Civil Engineers.)

Latitude (° North)	Jan	Feb	Mar	Apr	May	Jun	Jul	Aug	Sep	Oct	Nov	Dec
64	3.81	5.27	8.00	9.92	12.50	13.63	13.26	11.08	8.56	6.63	4.32	3.02
62	4.31	5.49	8.07	9.80	12.11	12.92	12.73	10.87	8.55	6.80	4.70	3.65
60	4.70	5.67	8.11	9.69	11.78	12.41	12.31	10.68	8.54	6.95	5.02	4.14
58	5.02	5.84	8.14	9.59	11.50	12.00	11.96	10.52	8.53	7.06	5.30	4.54
56	5.31	5.98	8.17	9.48	11.26	11.68	11.67	10.36	8.52	7.18	5.52	4.87
54	5.56	6.10	8.19	9.40	11.04	11.39	11.42	10.22	8.50	7.28	5.74	5.16
52	5.79	6.22	8.21	9.32	10.85	11.14	11.19	10.10	8.48	7.36	5.92	5.42
50	5.99	6.32	8.24	9.24	10.68	10.92	10.99	9.99	8.46	7.44	6.08	5.65
48	6.17	6.41	8.26	9.17	10.52	10.72	10.81	9.89	8.45	7.51	6.24	5.85
46	6.33	6.50	8.28	9.11	10.38	10.53	10.65	9.79	8.43	7.58	6.37	6.05
44	6.48	6.57	8.29	9.05	10.25	10.39	10.49	9.71	8.41	7.64	6.50	6.22
42	6.61	6.65	8.30	8.99	10.13	10.24	10.35	9.62	8.40	7.70	6.62	6.39
40	6.75	6.72	8.32	8.93	10.01	10.09	10.22	9.55	8.39	7.75	6.73	6.54
38	6.87	6.79	8.33	8.89	9.90	9.96	10.11	9.47	8.37	7.80	6.83	6.68
36	6.98	6.85	8.35	8.85	9.80	9.82	9.99	9.41	8.36	7.85	6.93	6.81
34	7.10	6.91	8.35	8.80	9.71	9.71	9.88	9.34	8.35	7.90	7.02	6.93
32	7.20	6.97	8.36	8.75	9.62	9.60	9.77	9.28	8.34	7.95	7.11	7.05
30	7.31	7.02	8.37	8.71	9.54	9.49	9.67	9.21	8.33	7.99	7.20	7.16
28	7.40	7.07	8.37	8.67	9.46	9.39	9.58	9.17	8.32	8.02	7.28	7.27
26	7.49	7.12	8.38	8.64	9.37	9.29	9.49	9.11	8.32	8.06	7.36	7.37
24	7.58	7.16	8.39	8.60	9.30	9.19	9.40	9.06	8.31	8.10	7.44	7.47
22	7.67	7.21	8.40	8.56	9.22	9.11	9.32	9.01	8.30	8.13	7.51	7.56
20	7.75	7.26	8.41	8.53	9.15	9.02	9.24	8.95	8.29	8.17	7.58	7.65
18	7.83	7.31	8.41	8.50	9.08	8.93	9.16	8.90	8.29	8.20	7.65	7.74
16	7.91	7.35	8.42	8.47	9.01	8.85	9.08	8.85	8.28	8.23	7.72	7.83
14	7.98	7.39	8.43	8.43	8.94	8.77	9.00	8.80	8.27	8.27	7.76	7.93
12	8.06	7.43	8.44	8.40	8.87	8.69	8.92	8.76	8.26	8.31	7.85	8.01
10	8.14	7.47	8.45	8.37	8.81	8.61	8.85	8.71	8.25	8.34	7.91	8.09
8	8.21	7.51	8.45	8.34	8.74	8.53	8.78	8.66	8.24	8.37	7.98	8.18
6	8.28	7.55	8.46	8.31	8.68	8.45	8.71	8.62	8.23	8.40	8.04	8.26
4	8.36	7.59	8.47	8.28	8.62	8.37	8.64	8.57	8.22	8.43	8.10	8.34
2	8.43	7.63	8.49	8.25	8.55	8.29	8.57	8.53	8.21	8.46	8.16	8.42
0	8.50	7.67	8.49	8.22	8.49	8.22	8.50	8.49	8.21	8.49	8.22	8.50

Table 4.7. Seasonal consumptive use coefficients, K, for irrigated crops in Western United States. (Source: *Evaporation and Irrigation Water Requirements*, Eds. M.E. Jensen, R.D. Burman, and R.G. Allen. Copyright 1990 by the American Society of Civil Engineers. Reproduced with permission of the American Society of Civil Engineers.)

Crop	Length of normal growing season**	Coefficient (K)***
Alfalfa	Between frosts	0.80 to 0.90
Bananas	Full Year	0.80 to 1.00
Beans	3 months	0.60 to 0.70
Cocoa	Full year	0.70 to 0.80
Coffee	Full year	0.70 to 0.80
Corn (maize)	4 months	0.75 to 0.85
Cotton	7 months	0.60 to 0.70
Dates	Full year	0.65 to 0.80
Flax	7 to 8 months	0.70 to 0.80
Grains, small	3 months	0.75 to 0.85
Grains, sorghum	4 to 5 months	0.70 to 0.80
Oil seeds	3 to 5 months	0.65 to 0.75
Orchard crops		
•Avocado	Full year	0.50 to 0.55
•Grapefruit	Full year	0.55 to 0.65
•Orange and lemon	Full year	0.45 to 0.55
•Walnuts	Between frosts	0.60 to 0.70
•Deciduous	Between frosts	0.60 to 0.70
Pasture crops		
•Grass	Between frosts	0.75 to 0.85
•Ladino white clover	Between frosts	0.80 to 0.85
Potatoes	3 to 5 months	0.65 to 0.75
Rice	3 to 5 months	1.00 to 1.10
Soybeans	140 days	0.65 to 0.70
Sugar beets	6 months	0.65 to 0.75
Sugar cane	Full year	0.80 to 0.90
Tobacco	4 months	0.70 to 0.80
Tomatoes	4 months	0.65 to 0.70
Vineyard	5 to 7 months	0.50 to 0.60

* From USDA (1970); ** Length of Season depends largely on variety and time of year when the crop is grown. Annual crops grown during the winter period may take much longer than if grown in the summer; *** The lower values of K for use in the Blaney-Criddle formula are for the more humid areas, the higher values are for the more arid climates.

Out of curiosity, let's compare that prediction with the measured pan evaporation for those four months. The pan evaporation recorded at Hoytville was 24.22 inches/season, but we need to adjust the pan evaporation by a coefficient to estimate the ET from alfalfa. From Table 4.2, to adjust E_{pan} to E_{tr} (alfalfa), multiply by the coefficient 0.92. The estimated ET from alfalfa using the pan method would be 22.28 inches, or a 9.3% difference between estimation methods.

If you have monthly consumptive use coefficients available for the specific crop and location such as those given in Table 4.8, then monthly consumptive use (u) can be computed as follows:

Table 4.8. Blaney-Criddle monthly consumptive use factors (k)

Crop and Location	Jan	Feb	Mar	Apr	May	Jun	Jul	Aug	Sep	Oct	Nov	Dec
Alfalfa												
Mesa, AZ	0.35	0.55	0.75	0.90	1.05	1.15	1.15	1.10	1.00	0.85	0.65	0.45
Los Angeles, CA	0.35	0.45	0.60	0.70	0.85	0.95	1.00	1.00	0.95	0.80	0.55	0.30
Davis, CA				0.70	0.80	0.90	1.10	1.00	0.80	0.70		
Logan, UT				0.55	0.80	0.95	1.00	0.95	0.80	0.50		
Corn												
Mandon, ND					0.50	0.65	0.75	0.80	0.70			
Cotton												
Phoenix, AZ				0.20	0.40	0.60	0.90	1.00	0.95	0.75		
Bakersfield, CA					0.30	0.45	0.90	1.00	1.00	0.75		
Weslaco, TX			0.20	0.45	0.70	0.85	0.85	0.80	0.55			
Grapefruit												
Phoenix, AZ	0.40	0.50	0.60	0.65	0.70	0.75	0.75	0.75	0.75	0.70		
Oranges												
Los Angeles, CA	0.30	0.35	0.40	0.45	0.50	0.55	0.55	0.55	0.50	0.50	0.45	0.30
Potatoes												
Davis, CA				0.45	0.80	0.95	0.90	0.85				
Logan, UT						0.40	0.65	0.85	0.80			
ND					0.45	0.75	0.90	0.80	0.40			
Grain, small (wheat)												
Phoenix, AZ	0.20	0.40	0.80	1.10	0.60							
Grain, small (oats)												
Scottsbluff, NB					0.50	0.90	0.85					
Sorghum												
Phoenix, AZ						0.40	1.00	0.85	0.70			
Great Plains Field Station, TX						0.30	0.75	1.10	0.85	0.50		

$$u = k \frac{tp}{100} \qquad (4.21)$$

where k is an empirical coefficient (Table 4.8) and u is the monthly consumptive use in inches/month.

Jensen-Haise Alfalfa-Reference Radiation Method

The Jensen-Haise method is termed a radiation method because solar radiation is needed in the equation to incorporate the recognized link between a source of energy and evapotranspiration. Jensen and Haise used over 3000 observations of actual evapotranspiration determined by soil sampling and statistically related R_s to E_{tr} as shown in Equation 4.22 (Jensen and Haise, 1963).

$$E_{tr} = \frac{C_T(T-T_x)R_s}{\lambda} \qquad (4.22)$$

where E_{tr} = reference evapotranspiration in mm/d
           C_T = temperature coefficient (Equation 4.23)
           λ = latent heat of vaporization in MJ/kg (Equation 4.27)
           R_s = solar radiation received at the earth's surface on a horizontal sur-
               face, $MJ/m^2 \cdot d$
           T = mean temperature for a 5-day period, °C
           T_x = intercept of the temperature axis (Equation 4.26), °C

The temperature coefficient can be calculated as follows:

$$C_T = \frac{1}{C_1 + 7.3\ C_H} \qquad (4.23)$$

and C_1, which is needed to calculate C_T, can be calculated from:

$$C_1 = 38 - \frac{2(H)}{305} \qquad (4.24)$$

where H is the elevation above sea level in meters. C_H, which is also needed for Equation 4.23, is calculated as follows:

$$C_H = \frac{5.0\ kPa}{(e_2 - e_1)} \qquad (4.25)$$

where e_2 and e_1 are the saturation vapor pressures in kPa at the mean maximum and mean minimum temperatures, respectively, for the warmest month of the year in an area.

$$T_x = -2.5 - 1.4\ (e_2 - e_1) - \frac{H}{550} \qquad (4.26)$$

$$\lambda = 2.501 - 2.361 \times 10^{-3}\ T \qquad (4.27)$$

where λ is the latent heat of vaporization (MJ/kg), and T is temperature in °C (Harrison, 1963).

Example 4.10: Estimate the evapotranspiration from alfalfa grown in Hoytville, Ohio for August 7–11, 1992 using the Jensen-Haise estimating method. Hoytville's elevation is 700 feet above sea level. The climatic data for the period of interest are shown below.

August	7	8	9	10	11	5-Day Average
Max T (°F)	83	85	86	90	93	87.4
Min T (°F)	52	57	61	62	65	59.4
R_s (Ly/d)	485	544	609	543	631	562.4
Pan (inches)	0.23	0.16	0.29	0.13	0.34	

Solution:

Step 1: Estimate λ from Equation 4.27, using an average temperature for the 5-day period of 73.4°F {(87.2 + 59.4)/2} which is 23.0°C. λ = 2.447 MJ/kg

Step 2: Estimate T_x from Equation 4.26. To use this equation you need the elevation in meters (700 feet = 213.36 m), and the saturation vapor pressure at the mean maximum and mean minimum temperatures respectively for the warmest month of the year in an area. Checking the NOAA summaries for the summer months at Hoytville, Ohio, July was the warmest month of the year. The mean maximum temperature for July at Hoytville is 80.3°F, and the mean minimum temperature for July was 60.3°F. Computing e_s for both these temperatures (Equation 4.4) results in $e_2 = 3.532$ kPa, and $e_1 = 1.575$ kPa. Substituting into Equation 4.26, T_x = -5.63.

Step 3: Calculate C_H from Equation 4.25, using e_2 and e_1 calculated in Step 2. $C_H = 2.55$.

Step 4: Calculate C_1 knowing the elevation of Hoytville and using Equation 4.24. $C_1 = 36.6$.

Step 5: Calculate C_T from Equation 4.23, using C_1 from Step 4 and C_H from Step 3. $C_T = 0.0181$.

Step 6: Calculate E_{tr} from Equation 4.22. To use R_s you must first convert Ly (Langleys) to MJ/m^2d (1 Ly/d $= 0.0419$ MJ/m^2d). $E_{tr} = 5.0$ mm/d.

Answer: The estimated ET from alfalfa $= 5$ mm/d.

It is interesting to compare this estimate to the pan measurements for the same period. The pan evaporation was 1.15 inches for the five days or 5.84 mm/d. The Jensen-Haise method predicts evapotranspiration from alfalfa, so we need to multiply E_{pan} by a coefficient (Table 4.2) of 0.92 to compare the pan measurement with the Jensen-Haise estimate. The pan method estimated E_{tr} to be 5.37 mm/d, or a 7.3% difference between the pan and Jensen-Haise estimation methods.

Thornthwaite Method

Thornthwaite (1948) developed an equation to predict monthly evapotranspiration from mean monthly temperature and latitude data (Equation 4.28). The small amount of data needed is attractive because often, one needs to predict ET for sites where little weather data are available. Based on what you know about ET, you should be skeptical about the general applicability of such a simple equation. Thornthwaite (1948) himself was not satisfied with the proposed approach, saying "The mathematical development is far from satisfactory. It is empirical...The chief obstacle at present to the development of a rational equation is the lack of understanding of why potential ET corresponding to a given temperature is not the same everywhere."

Taylor and Ashcroft (1972), as cited in Skaggs (1980) provided insight into the answer to Thornthwaite's question. They said "...this equation, being based entirely upon a temperature relationship, has the disadvantage of a rather flimsy physical basis and has only weak theoretical justification. Since temperature and vapor pressure gradients are modified by the movement of air and by the heating of the soil and surroundings, the formula is not generally valid, but must be tested empirically whenever the climate is appreciably different from areas in which it has been tested. ... In spite of these shortcomings, the method has been widely used. Because it is based entirely on temperature data that are available in a large number of localities, it can be applied in situations where the basic data of the Penman method are not available." Jensen et al. (1990) warn that Thornthwaite's method is generally only applicable to areas that have climates similar to that of the east central U.S., and it is **not** applicable to arid and semiarid regions.

Thornthwaite found that evapotranspiration could be predicted from an equation of the form:

$$E_{tp} = 16 \left[\frac{10\ T}{I} \right]^a \tag{4.28}$$

where E_{tp} = monthly ET in mm
       T   = mean monthly temperature in °C
       a   = location dependent coefficient described by Equation 4.30
       I   = heat index described by Equation 4.29

In order to determine **a** and monthly *ET*, a heat index *I* must first be computed.

$$I = \sum_{j=1}^{j=12} \left[\frac{T_j}{5} \right]^{1.514} \qquad (4.29)$$

where T_j is the mean monthly temperature during month j (°C) for the location of interest.

Then, the coefficient **a** can be computed as follows:

$$a = 6.75 \times 10^{-7}I^3 - 7.71 \times 10^{-5}I^2 + 1.792 \times 10^{-2}I + 0.49239 \qquad (4.30)$$

Example 4.11: Compute the monthly potential ET for August, 1992 at Hoytville, Ohio using Thornthwaite's method. The average temperature for the month of August, 1992 at Hoytville was 18.72°C. Information from the NOAA summaries is given below.

Solution:

Step 1: Determine I using Equation 4.29. To use this equation, we first need to know the mean monthly temperature for all the months of the year at Hoytville, Ohio. These were obtained from the NOAA summaries, and are shown below:

1992	Jan	Feb	Mar	Apr	May	Jun	Jul	Aug	Sep	Oct	Nov	Dec
T °C	-2.7	-0.1	2.0	8.0	14.1	17.9	21.3	18.7	16.4	9.4	4.7	-0.2

Calculating I using Equation 4.29; I=39.9.

Step 2: Use Equation 4.30 to calculate **a**, knowing I from Step 1. This results in a=1.13.

Step 3: Substituting into Equation 4.28, E_{tp}=9.17 cm/month.

Answer: The potential ET for August, 1992 at Hoytville, Ohio is 9.17 cm.

The pan evaporation measured in Hoytville in August, 1992 was 5.83 inches or 14.8 cm/mo. Adjusting by a pan coefficient to obtain E_{tr} (coefficients for E_{tp} not available) the ET estimated from the pan was 13.6 cm/month or 32.6% more than the Thornthwaite method predicted.

*Penman's Method

Penman (1948) first combined factors to account for a supply of energy and a mechanism to remove the water vapor from the immediate vicinity of the evaporating surface. You should recognize these two factors as the essential ingredients for evaporation. Penman derived an equation for a well watered grass reference crop:

$$E_{tp} = \frac{\dfrac{\Delta}{\Delta+\gamma}(R_n - G) + \dfrac{\gamma}{\Delta+\gamma} \, 6.43 \, (1.0+0.53 \, u_2)(e_s-e_d)}{\lambda} \qquad (4.31)$$

where
$\begin{aligned}
E_{tp} &= \text{potential evapotranspiration in mm/day}\\
R_n &= \text{net radiation in MJ m}^{-2} \text{ d}^{-1}\\
G &= \text{heat flux density to the ground in MJ m}^{-2} \text{ d}^{-1}\\
\lambda &= \text{latent heat of vaporization computed by Eq. 4.28 in MJ/kg}\\
u_2 &= \text{wind speed measured 2 m above the ground in m s}^{-1}\\
\Delta &= \text{slope of the saturation vapor pressure-temperature curve, kPa °C}^{-1}\\
\gamma &= \text{psychrometric constant, kPa °C}^{-1}\\
e_s\text{-}e_d &= \text{vapor pressure deficit determined by Method 3; kPa}
\end{aligned}$

The slope of the saturation vapor pressure-temperature curve, Δ, can be computed knowing the mean temperature as follows:

$$\Delta = 0.200 \, [0.00738 \, T + 0.8072]^7 - 0.000116 \qquad (4.32)$$

where Δ is in kPa/°C, and T is the mean temperature in °C.

To calculate the psychrometric constant, you must first calculate P, the atmospheric pressure, which Doorenbos and Pruitt (1977) suggested could be calculated by Equation 4.33:

$$P = 101.3 - 0.01055 H \qquad (4.33)$$

where P is in kPa and H is the elevation above sea level in meters. Using P, λ calculated from Equation 4.27, and c_p, the specific heat of water at constant pressure [0.001013 kJ/(kg °C)], the psychrometric constant (in kPa/°C) can be calculated from Equation 4.34:

$$\gamma = \frac{c_p \, P}{0.622 \lambda} \qquad (4.34)$$

The remaining value to calculate is G, the heat flux density to the ground in MJ m^{-2} d^{-1}, and this can be determined from Equation 4.35, knowing the mean air temperature for the time period before and after the period of interest:

$$G = 4.2 \frac{(T_{i+1} - T_{i-1})}{\Delta t} \tag{4.35}$$

where T is the mean air temperature in °C for time period i+1 and i-1, and Δt is the time in days between the midpoints of time periods i+1 and i-1.

Penman (1963) developed Equation 4.31 using Method 1 for computing the vapor pressure deficit. Jensen et al. (1971) and Wright (1982) recommended using Method 3, because they found more accurate predictions when Method 3 was used with Penman's equation.

Example 4.12: Estimate the E_{tr} for August 2, 1992 from a field close to Hoytville, Ohio using Penman's method. Radiation data, wind speed and pan evaporation were measured at the site. The other weather data are taken from the Hoytville daily summaries published by NOAA and summarized below.

Given: T_{avg} (Aug. 2) = 18.6 °C
T_{max} (Aug. 2) = 24.4 °C
T_{min} (Aug. 2) = 12.8 °C
T_{avg} (Aug. 1) = 15.0 °C
T_{avg} (Aug. 3) = 20.6 °C
e_d (8 a.m.) = 1.1759 kPa
u_2 = 1.83 m/s
R_s = 19.33 MJ/m^{2}d
ELEV = 213.4 m
lat = 41°N

Solution:

Step 1: Calculate G from Equation 4.35.

$$G = \frac{4.2(20.6 - 15.0)}{2} = 11.76 \text{ MJ}/(m^2 d)$$

Step 2: Calculate λ from Equation 4.27.

$$\lambda = \left(2.501 - 2.361 \times 10^{-3} \times (18.6)\right) = 2.4571 \text{ MJ/kg}$$

Step 3: Calculate P from Equation 4.33.

$$P = 101.3 - 0.01055(213.4) = 99.05\,\text{kPa}$$

Step 4: Calculate γ from Equation 4.34, remembering $c_p = 0.001013$ kJ/(kg °C). $\gamma = 0.0657$.

Step 5: Calculate Δ from Equation 4.32. $\Delta = 0.1340$.

Step 6: Determine R_n from R_s as shown in the section on solar radiation. From these calculations, $\epsilon = .1795$, $R_{bo} = 6.363$ MJ/(m^{2}d), $R_{so} = 28.95$ from Table 4.5, $R_b = 4.25$ MJ/(m^{2}d), and $R_n = 10.63$ MJ/(m^{2}d).

Step 7: Determine $(e_s - e_d)$ by Method 3 shown in the vapor pressure deficit section. The calculated vapor pressure deficit $= 1.0927$ kPa.

Step 8: Substitute into Penman's Equation (4.31).

$$E_{tp} = \frac{\dfrac{.134}{0.134+0.0657}(10.63-11.76) + \dfrac{0.0657}{0.134+0.0657}(6.43)(1+0.53(1.83))(1.0927)}{2.4571}$$

$$= 1.54\,\frac{\text{mm}}{\text{d}}$$

Answer: The potential ET for a site near Hoytville for August 2, 1992 was 1.54 mm/d. The measured pan evaporation was 1.12 mm/d, and multiplying by the E_{tr}/E_{pan} coefficient of 0.92 results in a predicted potential ET of 1.03 mm/d.

*Penman-Monteith Combination Method

Penman (1948) developed an equation of great significance (Equ. 4.31) because it combined both aspects of evaporation, namely the required energy source, and a mechanism to move vapor away from the evaporating surface. Penman developed an empirical equation for wind, which in practice accounts for the wind removing the water vapor and allowing ET to continue. Penman did not, however, have much theoretical basis for this equation. It did not include an aerodynamic resistance function (to quantify boundary layer resistance), nor did it include surface resistance to vapor transfer (to account for stomatal resistance). Several investigators have proposed equations to remedy these omissions, but the one most often cited is Monteith (1981). The modified equation is called the Penman-Monteith equation (Jensen et al., 1990). The reader is referred to Jensen et al. (1990, pp. 92–97) for suggestions on applying the Penman-Monteith equation in practice.

4.6. CONVERTING POTENTIAL OR REFERENCE CROP ET USING CROP COEFFICIENTS

The methods described above enable the estimation of potential or reference crop ET. The equations account for meteorological conditions, but do not account for crop or soil water status. A more useful quantity is actual ET, which estimates the amount of water the crop used during a given time period.

Crop coefficients can be used to estimate actual ET if care and attention are used to ensure that the same procedures and methods are used in applying the crop coefficients as were used in developing the methods. The coefficients were obtained by relating potential or reference crop ET to values of actual ET, typically measured with lysimeters. With careful adherence to recommended procedures, estimates of actual ET within $\pm$ 10% can be obtained (Jensen et al., 1990). When procedures and coefficients are applied to a different climate (i.e., arid vs. humid), testing is advisable because the coefficients are likely to be invalid.

To estimate crop actual ET (E_t):

$$E_t = k_c \ E_{tr} \ \text{or} \ E_t = k_c \ E_{tp} \qquad (4.36)$$

where E_{tr} is reference crop ET, E_{tp} is potential ET, E_t is actual evapotranspiration and k_c is the experimentally derived crop coefficient. Typical reference crops used to develop the coefficients are alfalfa or grass. Table 4.9 shows which reference crop was used for each method discussed in this chapter.

Table 4.10 presents crop coefficients k_c for normal irrigation and precipitation conditions, for use with alfalfa reference, E_{tr} (Jensen et al., 1990). A word of caution is in order because these values were based on data from Kimberly, Idaho which is an arid environment. The coefficients should be usable in other areas if the procedure has been verified locally. To use Table 4.10, select the appropriate crop from the first column, then select the applicable percent canopy cover. If the crop is less than completely closed over, use the first set of rows in the table, but if 100% cover has been established, the second group of rows is appropriate.

Example 4.13: The alfalfa reference E_{tr} was calculated for June to be 5 mm/day. The grower is interested in corn, however, not alfalfa. Predict the E_t if 30% of the time from planting to effective canopy cover has elapsed at the time of ET estimation.

Solution: From Table 4.10, select corn from the first column, and look in the 30% PCT column. The crop coefficient is 0.20. Multiplying E_{tr} by k_c, 5 mm/d (0.2)= 1.0 mm/d.

Answer: The estimated actual ET expected from corn in June with a 30% canopy cover is 1.0 mm/d.

If the method used to predict ET did not compute alfalfa reference E_{tr} by the combination method, Table 4.2 can be used to convert the method's prediction

Table 4.9.   Summary of the type of estimate produced by each of the methods in the chapter for estimating ET. (Source: *Evaporation and Irrigation Water Requirements*, Eds. M.E. Jensen, R.D. Burman, and R.G. Allen. Copyright 1990 by the American Society of Civil Engineers. Reproduced with permission of the American Society of Civil Engineers.)

Method	Type of Estimate	Remarks
Penman	Potential Evapotranspiration	Short green crop completely shading the ground and never short of water.
Penman-Monteith	Reference crop evapotranspiration, either alfalfa or grass	Reference type is dependent on surface roughness and canopy or bulk stomatal resistances used.
Jensen-Haise	Alfalfa reference	Reference crop is alfalfa (lucerne) well-watered with 30–50 cm of growth. Coefficients derived from USA data.
SCS Blaney-Criddle	Evapotranspiration from a specific crop	Evaluated for alfalfa and grass references using SCS crop coefficients and adjustment to negate the effects of cuttings.
Thornthwaite	Potential evapotranspiration for a standard surface	Grass solid cover with no water deficiency for conditions similar to those in Eastern U.S.

to alfalfa E_{tr}. To use Table 4.2, select the appropriate month from the top of the table, and the appropriate method from the left hand column, then determine the adjustment coefficient. Multiply the estimated ET by this coefficient to obtain E_{tr}. Now Table 4.10 can be used to find k_c, and Equation 4.36 can be used to calculate actual evapotranspiration. If the method used to estimate E_{tr} or E_{tp} was not developed for alfalfa, you can use the coefficients in Table 4.2 to convert to an alfalfa reference crop. Multiply the estimated ET by this coefficient to obtain E_{tr}. Now Table 4.10 can be used to find k_c, and Equation 4.36 can be used to calculate E_t.

Example 4.14: The Penman method was used to estimate E_{tp} for July to be 12 cm/mo. Estimate the actual ET which could be expected from corn if it has been 20 days since effective cover was established.

Solution: From Table 4.2, the coefficient $[E_{tr}/E_{tp}(P)]$ needed to adjust E_{tp} for the month of July is 1.20. Multiplying 12 cm/mo by $1.2 = 14.4$ cm/mo $= E_{tr}$. From Table 4.10, $k_c = 0.95$. Multiplying E_{tr} by k_c (Equation 4.36) gives 13.7 cm/month for actual ET.

Answer: The actual ET expected from field corn in July is 13.7 cm/month.

Table 4.10. Mean E_t crop coefficients, k_c, for normal irrigation and precipitation conditions, for use with alfalfa reference E_{tr}. (Source: *Evaporation and Irrigation Water Requirements*, Eds. M.E. Jensen, R.D. Burman, and R.G. Allen. Copyright 1990 by the American Society of Civil Engineers. Reproduced with permission of the American Society of Civil Engineers.)

Mean ET Crop Coefficients k_c

Crop	PTC, Time From Planting to Effective Cover (%)										
	0	10	20	30	40	50	60	70	80	90	100
Spring grain[a]	0.20	0.20	0.21	0.26	0.39	0.55	0.66	0.78	0.92	1.00	1.00
Peas	0.20	0.20	0.21	0.26	0.36	0.43	0.51	0.62	0.73	0.85	0.93
Sugar beets	0.20	0.20	0.21	0.22	0.24	0.27	0.33	0.45	0.60	0.80	1.00
Potatoes	0.20	0.20	0.20	0.22	0.31	0.41	0.51	0.62	0.70	0.76	0.78
Corn	0.20	0.20	0.20	0.20	0.23	0.32	0.42	0.55	0.70	0.85	0.95
Beans	0.20	0.20	0.20	0.26	0.35	0.45	0.55	0.66	0.80	0.90	0.95
Winter wheat	0.30	0.30	0.30	0.50	0.75	0.90	0.98	1.00	1.00	1.00	1.00

Crop	DT, Time since Effective Cover was Established (in Elapsed Days)										
	0	10	20	30	40	50	60	70	80	90	100
Spring grain[a]	1.00	1.00	1.00	1.00	0.90	0.50	0.30	0.15	0.10	-	-
Peas	0.93	0.93	0.70	0.53	0.35	0.20	0.12	0.10	-	-	-
Sugar beets	1.00	1.00	1.00	1.00	0.98	0.94	0.89	0.85	0.80	0.75	0.71
Potatoes	0.78	0.78	0.76	0.74	0.71	0.67	0.63	0.59	0.36	0.25	0.20
Corn	0.95	0.96	0.95	0.94	0.90	0.85	0.79	0.74	0.35	0.25	-
Sweet corn	0.95	0.94	0.93	0.90	0.85	0.75	0.58	0.40	0.20	0.10	-
Beans	0.95	0.95	0.90	0.67	0.33	0.15	0.10	0.05	-	-	-
Winter wheat	1.00	1.00	1.00	1.00	0.95	0.55	0.25	0.15	0.10	-	-

[a] Spring grain includes wheat and barley.

REFERENCES

Allen, R.G., M.E. Jensen, J.L. Wright, and R.D. Burman (1989) Operational estimates of re*rence evapotranspiration. *Agron. J.* 81:650–662.

Amatya, D.M., R.W. Skaggs, and J.D. Gregory (1992) Comparison of methods for estimating potential evapotranspiration. ASAE Winter Meeting Paper No. 92-2630.

Biswas, A.K. (1969) A short history of hydrology. In *The Progress of Hydrology.* Proceedings of the First International Seminar for Hydrology Professors, Volume II, Specialized Hydrologic Subjects. p. 914–935.

Doorenbos, J. and W.O. Pruitt (1977) Guidelines for prediction of crop water requirements. FAO Irrigation and Drainage Paper No. 24, 2nd ed., FAO Rome, Italy. 156 pp.

Harrison, L.P. (1963) Fundamental concepts and definitions relating to humidity. In Wexler, A. (ed.). *Humidity and moisture, Vol. 3,* Reinhold Publishing Company, NY.

Harrold, L.L., G.O. Schwab, and B.L. Bondurant (1986) *Agricultural and Forest Hydrology.* Agricultural Engineering Department, Ohio State University, Columbus, OH. 271 pp.

Jensen, M.E., R.D. Burman, and R.G. Allen (1990) Evapotranspiration and Irrigation Water Requirements. American Society of Civil Engineers, New York. 332 pp.

Jensen, M.E., and H.R. Haise (1963) Estimating evapotranspiration from solar radiation. *J. Irrig. and Drain. Div., ASCE,* 89:15–41.

Jensen, M.E., J.L. Wright, and B.J. Pratt (1971) Estimating soil moisture depletion from climate, crop and soil data. *Trans. ASAE,* 14:954–959.

Kramer, P.J. (1983) Water Relations of Plants. Academic Press, Inc. Orlando, FL. 489 pp.

Monteith, J.L. (1981) Evaporation and surface temperature. *Quart. J. Roy. Meteorol. Soc.,* 107:1–27.

NOAA (1992) Climatological data. Ohio. National Oceanic and Atmospheric Administration. National Climatic Data Center, Asheville, NC.

Nobel, P.S. (1983) Biophysical Plant Physiology and Ecology. W.H. Freeman and Company, New York.

Penman, H.L. (1948) Natural evaporation from open water, bare soil and grass. *Proc. Roy. Soc. London.* A193:120–146.

Penman, H.L. (1956) Evaporation: An introductory survey. *Netherlands J. Agric. Sci.,* 1:9–29, 87–97, 151–153.

Roberts, W.J. (1969) Significance of evaporation in hydrologic education. In *The Progress of Hydrology.* Proceedings of The First International Seminar for Hydrology Professors, Volume II, Specialized Hydrologic Subjects. p. 672.

Skaggs, R.W. (1980) Drainmod Reference Report. Methods for design and evaluation of drainage-water management systems for soils with high water tables. North Carolina State University, Raleigh, NC.

Schwab, G.O., D.D. Fangmeier, W.J. Elliot, and R.K. Frevert (4th Edition). 1992. *Soil and Water Conservation Engineering, 4th Edition.* John Wiley & Sons, Inc., New York.

Taylor, S.A. and G.L. Ashcroft (1972) Physical Edaphology. W.H. Freeman and Co., San Francisco, CA.

Thornthwaite, C.W. (1948) An approach toward a rational classification of climate. *Geograph. Rev.* 38:55–94.

Wright, J.L. (1982) New evapotranspiration crop coefficients. *J. Irrig. Drain. Div., ASCE,* 108(IR2):57–74.

Problems

4.1 If the measured pan evaporation was 3.2 inches for May, what would the alfalfa reference crop equivalent evaporation be if computed by the combination equation?

4.2 If the reference ET for grass was computed to be 4.5" for July, what would the reference ET be for alfalfa computed by the combination equation?

4.3 a) Find the saturation vapor pressure of air at 90°F.

b) Find the actual vapor pressure of air at 90°F and 80% relative humidity.

c) Find the vapor pressure deficit.

d) The book lists three methods for finding vapor pressure deficit (starting on p. 102). Which method did you use in part c?

4.4 If the average temperature on June 3, 1994 was 25°C, assuming a humid area (Latitude 40° N) with R_s = 23 MJ m^{-2} d^{-1}, calculate the net radiation (assume albedo of 0.23).

4.5 Find the wind speed at 25 feet if the wind speed at 6.6 ft is 2.3 mi/hr. The vegetation is soybeans, which are 2.5 ft tall.

4.6 Compute the evaporation for the month of August from a pond which had an average water temperature of 65°F measured one foot below the water surface, and an average wind speed of 2 mi/hr measured 25 feet above the pond. The average air temperature was 78°F and the relative humidity was 52%.

4.7 Using the information presented in the table below, calculate the monthly ET from corn grown near Columbus, OH (Latitude 40°N) for the 1992 growing season using the SCS Blaney-Criddle method (Equation 4.21).

The average maximum and minimum temperatures for a location close to Columbus, Ohio.

Month	June	July	August	September
Max. T(°F)	77.6	82.7	79.4	75.3
Min. T(°F)	56.9	64.2	59.3	54.1

4.8 Using the information presented in the previous table, calculate the seasonal ET (June-September) for corn grown near Columbus, OH using the SCS Blaney-Criddle method (Equ. 4.20).

4.9   Using the Jensen-Haise method, calculate the reference ET in mm/d for the data given below, if the data were collected at an elevation 219.5 m above sea level.   July was the warmest month of the year, with a mean max T = 27.8°C, and a mean min T = 15.5°C.

June	1	2	3	4	5	Average
Max T (°C)	27	30	32	31	32	30.4
Min T (°C)	13	15	20	18	16	16.4
R_s (MJ/m^{2}d)	24	20	27	25	26	24.4

4.10   Using Thornthwaite's method, determine the ET for the month of July, if the average monthly temperature was 21°C, and if the heat index for the location of interest was previously determined to be 40.

4.11   Estimate the E_{tr} for July 15, 1994 from a field at an elevation of 200 m using Penman's method.   The average temperature for July 15 was 16°C, with a maximum T = 22°C, and a minimum of 10°C.   The previous day's avg. temp. was 14°C, and the following day's avg. temp was 21°C.   The early morning actual vapor pressure on July 15 was 1.3 kPa, the wind speed was 2 m/s, and R_n was 15 MJ/m^{2}d.

4.12   The alfalfa reference E_{tr} was calculated (by the combination equation) to be 3.2 mm/day.   Predict the E_t from beans that have had full canopy cover for 30 days.

4.13   The Jensen-Haise method was used to estimate E_{tr} for June to be 10 cm/mo. Estimate the actual ET which could be expected from soybeans where 50% of the time from planting to full cover has elapsed.

CHAPTER 5

Surface Runoff and Subsurface Drainage

Andy D. Ward

5.1. INTRODUCTION

Rivers, lakes, reservoirs, and dams provide more than 75% of the water used in the United States. Surface water systems are also used for recreational purposes and in many cases are important transportation conduits. The importance of surface water in the development of the United States can be seen by looking at a map of the nation. Virtually all cities with populations exceeding 150,000 people are located on rivers. Furthermore, many smaller communities are also located on rivers or lakes. There are about two million streams and rivers in the United States including the mighty Mississippi River, which is the fourth longest river in the world, with a length of 3,710 miles. Four of the world's eleven fresh water lakes with the largest surface area are located in the Great Lakes system (Superior, 1st; Huron, 5th; Michigan, 6th; and Erie, 11th). In addition to the Great Lakes we have many thousands of smaller lakes, dams, reservoirs and ponds. These include 25 lakes with surface areas greater than 100 square miles and more than 200 with surface areas larger than 10 square miles.

Unfortunately, surface water flows can cause extensive damage by eroding soils and stream banks; carrying off valuable agricultural nutrients and pollutants; destroying bridges, utilities, and urban developments; and causing flooding and sediment deposits in recreational, industrial, and residential areas along stream systems. Since the Second World War we have seen a dynamic reshaping of our landscape. Rapid urbanization, agriculture, silviculture, and surface mining have changed hydrologic responses and sediment production on our watersheds. For example, urbanization of farm land and forests causes more rapid runoff, higher peak discharges, and larger runoff volumes. Growing concern for the adverse effects of these changes has resulted in the establishment of extensive federal, state, and local hydraulic and hydrologic design regulations. Intensive research has resulted in the development of numerous empirical methods for simulating the response of watersheds during storm events. Yet, little information is available on how to select an appropriate method and the likely accuracy associated with using the method.

The main sources of water entering surface water systems will vary from one location to another, but might include precipitation falling directly on the water body; surface runoff from storm events; snowmelt; flows from groundwater aquifers; interflow; subsurface drainage; and return flows from water used for domestic, commercial, industrial, mining, irrigation, and thermoelectric power generation purposes. The need to have detailed knowledge of each of these potential contributions will depend on the geographic location, the size of the area which contributes to flows at the location of interest, and the type of study which is being conducted. For example, surface runoff will dominate if flooding associated with severe storm events on small watersheds is being evaluated. In contrast, surface runoff due to storm events might be of little importance for a study of baseline flow during severe drought.

This chapter focuses on flows associated with single storm events and presents (1) details on the factors which influence surface runoff; (2) descriptions and examples on the use of several commonly used surface runoff techniques; (3) a perspective on the usefulness of these techniques; and (4) information on subsurface drainage which discharges back into surface water systems shortly after a rainfall event. In this book we have used the term surface runoff as being synonymous with overland flow and the term streamflow to describe flow in channels, streams and rivers.

5.2. FACTORS AFFECTING SURFACE RUNOFF PROCESSES

The potential for surface runoff exists whenever the rate of water application to the ground surface exceeds the rate of infiltration into the soil. Initially when water is applied to a dry soil, the application rate and infiltration rates might be similar. However, as the soil becomes wetter, the rate of infiltration will decrease and surface depressions will begin to fill. If all the surface depressions have been filled and the surface water application rate exceeds the infiltration rate, then surface runoff will be initiated. The soil profile near the soil surface need not be saturated for surface runoff to occur. The main components of the hydrologic cycle which are of importance in determining runoff from single events are illustrated in Figure 5.1 and include precipitation; infiltration; interception, surface storage, and detention; overland flow (surface runoff); interflow and subsurface drainage; and channel storage and streamflow.

The manner and degree to which these factors are accounted for varies from one runoff technique to another. Precipitation and infiltration have been discussed in Chapters 1 and 2. Further discussion on the specific aspects of precipitation and its relationship to runoff is presented in the next section. Interception, surface storage and detention influence both infiltration and surface runoff and are therefore not discussed as a separate topic. Channel storage and streamflow is the topic of Chapter 7 and, in many cases, will only have a small influence on overland flow processes.

5.2.1. Precipitation

When precipitation falls on plants, some of the precipitation is intercepted by the plant canopy and ultimately evaporates back to the atmosphere. The amount of interception will depend on the season of the year, the wind velocity, and the vegeta-

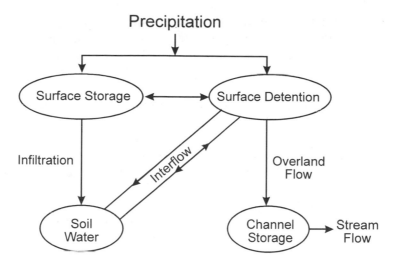

Figure 5.1. Storm runoff hydrologic cycle.

tion type and growth stage. Dense mature forests can intercept significant amounts of precipitation. Perhaps you have walked through a forest during a rain storm and stayed dry. The season of the year will influence the denseness of the plant canopy. In the autumn, plants which have shed their leaves have little capacity for intercepting precipitation. It is important to distinguish between precipitation that is retained on the canopy and that which is only temporarily delayed before reaching the ground. Water which trickles down a plant stem or drips from the canopy is not considered intercepted as it is still available for infiltration and surface runoff. Similarly, water that falls from the canopy due to wind shaking the leaves should be added back into that portion of the precipitation which directly reaches the ground. However, precipitation which is temporarily detained on the canopy sometimes reaches the ground surface long after the precipitation event has ended and might contribute little to runoff. Unfortunately, interception is difficult to measure or estimate.

Precipitation processes have been discussed in Chapter 2. Precipitation attributes which have the most influences on runoff processes are the precipitation type (rain, hail, sleet, etc.); the duration of the precipitation event; the amount of precipitation; how the precipitation intensity varied during the event; and spatial changes in the precipitation. Dams, runoff conveyance systems, flood control systems, earthworks and waste disposal systems are designed to control a specified design storm event. Commonly used design storm events are the Probable Maximum Flood (PMF) and events with return periods of 1, 10, 50, and 100 years. On the average, an event with a 10 year return period might be expected to occur once every 10 years. However, in the short term there might be several events with a 10 year return period in a 10 year period and even a chance of more than one of these events in a single year. It is also necessary to assign a rainfall duration to the return period of a rainfall event. For example, the 100-year, 1-hour rainfall depth might be 2 inches, while the 100-year, 24-hour rainfall depth could be 5 inches (see Chapter 2).

The PMF is defined as the flood which would result from the most severe combination of critical meteorologic and hydrologic conditions that might occur in a region. The rainfall associated with the PMF is known as the Probable Maximum

Precipitation (PMP) and is defined as the theoretically greatest depth of precipitation for a given duration that is physically possible over a watershed at a particular time of year. The PMF design event is the most stringent hydrologic design criteria which can be practically applied and is only used if the consequences of failure might be catastrophic or if the design life of the structure is long (at least 100 years).

The term design storm event is applied to both storm rainfall and storm runoff. It should not be expected that the 10-year rainfall event will produce the 10-year flood event. Storm runoff is very dependent on antecedent soil water conditions. A severe event following very dry conditions may produce only a small amount of runoff, while a small event following a wet period could produce a large volume of runoff. Although not strictly correct, it is generally assumed that flood flows for a given return period will be produced by storm events with the same return period—an assumption which is reasonable for extreme events. Therefore, the design hydrologist will use the rainfall depth associated with a 100-year rainfall event to determine the 100-year peak flood flow. The PMF is determined based on the PMP, and if high runoff conditions are assumed (i.e., wet antecedent soil water conditions or frozen ground) errors in associating the PMF to the PMP might be small. It should be noted that in regions where peak flows are caused by snowmelt, it is not appropriate to make flood estimations based on storm rainfall information.

The probability of exceeding a design storm event at least once can be determined from Equation 2.3 in Chapter 2. If a channel with a design life of 100 years is designed based on the 1:100 year event, the probability is about 0.63 (63%) that the design capacity will be exceeded at least once during the design life. If the same structure was designed based on the 1:200 year event, there would still be the very high probability of 39% that the design capacity would be exceeded at least once during the design life. A design return period in excess of 1:2000 years would need to be used to provide a 95% confidence that the design condition is not exceeded.

Rainfall depth, duration, time distribution, and areal distribution influence the rate and amount of runoff. If runoff information is determined from an actual storm event, then all of these rainfall characteristics will be known. However, in making design or environmental impact assessments, the rainfall characteristics need to be specified. Synthetic rainfall distributions which are commonly used are presented in Figure 5.2.

In the United States the most commonly used distributions are the SCS Type 1 and Type 2 curves. These curves should be used as follows:

Type 1:   Hawaii, coastal side of Sierra Nevada in southern California, and the interior regions of Alaska.

Type 1A: Storm distribution represents the coastal side of the Sierra Nevada and the Cascade Mountains in Oregon, Washington, and northern California, and the coastal regions of Alaska. Users requiring peak rates of discharge for these areas can obtain the graphs from the West Regional Technical Services Center, SCS, Portland, Oregon.

Type 2:   Remaining United States, Puerto Rico, and Virgin Islands.

The rainfall distribution used will have a major impact on the analysis. A steady rainfall throughout the storm duration would result in no runoff if the rainfall rate

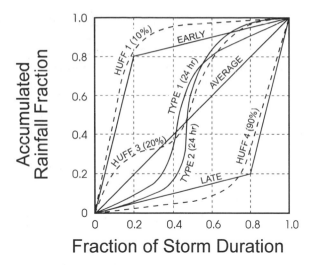

Figure 5.2. Commonly used rainfall distributions.

never exceeded the infiltration rate. However, a storm with the same duration and amount of rainfall might generate some runoff if there is a period of intense rainfall at some point during the event. The point in an event that the intense rainfall occurs is very important. At the start of an event, soils are the driest and infiltration rates are the highest, while near the end of an event soils are wetter and infiltration rates are lower. Therefore, the advanced-type rainstorm where the highest rainfall rates occur at the beginning is likely to result in less runoff than the delayed-type storm in which the highest rainfall rates occur later in the storm. The Huff 1 distribution is an example of an advanced-type distribution, while the Huff 4 distribution is a delayed-type storm (Huff, 1967). For storms in central Illinois, Huff divided the storms into four groups based on the time quartile most of the rain occurred. The frequency of occurrence of each group is as follows: Huff 1 events occur 30% of the time; Huff 2 events occur 36% of the time; Huff 3 events occur 19% of the time; Huff 4 events occur 15% of the time. Within each group there are families of curves with different probabilities of occurring (Barfield et al., 1981). For example, on Figure 5.2 the 90% Huff 4 storm is illustrated. What this signifies is that if a Huff 4 event occurs, then 90% of the time it will have this distribution. Note that of all storm events, there is only a 15% probability that it will be in the Huff 4 group. Therefore, out of all storm events the probability of the 90% Huff 4 storm is only 13.5% (0.90 × 0.15).

The areal (spatial) distribution of rainfall in a storm is often large. Methods for determining areal distribution have been discussed in Chapter 2. On small areas (less than a square mile) such as a farm field or the subdivision of an urban area, it is usually necessary to assume that the rain falls uniformly across the area of interest. In making environmental assessments or developing design details, this assumption will have little influence on the analysis. Due to a lack of information, or scattered climatic stations, the assumption of uniform rainfall coverage is often made on very large drainage basins. In this case the assumption of uniform coverage can result in significant errors. In many parts of the United States, the most severe events are associated with thunderstorms. These types of events result in different rainfall amounts and time distributions throughout a watershed. Fortunately, rainfall characteristics for regulatory assessments are usually specified. However, there are many applications where specifications on the rainfall characteristics are not adequate. For example, a small community might require that stormwater drains be designed to convey flow from a 1:10 year event. No information is provided on the duration, time distribution, or assumptions on the areal distribution.

5.3. WATERSHED FACTORS WHICH AFFECT RUNOFF

The watershed factors affecting runoff are size, topography, shape, geology, and perhaps most importantly, the soil and land use.

Size. The terms river basin, catchment, watershed, or subwatershed are used to delineate areas of different sizes which contribute runoff at their outlets. The terms are used interchangeably and no clear guidelines are available to identify when each term should be used. We suggest that the term river basin be used for large rivers which have several tributaries, while the term watershed be used for tributaries and small streams or creeks. Often it is desirable to subdivide a watershed into subwatersheds to reflect changes in land use, soil type, or topography. For convenience in this book, the term watershed will be used to reflect drainage areas of any size.

Determining the boundary of a watershed can be a challenging task and is normally done through use of topographic maps and field surveys. In the United States, a typical starting point might be to obtain a 1:24,000 scale (1 inch = 2,000 ft) USGS (United States Geological Survey) topographic map with 20 foot contour intervals. A practical exercise on this topic is presented in Chapter 12, Exercise 3.

A watershed boundary might consist of natural topographic features such as ridges or artificial features such as ditches along roads and storm water drains. No surface flow will occur across a watershed boundary. All flow within the boundary will drain to the watershed outlet, while flow outside of the boundary will be associated with a different watershed. The boundary of a watershed is often called the watershed divide because it divides the direction of flow. At a watershed divide, which is a natural feature (such as a ridge), flow is perpendicular to a tangent drawn anywhere along the watershed divide (Figure 5.3).

Topography. Common topographic features are illustrated in Figure 5.3. The only time that a natural watershed divide will cross a contour line will be when the divide runs along a ridge. Surface runoff will occur in the direction of the land slope and the flow direction will normally be perpendicular to the contours.

Topographic maps of a watershed show areas of steep and gentle slope, ridges, valleys, and stream systems. They also show the location of depressions with no surface outlet which contributes to runoff from the watershed. Overland flow slopes can be determined by measuring the lateral change in distance between contours (Figure 5.3). The land slope has little effect on infiltration rates or the depth of runoff. However, it has a significant influence on the velocity of water flow on land surfaces and in channels.

Shape. Circular or fan-shaped watersheds (Figure 5.4a) have high rates of runoff when compared to other shapes because runoff from different points in the watershed are more likely to reach the outlet at similar times. High rates of runoff on small catchments of this shape are short-lived. For long, narrow (elongated) watersheds (Figure 5.4b), tributaries join the main stream at intervals along its length. High

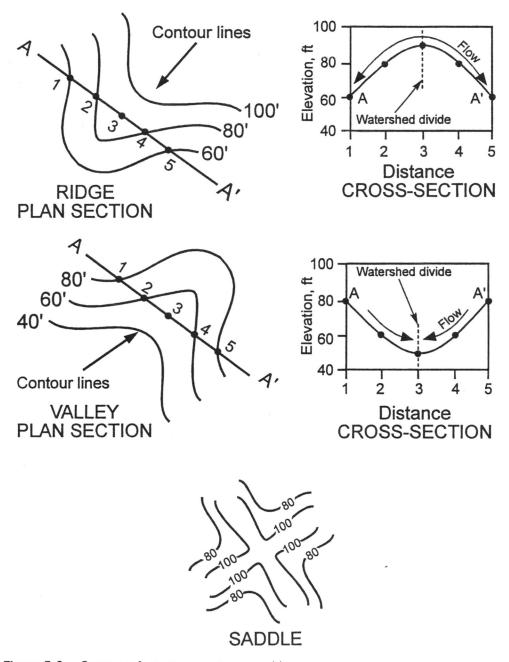

Figure 5.3. Common features on a topographic map.

flow rates from the downstream tributaries pass the gage before high flows from the upper tributaries arrive. Peak flow at the gage is less than for a fan-shaped catchment but persists for a longer time.

Orientation or aspect. The compass direction that the land surface faces is termed aspect or orientation. The aspect can be important where there is only one major slope face. On large areas, there are a multitude of aspects and the average tends to balance out to no specific direction. The aspect affects soil water content,

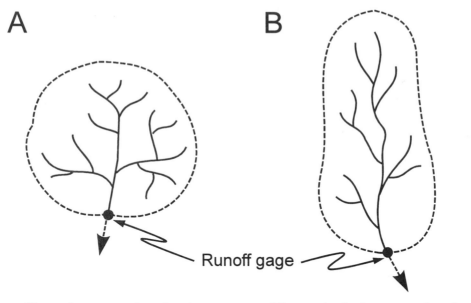

A

B

Runoff gage

Fan shape watershed   Elongated shape watershed

Figure 5.4. Catchment shapes: A. fan; and B. elongated.

vegetation, and is also important for soil freezing. In the northern hemisphere, north-facing slopes receive less solar energy, soil freezes to greater depth, and the period of frost is longer.

Geology. Geology, the science of the earth's crust and rocks, is a significant factor in the historic formation of soil and physical characteristics of watersheds and the establishment of surface and subsurface flow systems of the hydrologic cycle. Geological features (type of rock formations, faults, etc.) and processes (such as the movement of glaciers more than 10,000 years ago) have helped define the surface divide or ridge between watersheds, establish and control stream channel gradients, and formed the subsurface boundaries which control the movement of groundwater to surface streams. Groundwater flow boundaries do not always coincide with watershed divides and are more difficult to define.

Soil. Soil properties that influence runoff are the same as those that influence infiltration and have been discussed in detail in Chapter 3.

Land use. Compared to unprotected land surface, the vegetative cover of watersheds is likely to reduce storm runoff when soil water is significantly less than saturation. Vegetation improves the soil structure and depletes soil water, thus causing more infiltration and less surface runoff. Trees and deep-rooted crops usually consume more soil water by ET than shallow-rooted crops, especially in dry periods, and consequently their watersheds have less runoff (see Chapter 3).

Bare soil surfaces where infiltration rates diminish rapidly have more surface runoff. This is because bare soils tend to have poor structure and are less permeable. Even in permeable soils such as sand, a common problem known as surface sealing often occurs shortly after rainfall begins. Surface sealing can be due to both physical and chemical processes. For example, in an experiment with a sandstone spoil material from a surface mine, chemical bonding of particles occurred shortly after rainfall. This results in the formation of a very thin, almost impervious layer at the surface. A seal due to physical processes occurs when fine soil particles are washed into larger surface cracks and pores.

Soil and water conservation tillage practices that maintain high rates of infiltration could reduce surface runoff and those that reduce soil water evaporation could increase subsurface runoff. The effect of these practices on runoff depends somewhat on the soil water balance (the availability of excess soil water for surface or gravity flow).

5.4. SURFACE RUNOFF CHARACTERISTICS

A hydrograph is a plot of runoff rates against time (Figure 5.5). A rainfall excess (runoff depth) hyetograph for a single block of time is plotted in the upper left corner of the diagram. The runoff hydrograph is developed by routing blocks of rainfall excess to the watershed outlet. The volume of flow under the storm hydrograph will be equal to the sum of the rainfall excess in all the blocks of time associated with the storm event, multiplied by the watershed area. In many cases there will be flow prior to the start of the storm event. The flow associated with the storm event is added to this base flow. Base flow separation lines are drawn from the rising limb of the hydrograph to a point on the recession curve. If there is significant interflow, groundwater flow, subsurface drainage, or delayed runoff (such as from a forest) the receding limb will decay slowly and it is difficult to determine the point in time that flow from the rainfall event has ceased. A practical exercise on this topic is presented in Chapter 12, Exercise 4.

The hydrograph tells more about the hydrology of a small catchment than any other measurement. For example, Figure 5.6 shows that urban and forested watersheds had runoff for a few days after rain. However, the forested watershed continued to have streamflow more than 5 days after the rainfall on day 3, indicating the presence of soil or groundwater storage. The forested watershed's soil reservoir was recharged by storm water infiltration. The outflow into streams built up slowly and continued at a high rate after rainfall. Storm runoff rates from the urban watershed increased sharply during each rain period, then fell back rapidly after rain stopped. These sharp rises and falls indicate that much of the storm runoff came as surface or quick-return flow with little or no storm recharge to groundwater. The hydrograph of the urban watershed is termed "flashy." A flashy hydrograph has a peak runoff rate shortly after the most intense rainfall occurs. Conditions that can cause this type of rapid response are a small watershed area, impervious areas located adjacent to the stream channel, an impervious land use (such as a parking lot), and/or steep land slopes.

A stream acts as a temporary storage reservoir subduing (attenuating) the flashiness of runoff from small headwater watersheds (Figure 5.7). The shape of the headwater hydrograph reflects the individual rainstorm characteristics. But a point 50 miles or more downstream shows mostly the effect of temporary storage on runoff rates and practically none of the rainfall variations. Simple assumptions of this sort are usually adequate to help identify, on the hydrograph, sources of water flow from small headwater watersheds.

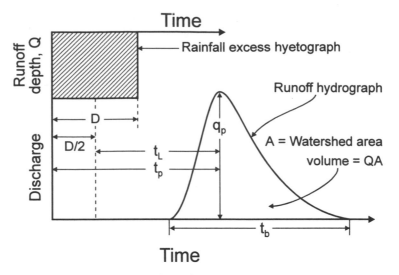

Figure 5.5. Storm hydrograph relationships.

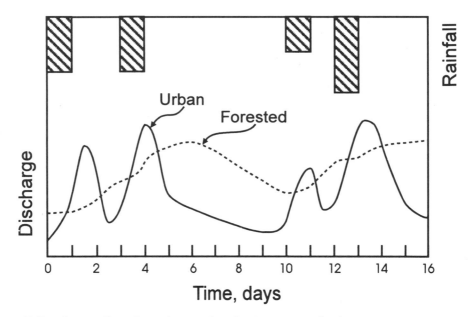

Figure 5.6. Streamflow from forested and urban watershed.

Runoff assessments often need to be made on watersheds where there is little or no streamflow information or in areas which undergo a land use modification such as an urban development or surface mining. Evaluation of observed runoff information is of limited value if the land features are going to be changed. Many surface runoff estimation techniques have been developed for these situations and used on ungaged watersheds. The most commonly used approaches are empirical equations, unit hydrograph techniques, the Time-Area Method, and kinematic methods. When assessing surface runoff, it is necessary to determine one or more of the following attributes: (1) the depth or volume of runoff; (2) the peak runoff rate; and (3) a

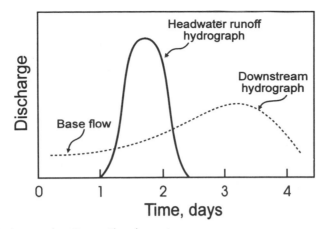

Figure 5.7. Hydrograph attenuation in a stream.

storm hydrograph. Common approaches to determining these attributes are presented in the next three sections.

5.5. VOLUME OF SURFACE RUNOFF

Rainfall excess (volume of runoff) could be determined by using one of the infiltration equations described in Chapter 3, but this approach has only been incorporated in a few hydrologic computer models. The most commonly used method is the SCS curve number procedure (SCS, 1972).

In this approach infiltration losses are combined with surface storage by the relationship:

$$Q = \frac{(P - I_a)^2}{(P - I_a + S)} \qquad (5.1)$$

where Q is the accumulated runoff or rainfall excess in inches, P is the rainfall depth in inches, and S is a parameter given by:

$$S = \frac{1000}{CN} - 10 \qquad (5.2)$$

where CN is known as the curve number. The term I_a is the initial abstractions in inches and includes surface storage, interception, and infiltration prior to runoff. The initial abstractions term I_a is commonly approximated as 0.2S and Equation 5.1 becomes:

$$Q = \frac{(P - 0.2S)^2}{(P + 0.8S)} \qquad (5.3)$$

A graphical solution of Equation 5.3 is presented in Figure 5.8.

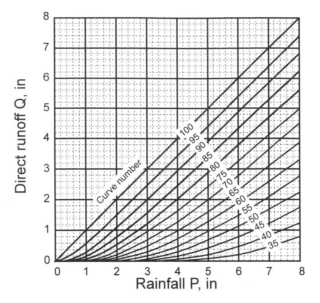

Figure 5.8. Relationships of runoff to rainfall based on SCS curve number method.

The SCS curve number is a function of the ability of soils to infiltrate water, land use, and the soil water conditions at the start of a rainfall event (antecedent soil water condition). To account for the infiltration characteristics of soils, the U.S. Soil Conservation Service has divided soils into four hydrologic soil groups which are defined as follows (SCS, 1984):

Group A: (Low runoff potential.)   Soils having high infiltration rates even when thoroughly wetted. These consist chiefly of deep, well-drained sands and gravels. These soils have a high rate of water transmission **(final infiltration rate greater than 0.3 inches/hr).**

Group B: Soils having moderate infiltration rates when thoroughly wetted. These consist chiefly of moderately deep to deep, moderately well to well-drained soils with moderately fine to moderately coarse textures. These soils have a moderate rate of water transmission **(final infiltration rate 0.15 to 0.30 inches/hr).**

Group C: Soils having slow infiltration rates when thoroughly wetted. These consist chiefly of soils with a layer that impedes downward movement of water or soils with moderately fine to fine texture. These soils have a slow rate of water transmission **(final infiltration rate 0.05 to 0.15 inches/hr).**

Group D: (High runoff potential.)   Soils having very slow infiltration rates when thoroughly wetted. These consist chiefly of clay soils with a high swelling potential, soils with a permanent high water table, soils with a claypan or clay layer at or near the surface, and shallow soils over nearly impervious materials. These soils have a very slow rate of water transmission **(final infiltration rate less than 0.05 inches/hr).**

Typical curve numbers for antecedent soil moisture condition II (AMC II) are shown in Table 5.1 (SCS, 1984). A summary of common U.S. soils and their hydrologic soil groups is presented in Appendix D.

Table 5.1.   Curve Numbers for Antecedent Soil Moisture Condition II (SCS, 1984).

Land Use Description	Hydrologic Soil Group			
	A	B	C	D
Commercial, row houses and townhouses	80	85	90	95
Fallow, poor condition	77	86	91	94
Cultivated with conventional tillage	72	81	88	91
Cultivated with conservation tillage	62	71	78	81
Lawns, poor condition	58	74	82	86
Lawns, good condition	39	61	74	80
Pasture or range, poor condition	68	79	86	89
Pasture or range, good condition	39	61	74	80
Meadow	30	58	71	78
Pavement and roofs	100	100	100	100
Woods or forest thin stand, poor cover	45	66	77	83
Woods or forest, good cover	25	55	70	77
Farmsteads	59	74	82	86
Residential 1/4 acre lot, poor condition	73	83	88	91
Residential 1/4 acre lot, good condition	61	75	83	87
Residential 1/2 acre lot, poor condition	67	80	86	89
Residential 1/2 acre lot, good condition	53	70	80	85
Residential 2 acre lot, poor condition	63	77	84	87
Residential 2 acre lot, good condition	47	66	77	81
Roads	74	84	90	92

Prior to estimating rainfall excess for a storm event, the curve numbers should be adjusted based on the season and 5-day antecedent precipitation. Antecedent soil moisture conditions are defined as follows:

AMC I: Dormant season antecedent soil moisture less than 0.5 inches. Growing season antecedent soil moisture less than 1.4 inches.

AMC II: Dormant season antecedent soil moisture between 0.5 and 1.1 inches. Growing season antecedent soil moisture between 1.4 and 2.1 inches.

AMC III: Dormant season antecedent soil moisture greater than 1.1 inches. Growing season antecedent soil moisture greater than 2.1 inches.

Curve number adjustments for antecedent soil moisture conditions can be made by using the information presented in Table 5.2. If Equation 5.3 is used to determine the runoff depth from an observed rainfall event, then the 5-day antecedent rainfall can be determined from rainfall records. However, when the procedure is used to determine runoff associated with a design storm event, an appropriate antecedent soil water condition will need to be specified. A conservative design approach is to assume wet conditions (AMC III) as they produce the most runoff. It should be recognized that in arid and semi-arid areas, the likelihood of these conditions occurring would be much lower than in semi-humid and humid areas.

Table 5.2.   Adjustments to runoff curve number (CN) for dry or wet antecedent soil moisture conditions.

Curve Number for Condition II	Factors to convert curve number for AMC II to AMC I or AMC III	
	AMC I (dry)	AMC III (wet)
10	0.40	2.22
20	0.45	1.85
30	0.50	1.67
40	0.55	1.50
50	0.62	1.40
60	0.67	1.30
70	0.73	1.21
80	0.79	1.14
90	0.87	1.07
100	1.00	1.00

Determining the depth or volume of runoff from watersheds with several land uses can be obtained by determining the volume of runoff from each land use and then summing these amount; or calculating an area-weighted curve number for the watershed and then using this single value with Equations 5.2 and 5.3. The first approach is preferred as it retains the runoff characteristics of the watershed, and small high runoff potential areas will not be dwarfed by large low runoff potential areas. However, for convenience the second approach is often included in hydrologic computer models.

Example 5.1: Estimate the depth and volume of runoff from 2 inches of rainfall on an eight acre, grassed area in an urban park. The grass is grown in Hoytville soils with good hydrologic conditions. Determine the runoff depth for AMC I, AMC II, and AMC III conditions.

Solution: From Appendix D it is determined that a Hoytville soil is in Hydrologic Soil Group D. From Table 5.1 the SCS curve number for grass (lawns) on soils in Hydrologic Soil Group D with good hydrologic conditions is 80.

For AMC II, substitute a curve number, CN, of 80 in Equation 5.2:

$$S = \frac{1000}{80} - 10 = 2.5 \text{ inches}$$

Substitute S = 2.5 inches and P = 2.0 inches in Equation 5.3:

$$Q = \frac{[2.0 - (0.2 \times 2.5)]^2}{[2.0 + (0.8 \times 2.5)]} = 0.5625 \text{ inches}$$

Therefore, $Q = 0.5625$ inches, or 28% of the rainfall for AMC II. (Note: an estimate of about 0.55 inches would be obtained by using Figure 5.8). Generally, Q would not be reported to more than two decimal places and curve numbers should not be determined to more than one decimal place.

The volume of runoff is determined by multiplying the runoff depth by the area contributing to runoff. Therefore, $Q = 0.5625$ inches $\times$ 8 acres = 4.5 acre-inches. Runoff volumes are normally expressed in acre-ft, so divide by 12 to convert inches to feet, and the runoff volume is 0.375 acre-ft.

For AMC I the curve number becomes $0.79 \times 80 = 63.2$ (factor of 0.79 from Table 5.2).

Substitute 63.2 in Equation 5.2:

$$S = \frac{1000}{63.2} - 10 = 5.8 \text{ inches}$$

Substitute $S = 5.8$ inches and $P = 2.0$ inches in Equation 5.3:

$$Q = \frac{[2.0 - (0.2 \times 5.8)]^2}{[2.0 + (0.8 \times 5.8)]} = 0.11 \text{ inches}$$

Therefore, $Q = 0.11$ inches or 5.5% of the rainfall for AMC I. Multiplying by 8 acres and dividing by 12 to convert inches to feet gives a runoff volume of 0.073 acre-ft.

For AMC III the curve number becomes $1.14 \times 80 = 91.2$ (factor of 1.14 from Table 5.2). Using the same approach, the runoff depth for AMC III is 1.18 inches, or 59% of the rainfall for AMC III. The runoff volume is 0.787 acre-ft.

Answer: AMC I, Q = 0.11 inches or 0.073 acre-ft
                      AMC II, Q = 0.56 inches or 0.375 acre-ft
                      AMC III, Q = 1.18 inches or 0.787 acre-ft

Example 5.2: Determine the depth and volume of runoff for a 1- in 25-year, 24 hr storm on an agricultural watershed in north central Ohio. The watershed has a 100 acre area of Hydrologic Group D soils which are cultivated with conventional tillage, and a 50 acre wooded area, with good cover, on soils in Hydrologic Group C.

Solution: From Table 5.1 the cultivated area with conventional tillage has a curve number of 91, while the woods have a curve number of 70. No information is provided on the AMC. However, in northern Ohio, severe storms could occur between March and October. Early spring storms would often occur during wet antecedent soil water conditions, particularly if there are heavy clay soils and no subsurface drainage. Therefore the solution will be developed for AMC III.

Solving for the cultivated area, from Table 5.2 the correction factor is 1.07 for a curve number of 90 and 1.0 for a curve number of 100. Use linear interpolation to determine the correction factor for a curve number of 91:

$$\frac{(X - 1.07)}{(1.00 - 1.07)} = \frac{(91 - 90)}{(100 - 90)}$$

and,

$$X = 1.063$$

Therefore, the AMC III curve number is $1.063 \times 91 = 96.7$. Substitute 96.7 in Equation 5.2:

$$S = \frac{1000}{96.7} - 10 = 0.34 \text{ inches}$$

From Appendix C the 1-in 25-year, 24 hour rainfall is 4.0 inches. Substitute $S = 0.34$ inches and $P = 4.0$ inches in Equation 5.3:

$$Q = \frac{[4.0 - (0.2 \times 0.34)]^2}{[4.0 + (0.8 \times 0.34)]} = 3.62 \text{ inches}$$

The runoff depth, Q, is 3.62 inches and the runoff volume is:

$$Q = 3.62 \text{ inches} \times 100 \text{ acres} \times \frac{1 \text{ ft}}{12 \text{ inches}}$$

Therefore, the runoff volume is 30.17 acre-ft.

Solving for the woods, from Table 5.2 the correction factor is 1.21 for a curve number of 70.

Therefore, the AMC III curve number is $1.21 \times 70 = 84.7$. Substitute 84.7 in Equation 5.2:

$$S = \frac{1000}{84.7} - 10 = 1.81 \text{ inches}$$

Substitute $S = 1.81$ inches and $P = 4.0$ inches in Equation 5.3:

$$Q = \frac{[4.0 - (0.2 \times 1.81)]^2}{[4.0 + (0.8 \times 1.81)]} = 2.43 \text{ inches}$$

The runoff depth, Q, is 2.43 inches and the runoff volume is:

$$Q = 2.43 \text{ inches} \times 50 \text{ acres} \times \frac{1}{12} \text{ ft}$$

Therefore, the runoff volume is 10.12 acre-ft.

The total runoff volume is 40.29 acre-ft (30.17 + 10.12). The average runoff depth is:

$$Q = 3.62 \text{ inches} \times \frac{100 \text{ acres}}{150 \text{ acres}} + 2.43 \text{ inches} \times \frac{50 \text{ acres}}{150 \text{ acres}}$$

Therefore, the average runoff depth is 3.22 inches.

Answer: The average runoff depth is 3.22 inches and the total runoff volume is 40.29 acre-ft.

5.6. PEAK RUNOFF RATE

Information on peak runoff rates is often used to design ditches, channels and storm water control systems. A practical exercise on this topic is presented in Chapter 12, Exercise 6. Several empirical methods have been developed, of which the Rational Equation, presented below, is the most widely used method:

$$q = 1.008CiA \tag{5.4}$$

where q is the peak flow (cfs), C is an empirical coefficient, i is the average rainfall intensity (in./hr) during the time of concentration, A is the catchment area (acres), and 1.008 is a unit conversion factor (1 in./hr $\times$ 1 ft/12 in $\times$ 1 hr/3600 sec x 1 acre $\times$ 43,560 ft^2/1 acre). The unit conversion factor of 1.008 is usually approximated as 1.0 because of the uncertainties associated with determining each of the other equation parameters. The time of concentration is the time it takes flow to move from the most remote point on a watershed to the outlet of the watershed. This longest flow path is called the hydraulic length.

The Rational Equation has been in use since 1851, is the simplest of the available methods, and has spawned numerous ways of computing the coefficient "C" and the time of concentration. The method has many limitations and is based on the following assumptions: (1) rainfall occurs uniformly over the drainage area; (2) peak rate of runoff can be reflected by the rainfall intensity, averaged over a time period equal to the time of concentration of the drainage area; (3) time of concentration is the time required for flow to reach the point in question from the hydraulically most remote point in the drainage area; and (4) frequency of runoff is the same as the frequency of the rainfall used in the equation.

Statistical fitting of data to Equation 5.4 results in the input parameters being dependent on each other. For example, if a particular approach is used to determine the time of concentration, then fitted C values will be a function of the time of concentration. This can result in erroneous predictions and inconsistences in tabulated C values. The empirical coefficient C can be determined from the information presented in Table 5.3. If the Rational Method is applied on a watershed with several different land uses, a single area weighted C value should be calculated for use with Equation 5.4.

The Graphical Peak Discharge Method (SCS, 1986) was developed for application in small rural and urban watersheds. It was developed from hydrograph analyses with TR-20 *Computer Program for Project Formulation - Hydrology* (SCS, 1973b) and has seen widespread application. The peak discharge equation is:

$$q = q_u \, A \, Q \, F \qquad (5.5)$$

where q is the peak discharge (runoff rate, cfs), q_u is the unit peak discharge (cfs per square mile per inch of runoff, csm/in), A is the drainage area (mi^2), Q is the runoff depth (in.) based on 24 hours, and F is an adjustment factor for ponds and swamps. The method depends on the SCS curve number method (Equations 5.1, 5.2, 5.3) to obtain Q and the necessary information to determine q_u. The unit peak discharge, q_u, is determined from Figure 5.9 and requires knowledge of the time of concentration and the initial abstraction, I_a, from Equations 5.1 and 5.3. Values of the adjustment factor, F, are presented in Table 5.4.

The method is intended for use in watersheds which are hydrologically homogeneous and have only one main stream or stream branches that have equal times of concentration. The F factor can only be applied to swamps and ponds which are not along the main flow path used to determine the time of concentration.

Numerous equations have been developed for determining the time of concentration and lag time. These methods may give results that are one or two orders of magnitude different from each other. Commonly used methods include those of the SCS (1972), Kirpich (1940), Overton and Crosby (1979) and Ward et al. (1980). In addition to these methods, several regional regression methods have been developed.

Commonly used U.S. methods are those developed by the SCS. The SCS lag equation is:

$$t_L = \frac{L^{0.8}(S+1)^{0.7}}{1900Y^{0.5}} \qquad (50 < CN < 95) \qquad (5.6)$$

where t_L is the lag time in hours, L is the hydraulic length of the watershed in feet, S is a function of the SCS curve number (see Equation 5.2), and Y is the average land slope in percent. The lag time is related to the time of concentration as follows

$$t_L = 0.6t_c \qquad (5.7)$$

The SCS lag time and time of concentration methods are used with the Graphical Peak Discharge Method and in the development of storm hydrographs rather than with the Rational Equation. Lag time is an estimate of the average flow time for all

Table 5.3.   Runoff coefficients, C, for use in the Rational Equation (Erie and Niagara Counties Regional Planning Board, 1981).

Land Use	Hydrologic Soil Group and Slope Range											
	A			B			C			D		
	0-2%	2-6%	6%+	0-2%	2-6%	6%+	0-2%	2-6%	6%+	0-2%	2-6%	6%+
Industrial	0.67[1]	0.68	0.68	0.68	0.68	0.69	0.68	0.69	0.69	0.69	0.69	0.70
	0.85[2]	0.85	0.86	0.85	0.86	0.86	0.86	0.86	0.87	0.86	0.86	0.88
Commercial	0.71	0.71	0.72	0.71	0.72	0.72	0.72	0.72	0.72	0.72	0.72	0.72
	0.88	0.89	0.89	0.89	0.89	0.89	0.89	0.89	0.90	0.89	0.89	0.90
High Density[3] Residential	0.47	0.49	0.50	0.48	0.50	0.52	0.49	0.51	0.54	0.51	0.53	0.56
	0.58	0.60	0.61	0.59	0.61	0.64	0.60	0.62	0.66	0.62	0.64	0.69
Medium Density[4] Residential	0.25	0.28	0.31	0.27	0.30	0.35	0.30	0.33	0.38	0.33	0.36	0.42
	0.33	0.37	0.40	0.35	0.39	0.44	0.38	0.42	0.49	0.41	0.45	0.54
Low Density[5] Residential	0.14	0.19	0.22	0.17	0.21	0.26	0.20	0.25	0.31	0.24	0.28	0.35
	0.22	0.26	0.29	0.24	0.28	0.34	0.28	0.32	0.40	0.31	0.35	0.46
Agricultural	0.08	0.13	0.16	0.11	0.15	0.21	0.14	0.19	0.26	0.18	0.23	0.31
	0.14	0.18	0.22	0.16	0.21	0.28	0.20	0.25	0.34	0.24	0.29	0.41
Open Space[6] (Grass/Forest)	0.05	0.10	0.14	0.08	0.13	0.19	0.12	0.17	0.24	0.16	0.21	0.28
	0.11	0.16	0.20	0.14	0.19	0.26	0.18	0.23	0.32	0.22	0.27	0.39
Freeways and Expressways	0.57	0.59	0.60	0.58	0.60	0.61	0.59	0.61	0.63	0.60	0.62	0.64
	0.70	0.71	0.72	0.71	0.72	0.74	0.72	0.73	0.76	0.73	0.75	0.78

1. Lower runoff coefficients for use with storm recurrence intervals less than 25 years.
2. Higher runoff coefficients for use with storm recurrence intervals of 25 years or more.
3. High density residential areas have more than 15 dwelling units per acre.
4. Medium density residential areas have 4 to 15 dwelling units per acre.
5. Low density residential areas have 1 to 4 dwelling units per acre.
6. For pastures and forests we recommend using the lower runoff coefficients which are listed for open spaces (our addition to original source).

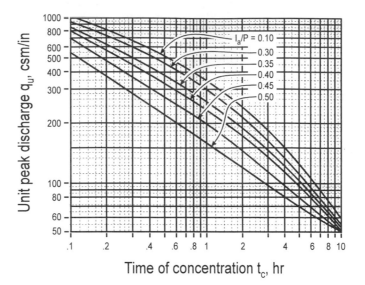

Figure 5.9. Unit peak discharge, q_u, for SCS type II rainfall distribution.

Table 5.4.   Adjustment factor, F, for ponds and swamps that are spread throughout the watershed.

Swamp and Pond Areas (%)	F
0.0	1.00
0.2	0.97
1.0	0.87
3.0	0.75
5.0	0.72

locations on a watershed. The coefficient of 0.6 in Equation 5.7 accounts for the fact that the average flow time will be 0.45 to 0.65 of the maximum flow time, depending on the watershed shape.

A commonly used time of concentration method which is used in conjunction with the Rational Method is the following equation which was developed by Kirpich (1940):

$$t_c = 0.00778L^{0.77}S^{-0.385} \qquad (5.8)$$

where L is the hydraulic length (maximum length) in feet, S is the mean slope along the hydraulic length expressed as a fraction, and t_c is the time of concentration in minutes. Solutions to Equation 5.8 are presented in Table 5.5.

Example 5.3: Use the Rational Method to determine the 1:25 year peak storm runoff rate (cfs) for a 100 acre agricultural catchment near Columbus, Ohio,

Table 5.5.   Small watershed time of concentration, t_c, determined with the Kirpich Method (Equation 5.8).

Mean Slope along the hydraulic length (S, ft/ft)	Time of concentration (t_c, min) Hydraulic length (L, ft)			
	500 ft	1,000 ft	2,000 ft	4,000 ft
0.0005	18	30	51	86
0.0010	13	23	39	66
0.0050	7	11	20	33
0.01	6	9	16	27
0.02	4	7	12	21
0.05	3	5	9	15

having a 4,000-ft hydraulic length with an average slope of 5%, hydrologic soil group B, and cropped in corn-conservation practice.

Solution: A = 100 acres; t_c = 15 min (Table 5.5); i = 4.88 inches/hr (Table 2.7 in Chapter 2 gives a 15-minute 1:25 yr rainfall of 1.22 inches or 1.22 × 60/15 inches/hr).

Use Table 5.3 to determine C, then C = 0.21 for an agricultural land use, slopes of 5%, and a 1:25 year storm event. Therefore, using Equation 5.4:

$$q = CiA = 0.21 \times 4.88 \times 100 = 102 \text{ cfs}$$

Answer: The peak runoff rate based on using the Kirpich equation and the Rational Method is 102 cfs.

Example 5.4: Use the Graphical Peak Discharge Method to determine the 1:25 year peak storm runoff rate (cfs) for a 100 acre agricultural catchment near Columbus, Ohio, having a 4,000-ft hydraulic length with an average slope of 5%, hydrologic soil group B, and cropped in corn-conservation practice.

Solution: The Graphical Peak Discharge Method is described by Equation 5.5 as follows:

$$q = q_u \text{ A Q F}$$

The time of concentration is determined by using the SCS lag time method described by Equation 5.6:

$$t_L = \frac{L^{0.8} (S + 1)^{0.7}}{1900 \ Y^{0.5}} \quad (50 < CN < 95)$$

From Table 5.1, the SCS curve number for cultivated land with conservation tillage and Hydrologic Soil Group B is 71. Therefore, S is equal to:

$$S = \frac{1000}{CN} - 10 = \frac{1000}{71} - 10 = 4.08 \text{ inches}$$

The average land slope was not provided, so it will be assumed to be equal to the slope along the hydraulic length. This assumption will underestimate the slope, as overland flow slopes to a stream are usually larger than the bed slope of the drainage system itself. Substituting:

$$t_L = \frac{4000^{0.8} (4.08 + 1)^{0.7}}{1900 \times 5^{0.5}}$$

and,

$$t_L = 0.55 \text{ hrs}$$

Using Equation 5.7:

$$t_L = 0.6 \ t_c$$

Therefore:

$$0.55 = 0.6 \ t_c$$

and, t_c is equal to 0.92 hours.

Using Table 2.7 in Chapter 2, the rainfall amount, P, for a 25-year, 24-hour storm event is 3.75 inches (note that for this method, the time of concentration is not used to calculate the rainfall amount, P). The initial abstraction, I_a, is 0.2S or 0.82 inches (0.2 × 4.08 inches). Therefore, the ratio of I_a/P is 0.82/3.75 or 0.21. From Figure 5.9, the unit peak discharge, q_u, is about 340 csm/in.

The runoff volume, Q, can be estimated from Figure 5.8 or calculated from Equation 5.3:

$$Q = \frac{(P - 0.2S)^2}{(P + 0.8S)}$$

and,

$$Q = \frac{(3.75 - 0.82)^2}{(3.75 + 3.26)} = 1.22 \text{ inches}$$

Therefore, the peak discharge is:

$$q = q_u \ A \ Q \ F_p = 340 \times \frac{100}{640} \times 1.22 \times 1.0 = 65 \text{ cfs}$$

Answer: The peak discharge based on the Graphical Peak Discharge Method is about 65 cfs.

In Examples 5.3 and 5.4 the Rational Method and the Graphical Peak Discharge Method gave very different answers for the same set of hydrologic conditions. The main reason for the difference is in the method used to compute time of concentration. Many studies have shown that the Kirpich equation estimates time of concentration values which are too rapid. For example, on 48 watersheds located in 16 states, McCuen et al. (1983) compared 14 different time of concentration methods with the SCS Upland Velocity method (SCS, 1986). The SCS Upland Velocity method is illustrated in Figure 5.10. It was found that the Kirpich Method (Equation 5.8), SCS Lag Method (Equation 5.6), and the SCS Velocity method (Figure 5.10) had mean time of concentration values of 0.81, 1.81, and 3.09 hrs, respectively. Also, of all methods evaluated, the SCS Lag Method had one of the lowest biases, while the Kirpich Method had one of the largest biases.

*5.7. STORM WATER HYDROGRAPHS

Storm water hydrographs provide information on the change in runoff rates with time, the peak runoff rate, and the volume of runoff. Storm hydrograph methods should be used wherever possible as they provide the most comprehensive information and the best estimates of runoff rates. However, these methods are also more complex, require more information to use, and are best solved by using a computerized procedure.

5.7.1. Unit Hydrograph Methods

One approach to developing a storm runoff hydrograph is the unit hydrograph method (Sherman, 1932). The unit hydrograph method is about 60 years old and was so named because the area under the hydrograph is equal to 1 inch, or 1 mm of runoff (Q). Many hydrologists over the world have studied, developed, reported on, and applied unit hydrographs to practical problems for over 50 years. The approach is empirical and based on the following assumptions: (1) uniform distribution of rainfall excess over the watershed; (2) uniform rainfall excess rate; and (3) the

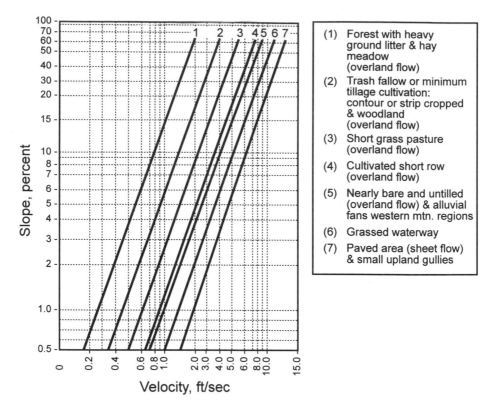

(1) Forest with heavy ground litter & hay meadow (overland flow)

(2) Trash fallow or minimum tillage cultivation: contour or strip cropped & woodland (overland flow)

(3) Short grass pasture (overland flow)

(4) Cultivated short row (overland flow)

(5) Nearly bare and untilled (overland flow) & alluvial fans western mtn. regions

(6) Grassed waterway

(7) Paved area (sheet flow) & small upland gullies

Figure 5.10.   SCS Upland Velocity Method (SCS, 1986).

runoff rate is proportional to the runoff volume for a rainfall excess of a given duration.

The basic factors which need to be determined in order to develop a unit hydrograph are the time to peak, the peak flow and the shape of the hydrograph. The time to peak is defined as:

$$t_p = t_L + D/2 \qquad (5.9)$$

where t_L is the lag time and D is the time increment of the rainfall excess. A plot of the rainfall excess versus time for each time increment, D, is called a hyetograph.

The peak flow of the unit hydrograph can be estimated by the equation (SCS, 1972):

$$q_p = \frac{484\ A}{t_p} \qquad (5.10)$$

where q_p is the peak flow in cfs (per inch of runoff), A is the watershed area in square miles and t_p is the time to peak in hours.

The shape of the unit hydrograph is usually modeled as curvilinear or triangular. The SCS has developed single triangle and curvilinear unit hydrograph procedures (Figure 5.11). For a practical exercise on SCS methods see Chapter 12, Exercise 5. Curvilinear procedures have also been developed by Haan (1970) and DeCoursey (1966). Double triangle shapes for use in rural and forested watersheds have been proposed by Overton and Troxler (1978) and Ward et al. (1980).

A single triangle can be developed by approximating the base time as 2.67 times the time to peak (t_p); using Equation 5.9 and a lag time equation (Equation 5.6); and using Equation 5.10 to determine the peak flow. Remember, the depth of rainfall excess represented by the unit hydrograph is 1 inch or 1 mm. A simple application of a single triangle method is illustrated in Example 5.5.

Example 5.5: Use a single triangle unit hydrograph procedure to develop a storm hydrograph for a 1 hour storm event on a proposed 500 acre commercial and business watershed with soils in Hydrologic Soil Group D. The watershed has an average overland slope of 1% and a hydraulic length of 6,000 feet. The rainfall depth during the one hour storm event was 2.5 inches.

Solution: Use the SCS curve number method (Equations 5.2 and 5.3) to develop a rainfall excess (runoff) hyetograph:

$$S = \frac{1000}{CN} - 10$$

From Table 5.1, CN = 95 (Hydrologic Soil Group D, and commercial land use). Assume antecedent moisture condition II, then:

$$S = \frac{1000}{95} - 10 = 0.53 \text{ inches}$$

and,

$$Q = \frac{[2.5 - (0.2 \times 0.53)]^2}{[2.5 + (0.8 \times 0.53)]} = 1.96 \text{ inches}$$

Assume a triangular unit hydrograph and determine the lag time and time to peak using Equations 5.6 and 5.9:

$$t_L = \frac{L^{0.8} (S + 1)^{0.7}}{1900 \times Y^{0.5}}$$

L = 6,000 feet, Y = 1%, and S = 0.53 inches:

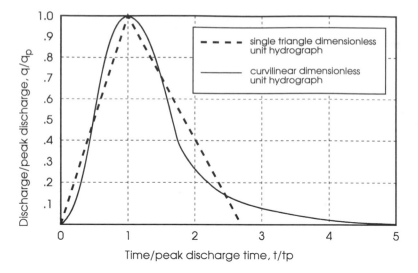

Figure 5.11.   SCS single and curvilinear unit hydrographs.

$$t_L = \frac{6000^{0.8} \, (0.53 + 1)^{0.7}}{1900 \times 1^{0.5}}$$

therefore,

$$t_L = 0.75 \text{ hrs} \quad (45 \text{ minutes})$$

and,

$$t_p = t_L + \frac{D}{2} = 45 + \frac{60}{2} = 75 \text{ minutes}$$

The peak flow of the unit hydrograph is determined from Equation 5.10:

$$q_p = \frac{484 \, A}{t_p}$$

and,

$$q_p = 484 \times \frac{500 \text{ acres}}{640} \times \frac{60}{75 \text{ minutes}} = 302.5 \text{ cfs (per inch of runoff)}$$

Note that the 500 acres are divided by 640 to convert them to square miles and the 75 minutes are divided by 60 to convert them to hours.

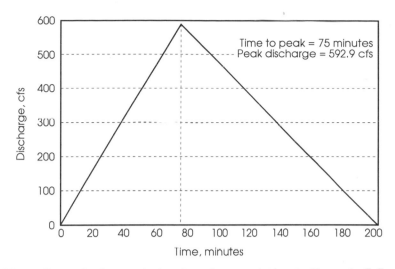

Figure 5.12.   Storm hydrograph developed as a solution to Example 5.5.

The peak flow is simply 302.5 cfs times 1.96 inches of runoff, which gives 592.9 cfs. The base time of the unit hydrograph is $2.67t_p = 2.67 \times 75 = 200$ minutes.

Answer: The hydrograph contains a runoff depth of 1.96 inches, has a peak flow of 592.9 cfs at a time of 75 minutes, and has a total runoff time of 200 minutes. The hydrograph is illustrated in Figure 5.12.

In Example 5.5 the storm duration was longer than the lag time (60 minutes versus 45 minutes). This causes an error in the solution as the shape of the hydrograph is primarily a function of the storm duration rather than the watershed characteristics. This problem can be prevented if knowledge is available of the rainfall time distribution. The design storm rainfall depth is generally associated with a synthetic rainfall time distribution (refer back to Figure 5.2).

Rainfall excess can be determined by using an infiltration equation or a procedure such as the SCS curve number method (Equation 5.3). The rainfall event should be divided into blocks of time which have a duration, D, that is not longer than one-third the time to peak of the unit hydrograph. Incremental storm hydrographs are then developed for each block of rainfall excess (Figure 5.13). The incremental runoff hydrograph ordinates are equal to the volume of runoff, for each block of rainfall excess, times the unit hydrograph ordinates. The incremental storm hydrographs each start at the beginning of the block of rainfall excess they are associated with.

When all the incremental storm hydrographs have been established the storm runoff hydrograph is obtained by adding the ordinates of the incremental hydrographs at each point in time. This is a time consuming activity which is subject to mathematical errors and is best performed by a computer program. The approach presented in Example 5.5 is commonly used and consists of approximating the storm

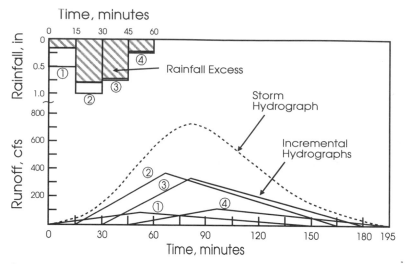

Figure 5.13.   Development of a storm hydrograph based on knowledge of the rainfall distribution.

hydrograph by using a single time block for the whole event; determining a time to peak and peak runoff rate for the unit hydrograph based on the single time block; and simply multiplying the ordinates of the unit hydrograph by the rainfall excess for the single block. This approach will underestimate the peak flow rate and will result in a base time for the storm hydrograph which is longer than that obtained if the hydrograph is developed from several incremental time blocks. The best approach is to subdivide the storm event into as many time blocks as is practical rather than using as few as possible.

Example 5.6: For the same watershed and storm that was evaluated in Example 5.5, use a unit hydrograph procedure to develop a storm hydrograph. In this case it is known that the rainfall during each 15 minutes of the event was 0.5, 1.0, 0.75, and 0.25 inches.

From Example 5.5, a 500 acre watershed has soils in Hydrologic Soil Group D, an average overland slope of 1%, a hydraulic length of 6,000 feet, and a proposed business and commercial land use. The lag time was calculated to be 45 minutes, therefore substituting in Equation 5.9:

$$t_p = t_L + \frac{D}{2} = 45 + \frac{15}{2} = 52.5 \text{ minutes}$$

The 15 minute time blocks that were provided can be used as they are less than one-third of the time to peak. The peak flow of the unit hydrograph is determined from Equation 5.10:

$$q_p = \frac{484 \ A}{t_p}$$

and,

$$q_p = 484 \times \frac{500 \ acre}{640} \times \frac{60}{52.5 \ minutes} = 432 \ cfs \ (per \ inch \ of \ runoff)$$

Use the SCS curve number method (Equations 5.2 and 5.3) to develop a rainfall excess (runoff) hyetograph.

$$Q = \frac{(P - 0.2 \ S)^2}{(P + 0.8 \ S)}$$

Solve for the first time block:

$$Q = \frac{(0.5 - 0.2 \times 0.53)^2}{(0.5 + 0.8 \times 0.53)} = 0.17 \ inches$$

Solve for the second time block:

$$Q_1 + Q_2 = \frac{(1.5 - 0.2 \times 0.53)^2}{(1.5 + 0.8 \times 0.53)} = 1.01 \ inches$$

and,

$$Q_2 = 1.01 - Q_1 = 1.01 - 0.17 = 0.84 \ inches$$

Note that it is necessary to determine the rainfall excess for the total time from the start of rainfall to the time at the end of the block being considered. The accumulated rainfall excess up to the start of the block being considered is then subtracted from the total rainfall excess for the time period up to the end of this block. This approach is used so that the curve number does not need to be adjusted for changing antecedent conditions during the event.

Solve for the third time block:

$$Q_1 + Q_2 + Q_3 = \frac{(2.25 - 0.2 \times 0.53)^2}{(2.25 + 0.8 \times 0.53)} = 1.72 \ inches$$

and,

$$Q_3 = 1.72 - Q_1 - Q_2 = 1.72 - 0.17 - 0.84 = 0.71 \ inches$$

Solve for the fourth time block:

$$Q_1 + Q_2 + Q_3 + Q_4 = \frac{(2.5 - 0.2 \times 0.53)^2}{(2.5 + 0.8 \times 0.53)} = 1.96 \text{ inches}$$

and,

$$Q_4 = 1.96 - Q_1 - Q_2 - Q_3 = 1.96 - 0.17 - 0.84 - 0.71 = 0.24 \text{ inches}$$

Answer: The solution is presented in Table 5.6 and Figure 5.13. The peak runoff rate is 729.7 cfs and occurs 82.5 minutes after the start of rainfall. Note that this answer is 136.8 cfs higher than the solution obtained in Example 5.5.

5.7.2. Time-Area Method

The time-area method is a process-based procedure to determine the volume of runoff and peak runoff rate from small watersheds (generally less than 500 acres). The approach does not consider interflow and assumes that surface and channel flow velocities do not change with time. If used on large watersheds, peak flows may be overestimated because the simple routing procedures do not consider attenuation. The most commonly used computer model which is based on time-area methods is the Illinois Urban Drainage Simulator, ILLUDAS, (Terstriep and Stall, 1974).

The approach is illustrated in Figure 5.14. The time-area diagram is obtained by drawing on the watershed isochrones of equal flow time, t, to the outlet. The value of time on each isochrone represents the travel time of water from the isochrone to the watershed outlet. Storage effects are ignored and the runoff hydrograph is due to a translation of rainfall excess to the outlet. Like the unit hydrograph procedure, incremental runoff hydrographs can then be developed for each increment of rainfall excess. Also a procedure such as the SCS curve number procedure can be used to develop a rainfall excess hyetograph. The storm hydrograph is then obtained by combining all the incremental hydrographs.

*5.7.3. The Kinematic Approach

This physically-based method is theoretically the most complete. Lighthill and Whitham (1955) imagined flow propagation in the form of kinematic waves for describing flood movement in rivers. They also developed kinematic equations for determining overland flow. However, the first application of kinematic equations to develop runoff hydrographs for steady rainfall excess conditions was by Henderson and Wooding (1964).

Complete details of the method may be found in hydrology texts such as Haan et al. (1982) and Bedient and Huber (1988). The method is based on a solution of the continuity and momentum equations, as well as relationships between the geometric characteristics of the watershed and drainage systems (such as slope, roughness, channel shape, etc.). The rainfall excess hyetograph is routed as overland flow to established channels and as streamflow to the watershed outlet.

Table 5.6.   Development of a Storm Hydrograph for Example 5.6 (Discharge, CFS).

Time (min)	Incremental Hydrograph 1	Incremental Hydrograph 2	Incremental Hydrograph 3	Incremental Hydrograph 4	Storm Water Hydrograph *
0	0.0	0.0	0.0	0.0	0.0
15	21.0	0.0	0.0	0.0	21.0
30	42.0	103.7	0.0	0.0	145.7
45	62.9	207.4	87.6	0.0	357.9
52.5	73.4	259.3	131.5	14.8	479.0
60	67.2	311.1	175.3	29.6	583.2
67.5	60.9	362.9	219.1	44.5	687.4
75	54.6	331.8	262.8	59.3	708.5
82.5	48.3	300.7	306.7	74.0	729.7
90	42.0	269.6	280.5	88.8	680.9
97.5	35.8	238.5	254.1	103.7	632.1
105	29.6	207.4	227.8	94.8	559.6
120	16.8	145.2	175.3	77.0	414.3
135	4.2	82.9	122.7	59.3	269.1
150	0.0	20.7	70.1	41.3	132.1
165	0.0	0.0	17.5	23.7	41.2
180	0.0	0.0	0.0	5.9	5.9

* Sum of columns 1, 2, 3, and 4.

Unfortunately, the method is data intensive and difficult to use. The approach has mainly been applied in urban areas where the watersheds are small and highly channelized. For most applications computer models are needed to develop a solution and it is necessary to make several assumptions and approximations (particularly in the geometry). The most commonly used computer model which incorporates kinematic equations is the Stormwater Management Model, SWMM (Huber et al., 1981).

5.8. ASSESSMENT OF FLOOD ESTIMATION TECHNIQUES

Hydrologists and engineers face two major problems on small catchment hydrology: (1) the availability of small catchment hydrological data; and (2) guidelines on the selection and expected accuracy of the different flood estimation techniques.

In general, little information is available on the suitability of the various flood estimation techniques as design tools. Hydrological control structures are frequently designed based on limited data and pragmatic modeling approaches. The success of the design depends largely on experience and judgement. In 1983, a study was conducted in South Africa to evaluate the suitability of commonly used small catchment surface runoff techniques as design methods (Ward et al., 1989). It is probable that the conclusions drawn from this study would be applicable to most countries in the world including the United States. The questionnaire survey indicated that a wide variety of flood estimation methods are currently being used in South Africa.

The results indicated that the Rational Method was the most commonly used method, but use of such methods as the Time-Area Method, SCS methods, and kinematic procedures was significant (Table 5.7). In general, hydrologists found all

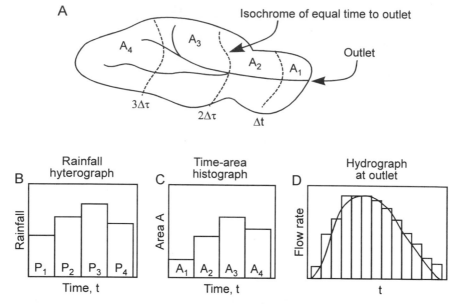

Figure 5.14.   Illustration of the time-area method.

the techniques fairly easy to use, but only 50% of the respondents felt the techniques were sufficiently accurate. In addition, flood estimation techniques are primarily being used to determine peak flows and volumes of runoff. Storm hydrograph generation was required less frequently but there was a significant need for this type of data. Despite being a peak flow technique, the Rational Method was used to generate hydrographs by 12% of the respondents. Flood estimation techniques were primarily being applied in urban, agricultural, rural and afforested areas smaller than 10 square kilometers. Respondents perceived that the main problem associated with the use of flood estimation techniques is the availability of adequate data. The survey also identified a lack of familiarity with the different techniques and no perception of accuracy.

As part of the same study, an evaluation of seven hydrological methods was made on 102 storm events on 26 diverse catchments. Catchment parameters were obtained from maps and design manuals. The analysis was conducted with both the measured rainfall distribution and with synthetic distributions. This was done to separate the effects of method limitations and parameter inaccuracies from the effects of simplified rainfall. The analysis of the results was based on two measures of method performance, namely peak flow and volume of runoff.

The results indicate that while urban catchments could be adequately modeled by most of the methods, flood predictions on rural catchments was far less reliable. In general, the ability of the methods to predict volume of runoff was found to be better than their ability to determine peak flow rates. The Rational Method was found to be very conservative and gave excessively high peak flow estimates for rural catchments. The SCS methods gave reasonable peak flow and runoff volume estimates for most of the watersheds and performed slightly better on rural catchments. The time-area and kinematic models gave reasonable estimates for urban watersheds but performed poorly on rural catchments.

Table 5.7.   Usage of surface runoff methods in South Africa (Ward et al., 1989).

Question	Rational method (%)	Time-area (%)	SCS (%)	Kinematic (%)	Other (%)
Use of Technique*	90	34	30	10	26
Reason for Use**					
• Easy to use	85	61	56	7	57
• Not familiar with other techniques	18	17	9	9	7
• Technique sufficiently accurate	55	50	38	18	46

* Percentage of respondents using flood estimation techniques. Some respondents use several methods.

** Percentage of respondents using this technique.

5.9. INTERFLOW

Baseflow and interflow (sometimes called throughflow) are two important flow processes which contribute to streamflows. Baseflow is generally considered to be groundwater flow from water tables which are above the level of the streambed and result in a hydraulic gradient towards the stream (see Chapter 9).

Interflow is usually due to lateral flow along a subsurface sloping layer which has impeded downward movement of infiltration or percolation. Interflow is considered to occur if the lateral flow returns to the land or a stream without first reaching groundwater (Kirkby, 1978). In steep forested areas with shallow soils, interflow might be the primary runoff process.

Measuring or estimating interflow is difficult, and because the flow has the potential to mix with surface runoff, it is often quantified as part of the runoff. The double triangle unit hydrograph methods discussed earlier are often used to account for interflow. In a steep forested watershed, the peak initial runoff response might be due to runoff from rock outcrops, bare rock areas along the streams, precipitation directly on the streams, and/or rapid interflow. The delay response is conceptualized as being due to steamflow, surface runoff delayed by bed litter (duff), and/or delayed interflow. This topic is discussed in further detail in Chapter 8.

5.10. AGRICULTURAL LAND DRAINAGE MODIFICATIONS

In the United States, drainage improvements are required on more than 20% of our cropland (110 out of 421 million acres). Egyptian and Greek use of surface ditches to drain land for agriculture dates back to about 400 BC (USDA, 1987). In the United States, extensive drainage of swamps was initiated in the mid-1700s. The first important federal legislation on land drainage were the Swamp Land Acts of 1849 and 1850 (USDA, 1987). Early subsurface drainage materials were wood, concrete, clay, and bituminized fiber pipe. In 1967, corrugated plastic tubing was commercially produced in the United States and by 1973 had largely replaced concrete and clay tile for small size drains (Schwab et al. 1993). Today, subsurface drains are still commonly called tile drains, even though nearly all the drains installed in the last two decades are made from perforated corrugated plastic tubing. In the United States, much of the tubing used in agriculture has a diameter of 4 inches and is made from high density polyethylene (HDPE). In Europe, field drains often have diameters of only 2 or 3 inches and are made from polyvinylchloride (PVC).

Agricultural drainage is the removal of excess water from the soil surface and/or the soil profile of cropland, by either gravity or artificial means. The two main reasons for improving the drainage on agricultural land are for soil conservation and enhancing crop production. Research conducted throughout the Midwestern U.S. has documented many benefits of agricultural drainage improvements. Typically, most agricultural producers improve the drainage on their land to help create a healthier environment for plant growth and to provide drier field conditions, allowing farm equipment to access the farm field throughout the crop production season. Healthy, productive plants have the potential to produce greater yields, and more food. Also, research in Ohio has shown that agricultural drainage improvements can help reduce the year-to-year variability in crop yields. This helps reduce the risks associated with the production of abundant, high quality, affordable food. Improved access of farm equipment to the field provides more time for field activities, can help extend the crop production season, and helps reduce crop damage at harvest.

The two primary types of agricultural drainage improvements are surface and subsurface systems (Figure 5.15) used individually or in combination with each other.

5.10.1. Surface Improvements

Surface drainage improvements are designed to minimize crop damage resulting from water ponding on the soil surface following a rainfall event and to control surface water runoff without causing erosion. Surface drainage can affect the water table by reducing the volume of water entering the soil profile. This type of improvement includes land leveling and smoothing; the construction of surface water inlets to subsurface drainage improvements; and the construction of shallow ditches and grass waterways, which empty into open ditches, streams, and rivers.

Land smoothing or leveling is a water management practice designed to remove soil from high spots in a field, and/or fill low spots and depressions where water may pond. Shallow ditches may be constructed to divert excess water to grass waterways and open ditches, which often empty into existing surface water bodies.

We should also be aware of the disadvantages of surface drainage improvements. First, these improvements require annual maintenance and must be carefully designed to ensure that erosion is decreased. Second, extensive earthmoving activities are expensive, and land grading might expose less fertile and less productive subsoils. Further, open ditches may interfere with moving farming equipment across a field.

5.10.2. Subsurface Improvements

The objective of subsurface drainage is to drain excess water from the plant root zone of the soil profile by artificially lowering the water table level. Subsurface

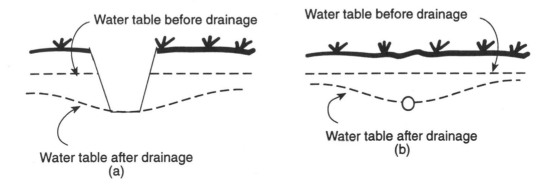

Figure 5.15.   Types of agricultural drainage improvements.

drainage improvements are designed to control the water table level through a series of drainage pipes that are installed below the soil surface, usually just below the root zone. In the Midwest region of the United States, the subsurface drains typically are installed at a depth of 30 to 40 inches, and at a spacing of 20 to 80 feet. The subsurface drainage network generally outlets to an open ditch, stream, or river. Subsurface drainage improvements require some minor maintenance of the outlets and outlet ditches. For the same amount of treated acreage, subsurface drainage improvements are generally more expensive to construct than surface drainage improvements.

Whether the drainage improvement is surface, subsurface, or a combination of both, the main objective is to remove excess water quickly and to safely reduce the potential for crop damage. In a situation where water is ponded on the soil surface immediately following a rainfall event, a general rule of thumb for most crops is to lower the water table 10 to 12 inches below the soil surface within a 24-hour period, and 12 to 18 inches below the soil surface within a 48-hour period. Properly draining excess water from the soil profile where plant roots grow helps aerate the soil and reduces the potential for damage to the crop roots. Further, it will produce soil conditions more favorable for conducting farming operations. In states which depend heavily on irrigation, subsurface drainage is often used to prevent harmful buildups of salt in the soil.

5.10.3. Perceptions

In recent years, public concern has increased about the nature of agricultural drainage and the impact of agricultural drainage improvements on the quality of water resources and environment. Few individuals not directly connected with agriculture realize that improved drainage is necessary for sustained agricultural production. In fact, controlled subsurface drainage, an agricultural drainage practice, is considered a sound water quality management and enhancement practice in North Carolina.

Land drainage activities have impacted much of the Midwest region's environment and water resources. For example, early settlers began draining Ohio's swamps in the 1850s and today, approximately 90% of Ohio's wetlands have been converted to other uses. This loss of wetlands is attributed to public health considerations, rural, urban and industrial development, and agriculture. While the loss of these wetlands has provided many benefits, wetlands also provide many benefits for the environment including wildlife habitat and enhanced water quality. An important water quality function of wetlands is the trapping and filtering of sediment, nutrients, and other pollutants from agricultural, construction, and other rural and urban sources. Interestingly, subsurface drainage improvements, in a more limited capacity, provide some of these same water quality benefits while providing a necessary element for sustained agricultural production on many soils.

Present agricultural trends are toward more intensive use of existing cropland with much of the emphasis on management. Maintaining existing agricultural drainage improvements and improving the drainage on wet agricultural soils presently in agricultural production helps minimize the need for landowners to convert additional land to agricultural production. In many cases, restoration of previously converted wetlands would be impossible because of large-scale channel improvements, urbanization, and shoreline modification in the Great Lakes Region. Focus should be placed on protecting existing wetlands and establishing new wetland areas, while maintaining our highly productive agricultural areas.

The use of surface and subsurface drainage improvements is not limited to agricultural lands. Many residential homes use subsurface drainage systems, similar to those used in agriculture, to prevent water damage to foundations and basements. Golf courses make extensive use of both surface and subsurface drains. Houses, streets, and buildings in urban areas depend heavily on surface and subsurface drainage systems for protection. These are generally a combination of plastic or metal gutters and concrete pipes or channels.

*5.11. DETERMINING SUBSURFACE DRAINAGE FLOWS

Details on the design and installation of subsurface drainage systems are presented in several texts and reports (USDA, 1987, Schwab et al., 1993, SCS, 1984). A common problem in irrigated agriculture is a build up of salts in the root zone. A practice to alleviate this problem is to apply additional irrigation water which will "flush" some of the salt from the root zone. Subsurface drains are then used to intercept leachate from the bottom of the root zone. In non-irrigated agriculture, subsurface drains are used to ensure that elevated water tables are lowered below the root zone within 1–4 days. Typically the drains are located at least 3 feet below the ground surface. The depth and spacing of the drains will depend on the soil characteristics, the crop, agricultural management practices, topography and surface drainage systems, and the depth to an impeding layer. For steady state conditions where the water table and rainfall or irrigation rate does not change with time, the ellipse

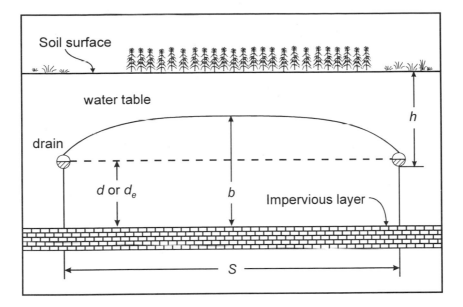

Soil surface

water table

h

drain

d or d_e

b

Impervious layer

S

Figure 5.16.   Subsurface drainage geometry for Equations 5.11 and 5.12.

equation can be used to determine a drain spacing and flow rate. The geometry of the system is presented in Figure 5.16 and the ellipse equation is:

$$q = \frac{4K(b^2-d^2)}{S} \qquad (5.11)$$

where K is the saturated hydraulic conductivity (ft/day), q is the flow rate (ft^2/day) into the drain per foot length of drain, S is the drain spacing in feet, b is the height (feet) at the midpoint between the drains of the water table above an impeding layer, and d is the height (feet) of the drains above the impeding layer.

If the flow rate, q, is set equal to Si, where i is the drainage rate in ft/day, then Equation 5.11 can be written as:

$$S = \left[\frac{4K(b^2-d^2)}{i} \right]^{1/2} \qquad (5.12)$$

An assumption in the development of Equations 5.11 and 5.12 was that resistance to flow resulting from flow lines converging near the drains could be ignored. However, reasonable results are only obtained if the drain spacing is large compared to the depth to the impeding layer. Van Schilfgaarde (1963) and Moody (1966) have developed modifications to the ellipse equations for conditions where the drain spacing, S, is small relative to the depth of the impeding layer. We would suggest using an effective depth, d_e, which is 0.2 to 0.5 of the height of the drains above the impeding layer in place of the depth, d, in Equation 5.11 or 5.12 in the development of provisional assessments, or if detailed knowledge of the subsurface soil profile is not available. If an effective depth is used, then the height, b, of the water table at the midpoint between the drain must be reduced by the difference between d and d_e. Information on recommended drainage rates for different crops and soils is provided by Schwab et al. (1993). Depending on the soil and crop, these rates will usually range between about 0.4 to 2.0 inches per day.

Example 5.7: Determine the drain spacing if drains are located 4 feet below the ground surface, the depth of the impeding layer is 12 feet below the drains (16 feet below the ground surface) and the saturated hydraulic conductivity is 0.75 ft/day. For good plant growth it is required that the water table be at least 2 feet below the ground surface (14 feet above the impeding layer) and it will be necessary to remove irrigation water at a rate of 0.60 in./day.

Solution: The drainage rate needs to be changed to ft/day, therefore:

$$i = \frac{0.6 \text{ inches/day}}{12} = 0.05 \text{ ft/day}$$

then using Equation 5.12:

$$S = \left[\frac{4K(b^2 - d^2)}{i} \right]^{1/2}$$

and substituting for K, b, d, and i:

$$S = \left[\frac{4 \times 0.75 \ (14^2 - 12^2)}{0.05} \right]^{0.5}$$

and,

$$S = 55.9 \text{ feet.}$$

The diameter of the drains and the flow depth are unknown and have not been considered when determining b and d.

Answer: The maximum drainage spacing should be 55.9 feet. If an effective depth, d_e, of 0.25d had been used instead of d, the height of the water table at the midpoint, b, would reduce by 0.75d and the drainage spacing would be 31.0 ft.

5.12. WATER YIELD

Information on minimum water yield (streamflow) for different return periods and catchments is useful and important in water management programs; telling the planner that in very dry periods, one must plan to get along with only so much flow, or develop storage systems to supply extra water. This data will tell how often to expect these low flows (Table 5.8).

Table 5.8.   Minimum 12-month runoff for catchments of 29 to 17,400 acres for return period of 2 to 50 years, Coshocton, Ohio (Harrold, 1957).

Catchment area (acres)	Minimum 12-month water yield (inches) for return period of 2–50 years				
	2 yr	5 yr	10 yr	25 yr	50 yr
29	8.7	6.4	5.2	4.2	3.4
76	9.6	7.4	6.2	5.1	4.3
122	11.2	8.8	7.2	5.9	4.9
349	13.0	10.1	8.7	7.0	6.0
2,570	14.0	10.8	9.2	7.3	6.3
4,580	14.5	11.5	9.9	8.0	6.8
17,400	15.5	12.0	10.5	8.5	7.2

The mass curve (accumulation) of monthly runoff given in Figure 5.17 shows low flow for the period June, 1953 through February, 1954 and again June, 1954 through February, 1955. This type of information is necessary for determining the amount of reservoir storage needed to provide water for a uniform daily demand of a municipality or industry.

Example 5.8: Determine the amount of reservoir storage needed to provide an urban development with an assured supply for a daily water use rate of 97,200 gal/day, given the mass curve of monthly stream flow for the continuous 28-month period January 1953 through April 1955 (Figure 5.17).

Solution: Calculate the uniform stream flow (cfs) needed to meet the uniform water use demand rate of 97,200 gal/day (7.48 gal $= 1$ ft³).

$$\frac{97,200 \text{ gal}}{\text{day}} \times \frac{1 \text{ day}}{24 \text{ hr}} \times \frac{1 \text{ hr}}{60 \text{ min}} \times \frac{1 \text{ min}}{60 \text{ sec}} \times \frac{1 \text{ ft}^3}{7.48 \text{ gal}} = 0.15 \text{ ft}^3/\text{sec}$$

Total volume of runoff for a year at a uniform stream flow of 0.15 cfs for this catchment of 303 acres is:

$$\frac{0.15 \text{ ft}^3}{\text{sec}} \times \frac{60 \text{ sec}}{1 \text{ min}} \times \frac{60 \text{ min}}{1 \text{ hr}} \times \frac{24 \text{ hr}}{1 \text{ day}} \times \frac{365 \text{ day}}{1 \text{ yr}} \times \frac{1}{303 \text{ acre}}$$

$$\times \frac{1 \text{ acre}}{43,560 \text{ ft}^2} \times \frac{12 \text{ in.}}{1 \text{ ft}} = 4.3 \text{ in.}$$

Start the uniform flow line at zero-point (*a*) (Figure 5.17) and at the end of the year the accumulation is 4.3 inches. The slope of this straight line is 0.15 cfs or 97,200 gal/day. Prior to point (*b*) on the mass curve (Figure 5.17), stream flow accumulated faster than the required uniform use rate of 97,200 gal/day and no help from storage is needed. From point *b* to *c*, the flow rate is less than the use rate and water from storage is needed to augment the natural stream flow rate, and the dotted vertical line from point *c* to the dashed line *b-d* indicates that 60 ac-ft of water from storage is required. After

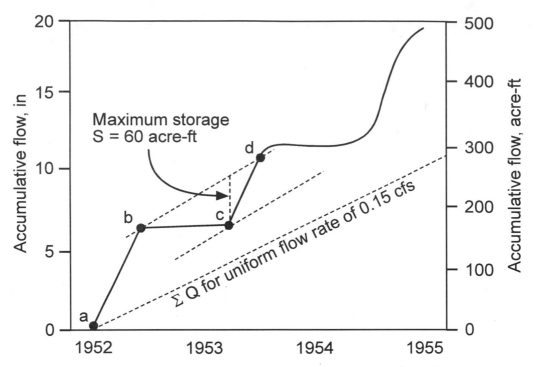

Figure 5.17. Mass curve of monthly streamflow from 303 acres at Coshocton, OH (Harrold, 1957).

point *c,* the mass curve of stream flow shows the flow rate exceeds the use rate. If the reservoir had been full at point *b,* its content would have decreased to a minimum at time *c* and then would be full again by time *d.*

Answer: The amount of reservoir storage needed is 60 acre-ft.

REFERENCES

Barfield, B.J., R.C. Warner, and C.T. Haan. 1981. Applied Hydrology and Sedimentology for Disturbed Areas. Oklahoma Technical Press, Stillwater, OK.

Bedient, P.B. and W.C. Huber. 1988. Hydrology and Floodplain Analysis. Addison-Wesley Publishing Company.

DeCoursey, D.G. 1966. A Runoff Hydrograph Equation. Publication ARS 41-116, U.S. Department of Agriculture, Washington, D.C.

Erie and Niagara Counties Regional Planning Board. 1981. Storm Drainage Design Manual. Erie and Niagara Counties Regional Planning Board, Grand Island, New York.

Harrold, L.L. 1957. Minimum Water Yield from Small Agricultural Watersheds. Trans. Amer. Geo. Union 38(2):201–208.

Haan, C.T. 1970. A Dimensionless Hydrograph Equation. File Report, Agricultural Engineering Department, University of Kentucky, Lexington, KY.

Haan, C.T., H.P. Johnson, and D.L. Brakensiek (Editors). 1982. Hydrologic Modeling of Small Watersheds. ASAE Monograph No. 5, American Society of Agricultural Engineers, St. Joseph, MI.

Henderson, F.M. and R.A. Wooding. 1964. Overland Flow and Groundwater Flow from a Steady Rainfall of Finite Duration. Geophysical Research 69:1531–1540.

Huber, W.C., J.P. Heaney, S.J. Nix, R.E. Dickinson, and D.J. Polmann. 1981. Storm Water Management Model User's Manual Version III, EPA-600/2-84-109a, Environmental Protection Agency, Athens, GA.

Huff, F.A. 1967. Time Distribution of Rainfall in Heavy Storms. Water Resources Research 3(4):1007–1019.

Kirpich, P.Z. 1940. Time of Concentration of Small Agricultural Watersheds. Civil Eng. 10:362.

Kirkby, M.J. (editor). 1978. Hillslope Hydrology. John Wiley & Sons, New York.

Lighthill, F.R.S., and C.B. Whitham. 1955. On Kinematic Waves, Flood Movement in Long Rivers. Proc. Royal Society of London, Vol. 229 (1178):281–316.

McCuen, R.H., S.L. Wong, and W.J. Rawls. 1983. Estimating the time of concentration of urban watersheds. Proceedings of the ASCE Conference on Frontiers in Hydraulic Engineering, Cambridge, Massachusetts, August 9–12, pp 547–552.

Moody, W.T. 1966. Nonlinear differential equation of drain spacing. Proceeding ASCE 92(R2):1–9.

Overton, D.E., and W.L. Troxler. 1978. Regionalization of Stormwater Response. Paper presented at the American Geophysical Union Meeting, Miami, Florida, April 17–21.

Overton, D.E. and E.C. Crosby. 1979. Effects of Contour Coal Strip Mining on Stormwater Runoff and Quality. Report to the U. S. Department of Energy. Civil Engineering Department, University of Tennessee, Knoxville, Tennessee.

Schwab, G.O., D.D. Fangmeier, W.J. Elliot, and R.K. Frevert (4th Edition). 1993. Soil and Water Conservation Engineering, 4th Edition. John Wiley & Sons, Inc., New York.

Sherman, L.K. 1932. Stream flow from Rainfall by Unit Graph Method, Eng. News-Record 108:501–505.

Soil Conservation Service. 1972. National Engineering Handbook, Section 4, Hydrology, SCS.

Soil Conservation Service. 1973b. Computer Program for Project Formulation Hydrology. Technical Release No. 20, U.S. Department of Agriculture, Washington, D.C.

Soil Conservation Service. 1984. Engineering Field Manual (including Ohio Supplement). U.S. Department of Agriculture, Washington, D.C.

Soil Conservation Service. 1986. Urban Hydrology for Small Watersheds, 2nd Ed. Technical Release No. 55, U.S. Department of Agriculture, Washington, D.C.

Terstriep, M.L., and J.B. Stall. 1974. The Illinois Urban Drainage Area Simulator, ILLUDAS. Bull. 85, Illinois State Water Survey, Urbana, Illinois.

USDA. 1987. Farm Drainage in the United States: History, Status, and Prospects. Miscellaneous Publication No. 1455, Economic Research Service, U.S. Department of Agriculture, Washington, D.C.

Van Schilfgaarde, J. 1963. Tile Drainage Design Procedure for Falling Water Tables. Proceedings ASCE 89(1R2).

Ward, A.D., B.N. Wilson, T. Bridges, and B.J. Barfield. 1980. An Evaluation of Hydrologic Modelling Techniques for Determining A Design Storm Hydrograph. Proceedings, International Symposium on Urban Storm Runoff, University of Kentucky, Lexington, KY, July 28–31.

Problems

5.1   Convert 1 inch per hour of runoff from an area of 10 acres to cubic feet per second (cfs) and to gallons per minute (gpm). (7.48 gal = 1 ft^3)

5.2   Determine the volume of water (acre feet) that could be added to a reservoir in one day from a stream with an average flow rate of 10 cfs (cubic feet per second).

5.3   Determine the time (hours) required to fill a 20 acre-ft reservoir if the average stream flow was 2 cfs (cubic feet per second).

5.4   Determine the runoff volume in inches from a 50-year return period storm at your present location assuming antecedent moisture condition III, SCS curve number 75 (condition II) and storm duration of 4 hours.

5.5   Determine the volume of runoff in acre feet for a flood from a 500-acre watershed at your present location during the growing season. Critical duration of storm causing the flood was 3 hours and the return period of the event was 10 years. The antecedent rainfall 5 days prior to the storm was estimated as 1.5 inches. Soils, land use, and hydrologic conditions for subareas follows:

Subarea Acres	Soil Group	Land Use Treatment and Hydrologic Condition
100	C	residential 1 acre lots, good condition
300	D	cultivated conservation tillage
100	B	forest, good cover

5.6   Determine the runoff depth and runoff volume for Example 5.2 by calculating an area weighted curve number and then using it with Equations 5.2 and 5.3. Determine the percentage difference between your answers and the answers presented in Example 5.2.

5.7   Determine the time of concentration for a watershed with a hydraulic length of 1000 ft and mean slope of 0.05%.

5.8   If the watershed in Problem 5.7 is located in St. Louis, Missouri determine the rainfall rate associated with a storm with a 100-year return period and duration equal to the time of concentration (use Figure 2.17 in Chapter 2).

5.9   Compute the peak runoff rate (cfs) that would occur once in 10 years for a 50-acre catchment using the Rational Method. It is located near St. Louis, Missouri and has an average slope of 0.5%, a maximum flow length of 2,000 ft, meadow (in crop rotation), and soils in Hydrologic Soil Group C.

5.10   Solve Problem 5.9 using the Graphical Peak Discharge Method.

5.11 Compute the 25-year peak runoff rate from a 30-acre catchment in your area using the Graphical Peak Discharge Method. The watershed is a low density residential area with Hydrologic Soil Group D, an average slope of 1.0%, and a flow length of 1,140 feet. Compare answers obtained with AMC II and AMC III.

5.12 Compute the peak runoff rate from a 100-acre catchment using the Rational Method. The estimated 100-year rainfall rate for the catchment is 7.0 inches per hour. Corn (cp) is grown on 20 acres of soil B and remainder of catchment is in permanent pasture, soil Group C. (Use a weighted average for the runoff coefficient.)

5.13 Use a single triangle unit hydrograph procedure to develop a storm hydrograph for a 1-hour, 50-year storm event on a proposed 100-acre commercial and business watershed with soils in Hydrologic Soil Group C, and a proposed residential land use (1/2 acre lots). The watershed has average overland slopes of 1% and a hydraulic length of 6,000 feet. The rainfall depth during the one hour storm event was 2.0 inches.

5.14 Write a computer program or spreadsheet solution to develop a storm hydrograph using unit hydrograph procedures. Use your procedure to solve Problem 5.13 if there were 0.25, 1.00, 0.50, and 0.25 inches of rainfall during each 15 minutes of the rainfall event.

5.15 Determine the drain spacing if drains are located 4 feet below the ground surface, the depth of the impeding layer is 15 feet below the ground surface, and the saturated hydraulic conductivity is 0.25 ft^3/day. For good plant growth it is required that the water table be at least 3 feet below the ground surface and it will be necessary to remove excess rainfall at a rate of 0.72 in/day.

5.16 Compute the depth of water in inches from a 303-acre catchment to provide a park with a water usage of 0.1 cfs. From the mass curve in Figure 5.17 estimate graphically the required reservoir storage in acre-ft to assure an adequate water supply for this 1953–1954 drought period.

5.17 From Table 5.8 determine the minimum 12-month water yield, in acre-feet, for a 30-acre catchment to be expected once in 5 years and once in 50 years.

5.18 From Table 5.8 determine the minimum 12-month water yield in inches to be expected once in 10 years from a 30-acre catchment and from a 4600-acre catchment. Explain the greater flow from the larger catchment.

CHAPTER 6

Soil Erosion and Control Practices

William J. Elliot and Andy D. Ward

6.1. INTRODUCTION

Erosion is one of the most important and challenging problems to natural resource managers worldwide. It is the main source of sediment that pollutes streams and fills reservoirs. Some estimates of erosion rates in the 1970s were as high as 4 billion tons annually in the United States of America (Schwab et al., 1993). This amount dropped to 3 billion tons in 1982, and was estimated at 2.13 billion tons in 1993 due to recent advances in soil erosion control and reduced acres under cultivation (SCS, 1994).

In recent years, greater emphasis has been given to erosion as a contributor to nonpoint pollution. "Nonpoint" refers to erosion from the land surface rather than from industries, feedlots, or gullies. Eroded sediment can carry nutrients, particularly phosphates, to waterways, and contribute to eutrophication of lakes and streams. Adsorbed pesticides are also carried with eroded sediments, lowering surface water quality.

Soil erosion can also reduce the productivity of some soils (Lowdermilk, 1953; Shertz et al., 1989). Eroded sediments remove soil organic matter, degrading soil structure and reducing its fertility. On shallow soils, the loss of topsoil can lead to reduced availability of soil water to plants resulting in restricted growth because of drought stress (Figure 6.1).

The two major types of erosion are geological erosion and erosion from human or animal activities. Geological erosion includes soil-forming as well as soil-eroding processes that maintain the soil in a favorable balance, suitable for the growth of most plants. Geological erosion has contributed to the formation of our soils and caused many of our present topographic features, such as canyons, stream channels, and valleys. Conversely, human tillage or vegetation removal by animals or other natural events may cause accelerated erosion, which leads to a loss of soil productivity.

Water erosion is the detachment and transport of soil from the land by water, including rainfall and runoff from melted snow and ice. Types of water erosion

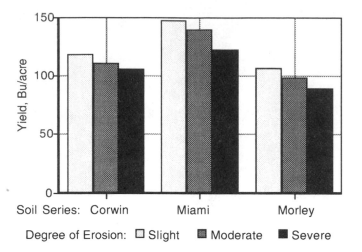

Figure 6.1.   Effect of erosion phase and soil series on crop yield. (Based on data by Shertz et al., 1989.)

include interrill (raindrop and sheet), rill, gully, and stream channel erosion. Water erosion is accelerated by farming, forestry, grazing, and construction activities.

6.2. FACTORS AFFECTING EROSION BY WATER

The major variables affecting soil erosion are climate, soil, vegetation, and topography. Of these, the vegetation, and to some extent the soil and topography, may be controlled. Climatic factors are beyond human control.

6.2.1. Climate

Climatic factors affecting erosion are temperature, humidity, solar radiation, wind, and precipitation. Temperature, humidity, solar radiation and wind are most evident through their effects on evaporation and transpiration. These processes reduce soil water content, and subsequently decrease surface runoff rates and erosion. Wind also changes raindrop velocities and the angle of impact, which may influence erosion rates. The relationship between precipitation characteristics, runoff and soil loss is complex, and a more complete discussion follows.

6.2.2. Soil

Soil properties affect the infiltration capacity and the extent to which soil particles can be detached and transported. For example, clay particles are more difficult to detach than sand, but clay is more easily transported. Texture is the dominant property determining erodibility, but soil structure, organic matter, water content, and density or compactness, as well as chemical and biological characteristics of the soil also influence erodibility (Elliot et al., 1993).

6.2.3. Vegetation

The major effects of vegetation in reducing erosion are (1) protecting the soil from raindrop impact, (2) reducing surface runoff velocity, (3) holding soil in place, (4) improving soil structure with roots, plant residue, and increased biological activity in the soil, and (5) increasing transpiration rates. These vegetative influences vary with the species, climate, season, soil type and degree of maturity of the vegetation, as well as the type of vegetative residue left from the previous crop, like roots, stems, or leaves.

6.2.4. Topography

Topographic features that influence erosion are slope steepness, length, and shape. On steep slopes, runoff water is more erosive, and can more easily transport detached sediment downslope. On longer slopes, an increased accumulation of overland flow results in increased rill erosion. Concave slopes, with less steep slopes at the foot of the hill, are less erosive than convex slopes. The relationship of the upland slopes to channels in a watershed, and the sediment transport capabilities and stability of the channel will also influence the total erosion within a watershed.

6.3. TYPES OF EROSION

6.3.1. Interrill Erosion

Interrill erosion includes raindrop splash and erosion from shallow overland flow (Figure 6.2). Splash erosion results from the impact of rain drops directly on soil particles or on thin water surfaces. Although the impact of raindrops on shallow water surfaces may not splash soil, it does increase turbulence, providing a greater sediment-carrying capacity.

Early erosion studies found that tremendous quantities of soil are splashed into the air, most particles more than once. The amount of soil splashed into the air was found to be 50 to 90 times greater than the runoff losses. Splashed particles may move more than 2 ft in height and more than 5 ft laterally on level surfaces (Ellison, 1947).

If raindrops fall on crop residue or growing plants, the raindrop energy is absorbed and soil splash is reduced. Raindrop impact on bare soil not only causes splash, but also leads to crusting and increased runoff. Raindrops detach the soil particles, and the detached sediment can reduce the infiltration rate by sealing the soil pores. Shallow overland flow detaches some sediment, but mainly transports sediment that was first detached by raindrops. Small overland flow rates may result in reduced interrill erosion because of low sediment transport capacity. Research has shown interrill erosion to be a function of soil properties, rainfall intensity, slope (Watson and Laflen, 1986; Liebenow et al., 1990), and in some cases, runoff rate (Kinnell and Cummings, 1993).

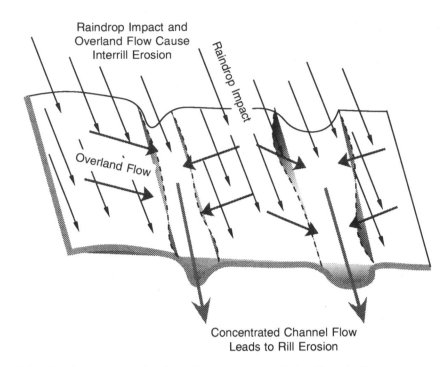

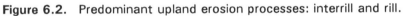

Figure 6.2. Predominant upland erosion processes: interrill and rill.

6.3.2. Rill Erosion

Rill erosion is the detachment and transport of soil by a concentrated flow of water. Rills are small enough to be removed by normal tillage operations. Rill erosion depends on runoff rate which is affected by rainfall intensity, soil infiltration rates, and length of slope contributing to overland flow.

Rill erosion is the dominant form of erosion on longer and steeper slopes, whereas interrill erosion is more dominant in shorter, flatter conditions. The interrill and rill erosion processes are incorporated in several physically-based erosion prediction computer models, including The Water Erosion Prediction Project (WEPP) (Lane and Nearing, 1989).

6.3.3. Gully Erosion

Gully erosion produces channels larger than rills. These channels carry water during and immediately after rains. Compared to rills, gullies cannot be obliterated by tillage. The amount of sediment from gully erosion is usually less than from rill and interrill erosion, but the nuisance from having fields divided by large gullies has been the greater problem.

The rate of gully erosion depends primarily on the runoff-producing characteristics of the watershed, the drainage area, soil characteristics, the alignment, size, and shape of the gully, and the slope in the channel (Bradford et al., 1973). Gully formation can be reduced or halted by diverting runoff away from the gully, or by installing engineering structures at the head of the gully (Schwab et al., 1993).

6.3.4. Stream Channel Erosion

Stream channel erosion consists of soil removal from stream banks or soil movements in the channel. Stream banks erode either by runoff flowing over the side of the stream bank, or by scouring and undercutting below the water surface. Stream bank erosion, less serious than scour erosion, is often increased by the removal of vegetation, overgrazing, tilling too near the banks, or straightening the channel. Scour erosion is influenced by the velocity and direction of flow, depth and width of the channel, and soil texture. Poor channel alignment and the presence of obstructions such as sandbars increase meandering. Meandering is the major cause of erosion along the bank. Straightening channels, however, may increase the rate of scour erosion. Additional discussion is presented in Chapter 7.

6.4. ESTIMATING SOIL LOSSES

Soil losses, or relative erosion rates for different management systems, are estimated to assist farmers, natural resource managers, and government agencies in evaluating existing management systems or in future planning to minimize soil losses. In the period from 1945 until 1965, a method of estimating losses based on statistical analyses of small field plot data from many states was developed, which resulted in the Universal Soil Loss Equation (USLE) (Wischmeier and Smith, 1978). A revised version of the USLE (RUSLE) has been developed for computer applications, allowing more detailed consideration of farming practices and topography for erosion prediction (Renard et al., 1991).

Since the mid-1960s, scientists have been developing process-based erosion computer programs which estimate soil loss by considering the processes of infiltration, runoff, detachment, transport, and deposition of sediment. Numerous research computer models have been developed, and some of these computer models are being improved for field use. The process-based WEPP model will be replacing the USLE in the SCS and other U.S. government agencies (Foster, 1988).

6.5. THE UNIVERSAL SOIL LOSS EQUATION

The USLE continues to be a widely accepted method of estimating sediment loss despite its simplification of the many variables involved in soil loss prediction. It is useful for determining the adequacy of conservation measures in resource planning, and for predicting nonpoint sediment losses in pollution control programs. The average annual soil loss, as determined by Wischmeier and Smith (1978) can be estimated from the equation

$$A = R \ K \ LS \ C \ P \tag{6.1}$$

where A = average annual soil loss in tons/acre,
        R = rainfall and runoff erosivity index for a geographic location (Figure 6.3),
        K = soil-erodibility factor (Figure 6.5),
        LS = slope steepness and length factor (Figure 6.6),
        C = cover management factor (Table 6.1 or Table 6.2),
        P = conservation practice factor (Table 6.3).

   In this text, English units are used for A, R, and K. Readers should ensure that if using other sources of information for the USLE, that units are consistent (Foster et al., 1981). Although developed for use in the United States, the procedure is used in many countries and has been the focus of considerable study during the past 30 years. Methods for determining each of the input parameters for the USLE and examples of their use follow.

6.5.1. Rainfall Erosivity, *R*

   The rainfall and runoff erosivity index, R varies with amount of rainfall and the individual storm precipitation patterns. For a given storm, a rainfall and runoff erosivity index (EI) is calculated. It is the product of the kinetic energy of the storm, E and the maximum 30-minute intensity for that storm, I. An example calculation was presented in Chapter 2. The EI values for all the storms occurring in a given year for a location are summed to give an annual erosivity index. The average annual rainfall and runoff erosivity index, R is shown in Figure 6.3 for the continental USA.

   Example 6.1: What is the R factor for a farm located in northeast Iowa?

   Solution: From Figure 6.3, determine R for northeast Iowa.

   Answer: $R = 170$.

For development projects or land uses which change throughout the year, it is sometimes desirable to determine the rainfall erosivity associated with different periods during the year. Figure 6.4 shows the cumulative distribution of rainfall erosivity for different locations within the U.S. Estimates of partial year erosivity can be determined by distributing the erosivity determined from Figure 6.3 with the percentages given in Figure 6.4.

   Example 6.2: Determine the rainfall erosivity for northeast Iowa for the period from April 1 to October 1.

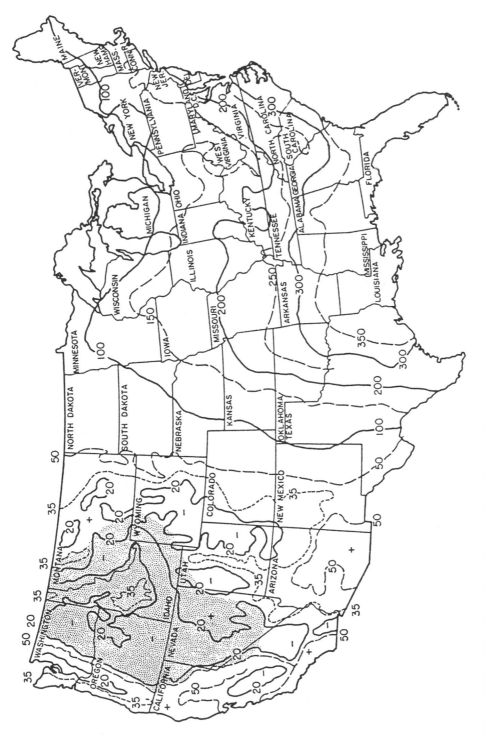

Figure 6.3. Rainfall and runoff erosivity index, *R* distribution in the United States   (Wischmeier and Smith, 1978).

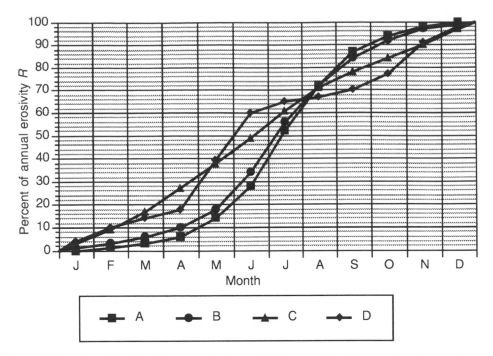

Figure 6.4   Monthly distribution of the rainfall erosivity index, *R*: Curve A: Iowa, Nebraska, South Dakota; Curve B, Missouri, Illinois, Indiana, and Ohio; Curve C, Louisiana, Mississippi, Tennessee, and Arkansas; Curve D, Atlantic Coastal Plains of Georgia and the Carolinas. (Based on Wischmeier and Smith, 1978.)

Solution: In Example 1 we determined that the annual erosivity was 170. On Figure 6.4, the climate in Iowa is described by Curve A. On April 1, curve A is at 4%, and on October 1, Curve A is at 90%. Therefore, the % erosivity that occurs during this period is 90 − 4 = 86%.

Answer: The erosivity that occurs between April 1 and October 1 in northeast Iowa is: $R = 170 \times 0.86 = 146$.

6.5.2. Soil Erodibility, *K*

The soil-erodibility factor *K* for a series of benchmark soils was obtained by direct soil loss measurements from fallow plots located in many U.S. states (Figure 6.7). Soils that have high silt contents tend to be the most erodible. The presence of organic matter, stronger subsoil structure, and greater permeabilities generally decrease erodibility. In RUSLE, *K* varies to account for seasonal variation in soil erodibility, with higher erodibility in the spring and/or after tillage.

Soil erodibility factors for the ten most common soils in each state are presented in Appendix D. A consideration with using the information in Appendix D is that

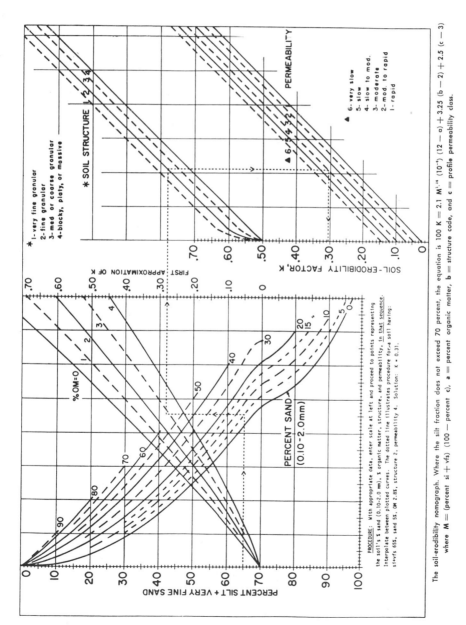

The soil-erodibility nomograph. Where the silt fraction does not exceed 70 percent, the equation is 100 K = 2.1 M$^{1.14}$ (10^{-4}) (12 − a) + 3.25 (b − 2) + 2.5 (c − 3) where M = (percent si + vfs) (100 − percent c), a = percent organic matter, b = structure code, and c = profile permeability class.

Figure 6.5. Nomograph to determine soil erodibility *K* factor (Wischmeier and Smith, 1978).

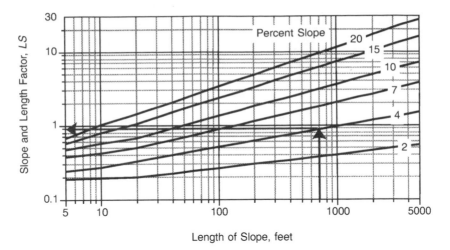

Figure 6.6.   Graph to determine slope length and steepness factor, *LS*.

land use activities might change the soil structure and/or permeability of a soil. This is particularly true if the land is used for agricultural purposes, or if it has been severely disturbed by construction activities. It can also be seen that some soils may exhibit a wide range of erodibilities. These problems can be addressed by applying the nomograph presented in Figure 6.5 to determine the erodibility of a given soil.

> **Example 6.3:** Determine the soil erodibility, *K* for a soil with the following properties: 65% silt and very fine sand; 5% sand; 2.8% organic matter content (OM); a fine granular soil structure; and a slow to moderate permeability.
>
> **Solution:** In Figure 6.5, enter the left graph at silt plus very fine sand = 65 percent; move right and intersect with 5% sand; move up to 2.8% organic matter; move right to soil structure = 2; move down to permeability = 4; and then move left to determine *K*.
>
> **Answer**: *K* = 0.31

6.5.3. Topographic Factor, *LS*

The topographic factor, *LS*, adjusts the predicted erosion rates to give greater erosion rates on longer and/or steeper slopes and lesser erosion rates on shorter and/or flatter slopes, compared to a USLE "standard" slope of 9% and slope length of 72.6 ft. The erosion plot in Figure 6.7 approximates these topographic conditions.

The slope length is measured from the point where surface flow originates (usually the top of the ridge) to the outlet channel or a point downslope where deposition begins. RUSLE considers nonuniform concave or convex slopes (Renard et al., 1991), as do most process-based erosion prediction programs For the USLE, the *LS* factor can be determined from Figure 6.6.

Table 6.1. Cover management C factors for permanent pasture, rangeland, and idle land[1] (Cooperative Extension Service and The Ohio State University, 1979).

Vegetal Canopy			Cover that Contacts the Surface					
Type and height of Raised Canopy[2]	Canopy Cover[3]	Type[4]	Percent Ground Cover					
			0	20	40	60	80	95–100
Column No:	2	3	4	5	6	7	8	9
No appreciable canopy		G	0.45	0.20	0.10	0.042	0.013	0.003
		W	0.45	0.24	0.15	0.090	0.043	0.011
Canopy of tall weeds or short brush (1.5 ft fall ht.)[2]	25	G	0.36	0.17	0.09	0.038	0.012	0.003
		W	0.36	0.20	0.13	0.082	0.041	0.011
	50	G	0.26	0.13	0.07	0.035	0.012	0.003
		W	0.26	0.16	0.11	0.075	0.039	0.011
	75	G	0.17	0.10	0.06	0.031	0.011	0.003
		W	0.17	0.12	0.09	0.067	0.038	0.011
Appreciable brush or bushes (6 ft fall ht.)[2]	25	G	0.40	0.18	0.09	0.040	0.013	0.003
		W	0.40	0.22	0.14	0.085	0.042	0.011
	50	G	0.34	0.16	0.085	0.038	0.012	0.003
		W	0.34	0.19	0.13	0.081	0.041	0.011
	75	G	0.28	0.14	0.08	0.036	0.012	0.003
		W	0.28	0.17	0.12	0.077	0.041	0.011
Trees but no appreciable low brush (12 ft. fall ht.)[2]	25	G	0.42	0.19	0.10	0.041	0.013	0.003
		W	0.42	0.23	0.14	0.087	0.042	0.011
	50	G	0.39	0.18	0.09	0.040	0.013	0.003
		W	0.39	0.21	0.14	0.085	0.042	0.011
	75	G	0.36	0.17	0.09	0.039	0.012	0.003
		W	0.36	0.20	0.13	0.083	0.041	0.011

[1] All values shown assume: (1) random distribution of mulch or vegetation, and (2) mulch of appreciable depth where it exists.

[2] Average fall height of waterdrops from canopy to soil surface.

[3] Percent of total-area surface that would be hidden from view by canopy in a vertical projection.

[4] G: Cover at surface is grass, grasslike plants, decaying compacted duff, or litter at least 2 in. deep.   W: Cover at surface is mostly broadleaf herbaceous plants (as weeds) with little lateral-root network near the surface, and/or undecayed residue.

Table 6.2.   Example of typical cover management *C* factors developed by state agencies for the Ohio climate and vegetation conditions (Cooperative Extension Service and The Ohio State University, 1979).

Vegetation	Tillage		Practice	
	Autumn Conventional	Spring Conventional	Spring Conservation	No-till
Agricultural Rotations				
Continuous Corn (Co)	0.40	0.36	0.27	0.10
Co and Soybeans (Sb)	0.42	0.37	0.24	0.10
Two-year rotation with corn grown in year 1 and soybeans and wheat in year 2	0.30	0.28	0.24	0.10
Co Co Oats (O) Meadow (M)	0.13	0.12	0.10	0.064
Co O M M	0.055	0.050	0.033	0.033
Co Co O M M	0.11	0.094	0.082	0.052
Co Sb O M M	0.13	0.12	0.082	0.052
Permanent Pasture				
Poor Condition				0.04
Good Condition				0.01

Table 6.3. Conservation Practice Factor P for the USLE (Wischmeier and Smith, 1978; and Bengston and Sabbagh, 1990).

Farming up and down slope		
All crops		$P = 1.0$

Contour Farming		
Land Slope Percent	Maximum Slope Length[a] (feet)	P factor
1 to 2	400	0.6
3 to 5	300	0.5
6 to 8	200	0.5
9 to 12	120	0.6
13 to 16	80	0.7
17 to 20	60	0.8

Strip Cropping	
With grass and row crop	Contour P x 0.5
With small grain and row crop	Contour P x 0.67

Terraces	
Loss from crop	Same as Contouring P
Loss from Terrace	
With graded channel outlet	Contour P x 0.2
With underground outlet	Contour P x 0.1

Subsurface Drainage	$P = 0.6$

[a] Maximum slope length for strip cropping can be twice contouring.

Figure 6.7. Fallow soil erosion plot located near Pullman, WA with a standard length of 72.6 ft and a slope steepness of about 9%.

Example 6.4: Determine the *LS* factor for the field with a slope steepness of 4% and a slope length of 700 ft.

Solution: From Figure 6.6, enter the x-axis at 700 ft. Move up until you intersect the 4% slope line. Move left to find the *LS* factor.

Answer: $LS = 0.9$.

6.5.4. Cover-Management Factor, *C*

The cover-management factor, *C,* includes the effects of vegetative cover, crop sequence, productivity level, length of growing season, tillage practices, residue management, and the expected time distribution of erosive events. For agricultural systems, *C* factors are generally based on crop rotations and tillage sequences. For forest, rangeland, and other non-agricultural conditions, *C* factors are generally estimated from the density of vegetation, and the amount of vegetative residue on the soil surface.

Table 6.1 provides *C* factors which are useful for many different land uses in much of the United States, provided that the land use exhibits some vegetative cover. For a disturbed bare soil, like Figure 6.7, a value of 1.0 or greater should be used. The table is particularly valuable when relative soil losses associated with land use changes are being determined. Erosion from permanent pasture, rangeland, and forest is generally much lower than from agricultural lands. Human and/or livestock activities that disturb the vegetation, like roads, grazing, or timber harvest are generally the source of most of the eroded sediment from rangeland and forests. Forest erosion processes are discussed in greater detail in Chapter 7.

Example 6.5: Logging operations in Pennsylvania result in part of a forest being clearcut. Immediately following the clearcutting, there is no appreciable canopy, but a 40% ground cover of decaying compacted duff. A year after the clearcutting, there is a 75% canopy of brush and short trees (4–8 ft high). Determine the change in the *C* factor during the first year following clearcutting.

Solution: From Table 6.1, the *C* factor immediately following clearcutting is 0.15. If we assume the percent ground cover does not change during the first year, the *C* factor at the end of this period will be 0.12. Therefore, based on the change in the *C* factor, soil losses at the end of the period will be 80% of that immediately following clearcutting.

Answer: The change in canopy cover during the first year following clearcutting will result in a 20% reduction in the *C* factor.

Table 6.2 gives some typical agricultural *C* factors for Ohio. These values should provide reasonable estimates for the central and eastern United States. However, these factors do not account for climate differences between locations which might influence seasonal changes in C factors. Also, the table does not fully consider differences in the amount of residue left in a field following harvest. Wherever possible, you should consult local SCS or other natural resource agency offices to obtain C factors for your region.

From Table 6.2, it can be seen that the C factor will vary based on the crop and tillage practice. The traditional functions of tillage were weed control and preparation of a seedbed in which traditional planters and drills would function. With the development of conservation tillage planters and drills, tillage management has become increasingly important as a conservation tool. Currently, the primary purpose of tillage operations is to provide an adequate soil and water environment for the seedling to germinate. The role of tillage in weed control has diminished with increased use of herbicides and improved timing of operations.

Conventional tillage tends to incorporate the majority of surface residue, leaving the surface bare and susceptible to erosion. Conservation tillage systems leave residue from previous crops on the surface, protecting the soil from raindrop impact and reducing the erosive shear of sheet and rill flow on the soil. The presence of residue significantly reduces erosion rates, as can be seen by the lower C factors for all the conservation systems listed in Table 6.2.

One of the major benefits of reduced tillage is the increased amount of residue left on the surface, which reduces runoff and erosion. Other methods for reducing runoff, such as putting checks in furrows in cropland or imprinting rangeland can also reduce erosion.

Example 6.6: As part of a conservation plan, a farmer in Indiana decided to place part of her land into pasture. Originally, the land was in continuous corn production and the farmer used conventional tillage in the autumn. Determine the relative soil reduction associated with this land use change, assuming the worst case condition for pasture.

Solution: From Table 6.2, it is determined that continuous corn production and conventional tillage in the autumn has a C factor of 0.40. The worst case for the pasture, poor condition, has a C factor of 0.04. 0.04 is 10% of 0.40.

Answer: Converting the land use into a pasture will reduce soil loss by 90 percent.

6.5.5. Erosion Control Practice, P

The vegetation cover or farming system selected can have a major effect on soil erosion rates. There are practices besides vegetation management that can be employed to control erosion. Erosion control practices include contouring, strip cropping, and terracing. For many applications, no erosion control practices will be used, and the P factor will be 1.0. Table 6.3 presents typical P factors which can be used to estimate the effect of conservation practices on predicted soil erosion.

6.5.5.1. Contouring

Contouring is the performing of field operations, such as tillage, planting, and harvesting, approximately on the contour. Contouring reduces surface runoff by impounding water in small depressions and decreasing the development of rills. The greatest concentration of contouring in the U.S. is in the eastern wheat belt where the benefits include both erosion reduction and conservation of water. The design of contours requires site surveying, and is discussed in detail in Schwab et al. (1993).

The relative effectiveness of contouring erosion on various slopes is shown by the conservation practice factor, P in Table 6.3. The benefits of contouring decrease as the slope increases because the water holding capacity of the rows diminishes with an increase in slope. Also, there are practical limitations on the slope length for which contouring is effective. On 2% slopes, the maximum slope length which should be contoured is about 400 feet, while for a 12% slope, the maximum length is only 120 feet. Contouring on steep slopes or under conditions of high rainfall intensity and soil erodibility will increase the risk of gullying because row breaks may release the stored water. Row breaks cause cumulative damage as the volume of water increases with each succeeding row.

6.5.5.2. Strip Cropping

Strip cropping is the practice of growing alternate strips of different crops in the same field. For controlling water erosion, the strips are on the contour (Figure 6.8). Rotations that provide strips of close-growing perennial grasses and legumes, alternating with grain and row crops, are the most effective practice to reduce erosion by water. Generally, strips that are farmed in row crops have parallel borders, and widths that are convenient for multiple row equipment operation. Maximum slope lengths for strip cropping can be twice those for contouring alone. For slope lengths longer than those recommended in Table 6.3, terracing is recommended if tillage operations are required for the management system.

The strips of grass or small grains act as sediment filters by slowing down the overland flow velocity and increasing infiltration, causing sediment to be deposited. Also, the velocity of runoff entering the next strip of crop is reduced, and therefore causes less soil erosion. The reason the grass strips act as filters is because they provide more near-surface vegetative cover than row crops such as the corn or soybeans. A short term economic impact of this practice is that some of the land is no longer in crop production, or is planted to a crop such as grass, which has less economic value than corn or soybeans.

Example 6.7: A farmer decides to switch from farming up and down the slope to contouring. The farmer's field has a slope of 3%. Determine the soil loss reduction benefit of this switch in practices. Also, determine any additional benefit that might be obtained if the farmer strip crops with grass.

Figure 6.8.   Contour stripcropping in southeast Wisconsin.

Solution: From Table 6.3, a *P* factor of 0.5 is obtained for contouring on a 3% slope. Without contouring, the *P* factor would be 1.0. Therefore, contouring reduces the soil loss by 50%. Also, from Table 6.3, we see that strip cropping with grass will further reduce the *P* factor to 0.25. This will result in a 75% reduction (1 minus 0.25 all times 100) in the predicted soil erosion. An additional reduction in erosion may also occur if the farmer has more years of grass (meadow) in his rotation because of a reduced *C* factor (Table 6.2).

Answer: A switch to contouring would reduce soil loss by 50 percent. Including strip cropping would reduce soil loss by 75 percent.

6.5.5.3. Terracing

A common erosion control practice is to construct terraces on eroding slopes (Figure 6.9). Terraces reduce the slope length and reduce runoff. The terrace channel may be level or graded to direct runoff to grassed waterways or subsurface drains that convey runoff away from the susceptible area. A level terrace restricts the lateral movement of water and is generally installed on soils with high infiltration rates, or in drier climates for both soil and water conservation. The details of terrace design and/or construction are presented in Schwab et al. (1993) and other engineering textbooks.

Terracing affects the slope length so that the *LS* factor in a terracing system is altered to determine the amount of sediment delivered to the terrace. In terrace systems, the *P* factor can be used to estimate the reduction in sediment yield leaving the terrace. From Table 6.3, it is possible to predict either i) sediment detached from the cropping area (the *P* factor is the same as for contouring); or ii) sediment yield leaving the field, and include the terrace factor. *P* for terrace sediment yield is 10 to 20% of the contour farming *P* factor.

Figure 6.9.   Grassed backslope terraces with subsurface drains in central Iowa.

Example 6.8: For a field with a total slope length of 700 feet, and an average steepness of 4%, find the P factor for a) Sediment delivered to terraces at 100-foot spacings; b) Sediment being transported from a graded terrace channel with a spacing of 100 feet leading to a surface channel outlet.

Solution: Refer to Table 6.3 to determine the appropriate P factors.

Answers:

a) For the sediment delivered to terraces at 100-foot spacings, P is the same as the contour farming value: $P = 0.50$. The slope length has been reduced from 700 to 100 feet so from Figure 6.6, the LS factor will reduce from 0.9 to 0.5.

b) If the terraces have a graded channel to a surface drain, then the P factor for sediment delivery is $0.5 \times 0.2 = 0.1$.

6.6. SOIL LOSS TOLERANCE, *T*

Natural resource managers sometimes find it helpful to determine a maximum level or tolerable level of erosion that can be allowed. By comparing the predicted erosion to a maximum or tolerable level, management systems can be evaluated. Both physical and economic factors, as well as social aspects, need to be considered in establishing soil-loss tolerances, sometimes called T values. These values vary with topsoil depth and subsoil properties and range from 3 to 20 tons/acre. They are the maximum rates of soil erosion that will permit a high level of crop productivity to be sustained economically and indefinitely. In some cases, criteria for control of

Table 6.4.   Details from a field in Iowa observed to be susceptible to soil erosion (USDA, 1982).

Feature	Value
Location: Clayton County in Northeast Iowa	
Soil Series and Texture: Kenyon Loam	
Erodibility K Factor	0.27
Stated T Value	5 t/ac/yr
Total Length of Slope	1200 ft
Length of Eroding Slope	720 ft
Average Steepness of Eroding Slope	3 percent
Current Farming System: Contour Strip Cropping with Grass, Spring Conventional Tillage, 1.7 t/a Spring Residue Rotation: Corn/Oats/Meadow/Meadow	

sediment pollution may dictate lower tolerance values. In other cases, it may be necessary to allow for higher levels of erosion to ensure the economic viability of a rural community.

6.7. APPLICATIONS OF THE USLE

In the previous section we have focused on determining the factors in the USLE, and determining soil loss changes associated with each factor. Often, we will be interested in determining the soil loss associated with a combination of all or most of the factors. This will require knowledge of the location, soil, land use, slope steepness, slope length, and any erosion control practices. We will present two examples of applying the USLE to estimate soil erosion. Table 6.4 describes an actual field that had been experiencing severe erosion prior to the implementation of conservation practices.

Example 6.9: Determine the average annual soil loss for the field described in Table 6.4.

Solution: From Example 6.1, $R = 170$

From Table 6.4, $K = 0.27$

From Figure 6.6 for length = 720 feet and steepness = 3%: $LS = 0.6$

Assume Iowa C factors similar to Ohio, so from Table 6.2 for Co O M M rotation, spring tillage, $C = 0.05$

From Table 6.3 for contour strip cropping, $P = 0.5 \times 0.5 = 0.25$

Answer: Applying Equation 6.1:

$$A = 170 \times 0.27 \times 0.6 \times 0.05 \times 0.25 = 0.34 \text{ tons/acre/year}$$

The above erosion rate is relatively low for an agricultural system, and is due to the combination of crop rotation and contour strip cropping, which are both very effective management techniques in reducing soil erosion. The T value for the field described in Table 6.4 is 5 t/ac/yr. The above example predicts an erosion rate of under 1 ton/acre/year, which will ensure that the productivity of that field is maintained indefinitely.

The USLE can be applied to many land uses, including agriculture, forests, pastures, rangeland, idle land, and land disturbances associated with surface mining or land development. However, when using the USLE, it is important to recognize that our knowledge of C and P is better for agricultural applications than for surface mining or other land disturbances such as urban development. Also, there is limited information available on LS factors for slopes greater than 20% and/or slope lengths greater than 2000 feet.

Example 6.10: Compare annual soil losses prior to and following reclamation of a surface mining activity in Eastern Kentucky. Disturbed areas are returned to the original contours, the average land slope is 20%, and typical slope lengths are 750 ft. Mining is being conducted in a Shelocta soil series. Prior to mining, the area was predominately a forest and disturbed areas are being reclaimed to permanent pasture. Poor soil fertility and acid conditions prevent establishment of more that 60% ground cover of grass and legumes in the pasture. Suggest conservation practices which might reduce the post-mining soil losses to less than a soil loss tolerance, T, value of 5 t/ac/yr.

Solution: From Figure 6.3, determine that for Eastern Kentucky, the R factor is 150.

From Appendix D, we find that a Shelocta soil has a K factor of 0.32.

From Figure 6.6, establish that the LS factor for a 20% slope and a 750 foot slope length is 10.

Using Table 6.1, the C factor for a forest is 0.003 (95–100% ground cover of decaying compacted duff), and the C factor for the permanent pasture will be about 0.042 (60% ground cover).

The only factor that changes when we move from the pre-mining to post-mining use is the C factor. The soil losses before and after reclamation are determined to be:

Condition	A	R	K	LS	C	P
Pre-Mining Forest	1.4	150	0.32	10	0.003	1.0
Post-Mining Grass	20	150	0.32	10	0.042	1.0

It will be difficult to reduce post-mining predicted losses from 20 to under 5 t/ac/yr, but a number of options are available. Improved management to increase the vegetative cover will reduce the C factor. If soil fertility can be improved to give 80% cover rather than 60%, the C factor will decrease from 0.042 to 0.013. Installing terraces at 250-ft spacings will reduce the LS factor from 10 to 5.5. The following table summarizes our options:

Condition	A	R	K	LS	C	P
Post-Mining Grass	20	150	0.32	10	0.042	1.0
Improved Grass	6.24	150	0.32	10	0.013	1.0
Improved Grass and Terraces	3.4	150	0.32	5.5	0.013	1.0

Answer: To ensure that predicted post-mining onsite erosion is reduced to under 5 t/ac/yr, it is necessary to improve the soil fertility to ensure an 80 percent grass cover, and to install terraces at 250-foot spacings.

In the above example, if the proposed terraces drain into a grassed waterway, the sediment leaving the site will be reduced even further with a P factor of 0.2, leading to a predicted net soil loss from the site of 0.7 t/ac/yr. Natural resource managers need to determine if their goal is to reduce on-site erosion to maintain site productivity, or to reduce off-site impacts to maintain water quality. See exercise 8 for additional examples of applications of the USLE.

6.8. DOWNSTREAM SEDIMENT YIELDS

Natural resource managers may need to know sediment delivery downstream in a watershed. One method for estimating sediment delivery is to use the USLE and a sediment delivery ratio. The USLE estimates gross sheet and rill erosion, but does not account for sediment deposited enroute to the place of measurement, nor for gully or channel erosion downstream. The sediment delivery ratio is defined as the ratio of sediment delivered at a location in the stream system to the gross erosion from the drainage area above that point. This ratio varies widely with size of area, steepness, density of drainage network, and many other factors. For watersheds of 6.4, 320, 3200, and 64,000 acres, ratios of 0.65, 0.33, 0.22, and 0.10, respectively, were suggested as average values by Roehl (1962). The development of new process-based erosion models in GIS environments will enhance the prediction of sediment delivery from watersheds.

Example 6.11: Assume that the erosion rate calculated for the field in Example 6.9 is typical of a 320-acre watershed draining into a small reservoir. Estimate the sediment delivery to the reservoir for a 20 year period.

Solution: The average annual erosion rate is 0.34 t/ac/yr. The sediment delivery ratio for a 320-acre watershed was given as 0.33.

Answer: The sediment yield from 20 years will be:

Sediment = 0.34 t/ac/yr × 320 acres × 20 years × 0.33 = 718 tons

Exercise 8 has additional applications of predicting offsite impacts of soil erosion.

*6.9. SINGLE EVENT SEDIMENT YIELDS

The USLE is useful for determining gross erosion from an area for seasonal, annual, and extended time periods. In the previous section, we have seen how a delivery ratio can be used to determine a downstream sediment yield estimate. However, applying a delivery ratio to gross erosion estimates based on the USLE cannot be used to obtain sediment yields for individual events, only long-term averages. A method called the Modified Universal Soil Loss Equation (MUSLE) (Williams, 1975) illustrates how sediment yields for individual storm events might be obtained:

$$Y = 95 \, (Q \, q)^{0.56} \, K \, LS \, C \, P \qquad (6.2)$$

where Y is the single storm sediment yield in tons, Q is the storm runoff volume in acre-ft, q is the peak discharge in cfs, and the other terms are the standard USLE factors discussed earlier. The approach has seen widespread application, but should be used with caution as it was developed empirically based on limited data for Texas and the southwestern United States. The procedure should only be used on small watersheds, and considerable judgement is required in selecting an appropriate slope length when determining the LS topographic factor. A delivery ratio should not be included if the sediment yield is determined at the outlet of the watershed used to obtain the runoff volume, Q, and the peak discharge, q. However, if a sediment yield estimate is required downstream of the watershed evaluated with the MUSLE, then a delivery ratio would also need to be used. Estimates of Q and q can be obtained by using the procedures presented in chapter 5.

Example 6.12: Determine the soil loss from the 320-acre watershed described in Examples 6.9 and 6.11 for a storm which produces 12 acre-feet of runoff and a peak discharge of 200 cfs from a 2-inch rainfall.

Solution: From Example 6.9, $K = 0.27$, $LS = 0.6$, $C = 0.05$, and $P = 0.25$

Answer: $Y = 95 \times (12 \times 200)^{0.56} \times 0.27 \times 0.6 \times 0.05 \times 0.25 = 15$ t

*6.10. ESTIMATING SEASONAL *C* FACTORS FOR AGRICULTURAL CROPS

Sometimes local *C* factors such as those presented in Tables 6.1 and 6.2 are not available. This section will demonstrate how to determine a *C* factor based on general crop properties (Table 6.5) in conjunction with a local climate (Figure 6.4).

Table 6.5 gives the seasonal losses from several typical farming systems as a percent of the soil loss for continuous fallow during the growing season. The annual distribution of erosive events varies with geographic location. Examples from different regions of the U.S. are given in Figure 6.4. The *C* factor for a given crop rotation is found by first multiplying the soil loss ratio for each growth period in a crop rotation (Table 6.5) by the percent of annual erosion during each respective period (Figure 6.4). These products are then summed and an annual average calculated to give the *C* factor (Example 6.13).

Example 6.13: Calculate the *C* factor for the field described in Table 6.4.

Solution: Table 6.6 contains the solution to this example. The rotation given in Table 6.4 is corn, oats or small grain, meadow, meadow.

1) The various stages of crop growth and their respective months are determined for each growth stage, and the months entered in column 1 of Table 6.6.

2) The *C* values from Table 6.5 for each period are entered into Table 6.6, Column 2.

3) The percent of annual erosivity for each of these periods is determined from Figure 6.6, Curve A, and also entered into Table 6.6, Column 3.

4) The product of the *C* value and the erodibility (both expressed as decimal fractions) is calculated and entered in Column 4.

5) The sum of these products is divided by the number of years in the rotation to obtain an average *C* factor for the rotation, which is 0.060.

Answer: The *C* factor for the field described in Table 6.4 is 0.060.

The calculated *C* factor of 0.06 from Table 6.4 is greater than the *C* factor of 0.05 given in Table 6.2 for Ohio conditions because the Iowa climate tends to experience more erosive storms during the critical crop period in June and July (compare the steepness of curves A and B in Figure 6.6), and different assumptions may have been made regarding the *C* factor in the Ohio data.

Table 6.5. Seasonal distribution of cover management factor expressed as percent of soil loss from crops to corresponding loss from continuous fallow for selected crop and tillage management systems (Wischmeier and Smith, 1978).

Line no, Cover, Sequence and Management[b]	Spring Planting Residue[c] (tons)	Cover[d] (%)	Soil loss ratio[a] for crop stage period and canopy cover[e] F (%)	SB (%)	1 (%)	2 (%)	3:80 (%)	90 (%)	96 (%)	4L (%)
Continuous										
1 Corn (Co), RdL, Spring, TP	2.3	-	31	55	48	38	-	-	20	23
2 As above, less residue								24	20	
6 Co, RdL, fall, TP	1.7	-	36	60	52	41	-	24	20	30
46 Co, Chisel or Cultivate Only	GP	-	49	70	57	41	-	12	9	-
63 Co Ridge Tillage on Contour	1.7	50	-	16	13	12	7	4	2	24
130 Small Grain	-	-	-	0.7	0.7	0.7	0.7	0.7	0.7	0.7
200 Meadow	1.7	60	-	16	14	12	0.8	0.6	0.4	17
Rotation										
110 Co after Beans (Bn), Spring, Conv	GP	-	47	78	65	51	-	30	25	37
124 Bn after Co, Spring, TP	GP	-	39	64	56	41	-	21	18	22
210 Rowcrop after Meadow, Rdl, Spring, TP		-	9	24	21	18	-	12	10	18
Conservation Tillage										
52 Bn or Co after Co	1.3	40	-	21	20	19	19	15	12	30
121 Co after Bn	GP	30	-	33	29	25	22	18	14	33
131 Small Grain after Co in disk residue	1.7	40	-	27	21	16	9	5	3	22

[a] The soil loss ratios, given as percentages of the loss that would occur from fallow, assume that the indicated crop sequence and practices are followed consistently.

[b] Symbols: RdL, Crop residue left in field; TP plowed with moldboard.

[c] Dry mass per acre, after winter loss and reductions by grazing or partial removal, 2 tons/acre represents a yield of 100 to 130 bu/acre, GP is good productivity level.

[d] Percentage of soil surface covered by plant residue mulch after crop seeding. The difference between spring residue and that on the surface after crop seeding is reflected in the soil loss ratios as residues mixed with the topsoil.

[e] Crop stage periods are F, Rough Fallow; SB, Seedbed until 10% canopy cover; 1, Establishment until 50% canopy cover; 2, Development until 75% canopy cover; 3, Maturing until harvest for three different levels of canopy cover; 4L, residue or stubble.

Table 6.6. Solution to Example 6.13, calculating the *C* factor for the conditions described on Table 6.4.

Months	Soil Loss in % of Continuous Fallow (Table 6.5)	Annual Erosion (Figure 6.4) Percent	*C* Factor Col 2 × Col 3
Column 1	2	3	4
Corn, first year, Line 210			
January - April	0.6	10	0.0006
May	24	10	0.024
June	21	20	0.042
July	18	20	0.036
August - September	12	30	0.036
October - December	18	10	0.018
Small Grain with Meadow Seeding, Lines 130 and 200			
January - March	18	4	0.0072
April	16	6	0.0096
May	14	10	0.014
June	12	20	0.024
July- August	4	40	0.016
September - December	0.6	20	0.0012
Meadow, Line 200			
January - December	0.6	100	0.006
January - December	0.6	100	0.006
Total		400 percent	0.242
Average		0.242 / 4 =	0.060

REFERENCES

Bengtson, R.L., and G. Sabbagh. 1990. USLE P factors for subsurface drainage on low slopes in a hot, humid climate. *J. Soil Water Cons.* 45(4):480–482.

Bertoni, J., and F. Lombardi Neto. 1993. Conservação do solo, 3rd Edição. Ìcone Editora Ltda. São Paulo, Brazil.

Bradford, J.M., D.A. Farrell, and W.E. Larson. 1973. Mathematical evaluation of factors affecting gully stability. *Soil Sci. Soc. Amer. Proc.* 37:103–107.

Cooperative Extension Service and The Ohio State University. 1979. Ohio erosion control and sediment pollution abatement guide. Columbus, OH. 24 pp.

Elliot, W.J., J.M. Laflen, and G.R. Foster. 1993. Soil erodibility nomographs for the WEPP Model. Paper No. 932046, American Society of Agricultural Engineers, St. Joseph, MI.

Ellison, W.D. 1947. Soil erosion studies, part I. **Agricultural Engineering**, April, 1947, pp 145–146.

Foster, G.R. 1988. User requirements, USDA-Water Erosion Prediction Project (WEPP). USDA-ARS National Soil Erosion Lab. W. Lafayette, IN.

Foster, G.R., D.K. McCool, K.G. Renard, and W.C. Moldenhauer. 1981. Conversion of the universal soil loss equation to SI metric units. *J. Soil Water Cons.* 36(6):355–359.

Kinnell, P.I.A., and D. Cummings. 1993. Soil/slope gradient interactions in erosion by rain-impacted flow. *Transactions of the ASAE* 36(2):318–387.

Lane, L.J., and M.A. Nearing. 1989. USDA-Water Erosion Prediction Project: Hillslope profile model documentation. NSERL Report No. 2. USDA-ARS National Soil Erosion Research Laboratory. West Lafayette, IN.

Liebenow, A.M., W.J. Elliot, J.M. Laflen, and K.D. Kohl. 1990. Interrill erodibility: Collection and analysis of data from cropland soils. *Transactions of the ASAE* 33(6):1882–1888.

Lowdermilk, W.C. 1953. Conquest of the land through 7,000 years. Agric. Information Bulletin No. 99. USDA SCS, Washington, D.C.

Renard, K.G., G.R. Foster, G.A. Weesies, and J.P. Porter. 1991. RUSLE Revised Universal Soil Loss Equation. *J. Soil Water Conserv. Soc.* 46(1):30–33.

Roehl, J.N. 1962. Sediment Source Areas, Delivery Ratios and Influencing Morphological Factors. Intern. Assoc. Scientific Hydrology, Commission of Land Erosion. Publ. No. 59.

Schertz, D.L., W.C. Moldenhauer, S.J. Livingston, G.A. Weesies, and E.A. Hintz. 1989. Effect of past soil erosion on crop productivity in Indiana. *J. Soil Water Conserv. Soc.* 44(6):604–608.

Schwab, G.O., D.D. Fangmeier, W.J. Elliot, and R.K. Frevert. 1993. *Soil and Water Conservation Engineering, Fourth Edition*. John Wiley & Sons, Inc., New York. 507 pp.

USDA SCS. 1982. Soil Survey of Clayton County, Iowa. Washington, D.C. 356 pp.

USDA SCS. 1994. National Resources Inventory. Washington, D.C.

Watson, D.A., and J.M. Laflen. 1986. "Soil Strength, Slope, and Rainfall Intensity Effects on Interrill Erosion." *Am. Soc. Agr. Eng. Trans.* 29(1):98–102.

Williams, J.L. 1975. Sediment-yield prediction with universal equation using runoff factor. In Present and Prospective Technology for Predicting Sediment Yields and Sources. USDA-ARS, West Lafayette, IN, Pub. 540. pp 244–251.

Wischmeier, W.H. and D.D. Smith. 1978. Predicting Rainfall Erosion Losses-A Guide to Conservation Planning. USDA Hdbk. 537, Washington, D.C.

Problems

6.1   a) For the following cities, determine the average annual rainfall (Figure 2.3) and the R Factor (Figure 6.3): New Orleans, Kansas City, Minneapolis, Denver, and Atlanta.

b) Plot R versus annual rainfall.

c) From your graph, estimate the R factor for Brasilia, Brazil, with an annual rainfall of 60 inches.

d) According to Bertoni and Lombardi (1993), the R factor for Brasilia is about 440. Comment on the reliability of this method for estimating R in 6.1c.

6.2   Determine the soil erodibility factor, K, for a soil that has the following properties: 60% silt plus very fine sand; 0% coarse sand; 3% organic matter; very fine granular soil structure; slow permeability.

6.3   If the predicted soil erosion for a given set of conditions is 10 t/ac for a 450-ft slope length, and three terraces are installed, what is the predicted soil erosion rate from the slope for a 150-ft slope length if all other conditions remain unchanged and the slope is 8%?

6.4   Determine the annual soil loss from an urban area which is being developed with the following characteristics: 1) average annual rainfall erosivity of 200; 2) slope length of 500 feet and steepness of 4%; 3) no vegetation; and 4) soil erodibility factor of 0.4.

6.5   a) Determine the soil loss for a field at your present location if $K = 0.30$, slope length = 300 ft, slope steepness = 10%, $C = 0.2$, and up- and down-slope farming is practiced.

b) What conservation practice should be adopted if the soil loss is to be reduced to 4 t/ac?

6.6   You work for the Tennessee Department of Natural Resources and are in charge of evaluating applications for permits for surface mining. A mine operator wishes to mine a steeply sloping forested watershed located in the Appalachian Mountains (Eastern Tennessee). The watershed currently exhibits the following characteristics: 1) soils with an erodibility factor of 0.4; 2) slope steepness of about 10% and length of about 2000 feet; 3) Vegetation is an unmanaged forest with a medium stocked stand (tree canopy area of 50% and forest litter cover of about 80%).

a) Determine the average annual erosion rates prior to mining.

b) Following mining, the operator will: 1) place terraces every 1000 feet; 2) replace the land at average slopes of 8%; 3) reclaim the area by planting grass. It is anticipated that there will be no appreciable canopy and a 40% ground cover. If the maximum allowable soil loss is 4 t/ac, would you give the mine operator a permit? State your reasons.

6.7   a) Determine the annual soil loss from a 200-acre farm in central Indiana. The farm has Blount soils and typical slopes are 1000 feet long with a 2% steepness. The farmer uses conventional tillage in the autumn and has a corn-soybean rotation.

b) What percentage change in the annual soil loss would you expect if the farmer switched to a no-till farming system?

6.8   Determine the sediment yield during a 10-year period from the 200-acre farm in Problem 6.7. The farm is located on a 5-square mile watershed.

6.9   From the principles presented in this chapter, discuss three management techniques to minimize soil erosion on the slopes of a popular ski resort.

6.10   A watershed in western Ohio was intensively farmed for almost 100 years, and experienced severe erosion. The farmers all abandoned their farms, and the predominant vegetation on the watershed reverted to native woodland. When the woodland was about 80 years old, a large recreation and flood control reservoir was constructed on the stream draining the watershed. The engineers were surprised to discover sedimentation rates from the watershed were similar to cropland rates.

a) What do you think was the source of the sediment?

b) Describe a method to reduce sedimentation in the lake.

6.11   A reservoir is constructed on a stream draining 5 square miles (1 square mile = 640 acres). Two thirds of the watershed has an estimated upland erosion rate of 4.5 tons/acre/year, and the other third a rate of 0.5 tons/acre/year. How much sedimentation will the reservoir experience during its 30-year design life?

6.12   Determine the sediment yield from a storm with 80 acre-ft of runoff and a peak discharge of 600 cfs. The USLE factors for this watershed are soil erodibility: 0.3; topographic factor: 0.5; cover-management factor: 0.25; and erosion control practice factor 1.0.

6.13   Calculate a *C* factor for a farming system in Ohio that has a rotation of Corn, Soybeans, Small Grain (oats), Meadow, Meadow, using conventional autumn tillage. State your assumptions. Compare your answer to the value given in Table 6.2.

CHAPTER 7

Flow in Channels, Rivers, and Impoundments

Andy D. Ward

7.1. INTRODUCTION

Seldom does flow move far before entering some kind of channel. Consider for example runoff from the roofs of houses in urban areas. The water soon enters gutters, flows into a downspout and through a buried pipe to the street curb. The water then flows along the side of a road and in many parts of the country will enter a manhole leading to a subsurface storm water system. Even in rural areas, runoff from agricultural fields frequently flows into grassed waterways or open ditches running alongside roads. Does this mean that the procedures described in Chapter 5 have little or no usefulness? Certainly not, for they are used in the design of gutters, pipes, storm water systems, waterways, and ditches. Also, flow in small conveyance systems is often approximated as being similar to overland flow because it is not practical to divide every watershed into subwatersheds representing individual houses and fields.

Of interest is how flow in channels, rivers, and impounds differs from overland flow and at what point we should start considering the influences of these conveyance systems on flow processes. The main difference between the two flow processes is that water storage and its influence on flow rates is considered in conveyance systems such as channels, rivers, and impoundments. There is no specific land size or land use where the importance of these influences needs to be considered. Very large areas may contain no impoundments and small conveyance systems which have similar storage characteristics to the adjacent land. In these cases, assuming that overland flow processes dominate might result in little error. On the other hand, it is often required to construct small ponds downstream of land development activities in order to trap sediment and to reduce post-development peak flows to pre-development levels. In these cases, consideration of pond influences on flow processes is of importance.

In this text, floods and flood-reduction considerations will be related to headwater watershed areas of less than 1,000 mi^2. A headwater flood in small rivers and their tributaries resulting from small-area storms, is termed a small-area flood. It is distinguished from a large-area flood found in downstream channels where flood-control programs are wholly of an engineering nature and are beyond the scope of this text. Headwater floods are typically flash floods of short duration. In smaller headwater areas, the average rainfall intensity is much higher because of the smaller area. Variations in climatic and physiographic factors which affect runoff are often more extreme than in larger watersheds. The effects of crops, soil, tillage practices, and conservation measures are more important in headwater areas since the surface condition of the entire area might change completely from season to season and from year to year. The effectiveness of headwater flood-control measures decreases rapidly with distance downstream. Through the Army Corps of Engineers, the federal government develops and supervises most flood-control programs on large river systems, while the Soil Conservation Service (SCS) and Forest Service (USFS) conduct programs in headwater areas.

7.2. TYPES OF STREAMS

This book focuses on how land use and land use changes influence the partition-ing of precipitation and surface applied water into runoff, infiltration, soil water storage, groundwater flow and storage, and flow in surface water systems such as rivers and impoundments. Often interactions between the hydrologic cycle and the land use are established within a short time ranging from a few minutes, hours, or days to several years. Increasingly, similar rapid changes occur in the shape and flow directions of rivers systems. However, even today the general characteristics of most rivers are associated with complex processes that have occurred during tens of thousands of years. Good accounts of these processes are presented by Knighton (1984), Ikeda and Parker (1989), Jarvis and Woldenberg (1984), Morisawa (1968), Richards (1982), and Schumm (1977). We will limit our discussion to how river systems are described and classified, and an overview of factors which influence the shape and flow directions of a river.

It is often confusing to students learning hydrology to understand the differences between channel, creek, stream, and river flow. You should not be alarmed as this same confusion exists with the authors because each of these descriptions relate to channelized flow of different magnitude, which is only described qualitatively. A stream is a flow of running water and a river is simply a large stream of water. The term channel flow can be used to describe flow in any size channel, stream or river, and the terms discharge, discharge rate, and flow rate (cfs or m^3/s) are synonymous. Ephemeral streams only contain water during and immediately after some rainfall events and are dry most of the year. Intermittent streams are also dry through part of the year but flow when the groundwater is high enough as well as during and immediately after some rainfall or snowmelt events. Perennial streams flow through-out the year.

7.3. STREAM ORDERS (also see Chapter 12, Exercise 8)

Systems for classifying drainage networks were first developed by Horton (1945) and a later modification by Strahler (1952) resulted in an approach which is commonly used today. The smallest fingertip tributaries are called first-order streams (Figure 7.1). Where two first-order streams join, a second-order stream is created; where two second-order streams join, a third-order stream is created; and so on. This approach is only useful if the order number, n, is proportional to the channel dimensions, size of the contributing watershed, and stream discharge at each point in the system. The ratio of the number of stream segments of a given order, N_n, to the number of segments of the next highest order, N_{n+1}, is called the bifurcation ratio, R_b:

$$R_b = \frac{N_n}{N_{n+1}} \qquad (7.1)$$

Within a watershed, the bifurcation ratio will change from one order to the next but will tend to be constant throughout the series. This observation forms the basis of the law of stream numbers (Horton, 1945) which states that the number of stream segments of each order form an inverse geometric sequence with order number, which is described mathematically as:

$$N_n = a_1 \exp^{-b_1 n} \qquad (7.2)$$

where a_1 is a constant, and b_1 is $\ln R_b$.

Two other drainage network laws have been developed. The law of stream lengths was also developed by Horton (1945) and is defined as follows:

$$L_n = a_2 \exp^{b_2 n} \qquad (7.3)$$

where a_2 is a constant, b_2 is $\ln R_L$, and R_L is the stream length ratio. The stream length ratio is defined as:

$$R_L = \frac{L_{n+1}}{L_n} \qquad (7.4)$$

where L_n and L_{n+1} are the average lengths of streams of order n and n+1, respectively.

The law of drainage areas was developed by Schumm (1956) and is defined as follows:

$$A_n = a_3 \exp^{b_3 n} \qquad (7.5)$$

where a_3 is a constant, b_3 is $\ln R_A$, and R_A is the drainage area ratio:

$$R_A = \frac{A_{n+1}}{A_n} \qquad (7.6)$$

where A_n and A_{n+1} are the average drainage areas of streams of order n and n+1, respectively.

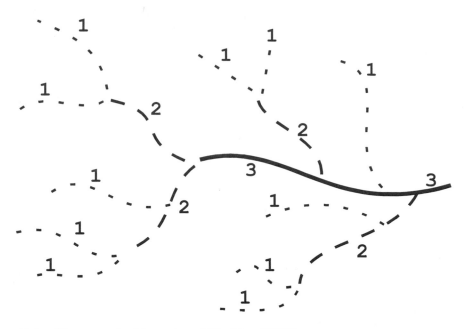

Figure 7.1. Stream order hierarchy of Strahler (1952).

The bifurcation ratio, stream length ratio, and drainage area ratio will in most cases fall within the ranges 3-5, 1.5-3.5, and 3-6, respectively. Also, Hack (1957) found that for many locations in the eastern U.S., the stream length was related to the drainage area as follows:

$$L = 1.4 \; A^{0.6} \qquad (7.7)$$

where L is the stream length in miles from the watershed boundary, and A is the watershed area in square miles.

There are several practical difficulties related to drainage network determination. Firstly, in developed countries such as the United States, many watersheds contain a large number of artificial channels such as storm water systems, vegetated waterways, and open drains alongside roads. These systems are often difficult to identify and do not usually lend themselves to inclusion in the steam order hierarchy associated with natural drainage systems. Secondly, maps with scales of 1:24,000 are often used to identify drainage networks because ground reconnaissance, or the use of remote sensing techniques, is too costly or time consuming. To adequately identify first-order and some second-order streams often requires use of detailed topographic maps with a scale in the order of 1:2,400.

The three drainage network laws and the stream ordering systems developed by Horton, Strahler, and Schumm are still commonly used. However, the Topologically Distinct Channel Networks approach developed by Shreve (1967) is also widely used. For complete details on this and other complex approaches, reference should be made to texts on rivers, river networks, or river morphology such as Knighton (1984), Jarvis and Woldenberg (1984), and Richards (1982).

Example 7.1: Use the drainage network laws to determine the bifurcation ratio, stream length ratio, and drainage area ratios for a 730-acre, fourth order watershed. The drainage network characteristics are summarized in Table 7.1.

Solution: Values of R_B, R_L, and R_A for each set of successive stream orders could be obtained by using the information in Table 7.1 and Equations 7.1, 7.4, and 7.6, respectively. For example, R_B for the ratio of stream orders 1 and 2 would simply be 80 divided by 14, or 5.7. However, what we need to obtain are values of R_B, R_L, and R_A which are representative of the whole network. This requires using Equations 7.2, 7.3, and 7.5. First we need to obtain natural logarithms of the N_n, L_n, and A_n values in Table 7.1. These values are reported in Table 7.2.

We now need to plot the set of ln N_n, ln L_n, and ln A_n values against stream order, n, and statistically or visually fit a straight line to each set of values. The slope of the line drawn through the set of ln N_n value will be b_1, and the slopes of the lines drawn through the sets of ln L_n, and ln A_n values will be b_2 and b_3, respectively. The plots are presented in Figure 7.2, and from the figure b_1, b_2, and b_3 are 1.46, 1.0, and 1.66, respectively. Then, by obtaining the anti-logarithm (e^{b1}, e^{b2}, and e^{b3}) of each of these values we find that R_B, R_L, and R_A are 4.3, 2.7, and 5.3, respectively.

Answer: The bifurcation ratio, R_B, is 4.3; the stream length ratio, R_L is 2.7; and the drainage area ratio, R_A is 5.3.

7.4. STREAM PATTERNS AND GEOMETRY

Factors influencing stream morphology include geology, topography, the size of the contributing watershed, flow velocity, discharge, sediment transport, sediment particle distribution, channel geometry, and other geomorphological controls on the system. Like all physical systems, a stream will attempt to be in equilibrium. European and Indian engineers often refer to the equilibrium between erosion and deposition in a stream as the regime theory which is described by a set of three empirical equations relating stream channel geometry to discharge.

Leopold et al. (1964) noted three different channel patterns, namely, sinuous, meandering, and braided. Attempts have been made to classify streams based on sinuosity which is the ratio between the stream length and the valley length. If the sinuosity is 1.0, the stream channel is straight; a condition which rarely occurs in natural channels except over short distances. If the sinuosity is greater than 1.0, the channel is sinuous. If the sinuosity is greater than 1.5, the river is said to meander, and if the sinuosity exceeds 2.1, the degree of meandering is tortuous (Schumm, 1977). Primarily because of the ability of the system to develop meanders, no exact analytical method exists to define the geomorphic behavior of a river system.

Features of a meandering stream are shown in Figure 7.3. Typically, sand bar ridges known as point bars will form on the concave inner side of stream bends where the flow velocity is slowest. On the outer apex of river bends the flow veloci-

Table 7.1.   Drainage network characteristics for Example 7.1.

Stream Order (n)	Number of Streams (N_n)	Average Stream Length (L_n, ft)	Average Drainage Area (A_n, acres)
1	80	400	5
2	14	1,100	27
3	4	3,000	140
4	1	8,000	730

Table 7.2.   Natural logarithms of selected drainage network characteristics for the solution to Example 7.1.

Stream Order (n)	ln N_n	ln L_n (ft)	ln L_n (acres)
1	4.38	5.99	1.61
2	2.64	7.00	3.29
3	1.39	8.01	4.94
4	0.00	8.99	6.59

ty is fastest and scour (channel erosion) will create deep pools. If equilibrium is not achieved, erosion on the outer bends will continue, the sinuosity will increase, and the stream flow might break through two bends to form a cutoff and the eventual formation of an oxbow lake.

A braided river does not have a single or well defined channel and consists of a network of interconnected streams (Figure 7.3). Braiding is considered an incipient form of meandering. Braided river reaches are generally steeper, wider, and shallower than undivided reaches carrying the same discharge. For any given discharge, meanders occur at smaller slopes.

Relationships between the slope of a stream and the stream discharge have been used by many scientists to distinguish between stream channel patterns (Knighton, 1984). The following relationship by Leopold and Wolman (1957) illustrates this approach:

$$s = 0.0576 \ Q^{-0.44} \tag{7.8}$$

where s is the streambed slope as a fraction, and Q is the bankfull stream discharge in ft^3/s (cfs). Points representing meandering channels lie below a plot of this equation, while braided channels plot above the line. Straight channels plot above or below the line and cannot be demarcated.

Horizontal and vertical instability might occur in a river. Horizontal instability will result in the river type changing. Vertical instability will result in a downcutting of the channel. Downcutting can cause a change in river type or erosion. Generally, downcutting is only significant in alluvial channels. The bed and banks of alluvial channels are formed from deposited material transported by the channel.

Often, knowledge of the type of channel materials and vegetation in the channel will be used in conjunction with information on flow rates to determine if scour will occur. The Soil Conservation Service (1984) determined maximum permissible

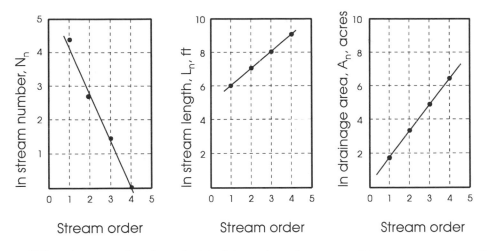

Figure 7.2.   Stream order plotted with: (a) number of streams; (b) mean stream length; and (c) mean drainage area for Example 7.1.

velocities for vegetated channels (Table 7.3).   If the maximum permissible velocity is exceeded, the channel will scour.

A problem with evaluating vegetated channels is that the hydraulic behavior of the vegetation will change as the flow velocity increases.   For example, as grass becomes submerged the roughness will initially increase, but when 20–40% submerged, it will start to decrease because the grass will begin to bend over and flatten.   At high flows, fully submerged grass may have a resistance to flow (roughness) which is only 10% of the unsubmerged value.   Grasses have been divided into four retardance classes based on their resistance to bending, and reference should be made to design texts for further details (Schwab et al., 1993).

7.5. FLOW IN CHANNELS AND RIVERS (also see Chapter 12, Exercise 7)

Flow in a channel or stream is a function of many factors including precipitation, surface runoff, interflow, groundwater flow, and pumped inflow and outflows; the cross sectional geometry and bed slope of the channel, the bed and side slope roughness; meandering, obstructions, and changes in shape; hydraulic control structures and impoundments; and sediment transport and channel stability.   Generally, flow in creeks, drainage ditches, streams, rivers, and impoundments is classified as open-channel flow because the surface of the flow is open to the atmosphere.   Open-channel flow can occur in many ways.   For example, it can be turbulent in steep rocky areas or following severe storm events.   Also, during severe storm events there might be rapid changes in the depth and amount of flow.   On other occasions it can be tranquil and difficult to detect that the water is flowing.   Flow may be classified into many types and a different mathematical equation is needed to describe each type.   Discussion in this text will primarily be limited to steady uniform flow.   Flow in an open channel is steady if the depth of flow does not change or is considered constant during the time interval under consideration.   The flow is uniform if the depths of flow are the same at every section along the channel.   Reference should be

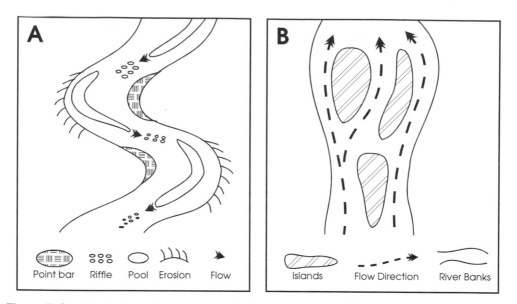

Figure 7.3. Illustration of: (a) a meandering stream; and (b) a braided stream.

made to Chow (1959) or texts such as Hwang and Hita (1987), or Bedient and Huber (1988) for comprehensive discussions on open-channel hydraulics.

The discharge of a channel or river is:

$$q = v \, a \tag{7.9}$$

where; q is the discharge (ft^3/sec or m^3/sec), a is the cross-sectional area of the stream (ft^2 or m^2), and v is the velocity of flowing water (ft/sec or m/sec).

If the discharge does not change along a section of a channel, called a reach, then:

$$q = v_1 \, a_1 = v_2 \, a_2 \tag{7.10}$$

where the subscripts designate the cross sectional areas and velocities at the upstream and downstream ends of a reach. This equation is the continuity equation for uniform flow in open channels. In Equations 7.9 and 7.10, v, v_1, and v_2 are the average velocity of flow within a cross-section. A typical flow distribution for a parabolic grass-lined channel is illustrated in Figure 7.4. Within a thin boundary layer at the interface between the flow and channel sides, the velocity of flow will reduce to zero.

For uniform flow in a channel, the average velocity, v, can be estimated by Manning's equation:

$$v = \frac{1.5}{n} R^{2/3} S^{1/2} \tag{7.11}$$

where v is the velocity of flow (ft/s); n is Manning's roughness coefficient of the channel; S is the channel bed slope (ft/ft); and R is the hydraulic radius of the chan-

Table 7.3.   Permissible velocities for vegetated channels (SCS, 1984).

| | Permissible Velocity for Erosion Resistant Soils (ft/s) | | |
| | Channel Bed Slope (%) | | |
Cover	0–5	5–10	Over 10
Bermuda grass	8[a]	7[a]	6[a]
Blue grama, Buffalo grass, Kentucky bluegrass, Smooth broom, Tall fescue	7[a]	6[a]	5[a]
Annual crops for temporary protection			
Alfalfa, Crabgrass, Kudzu, Lespedeza sericea, Weeping lovegrass	3.5[b]	NR[c]	NR
Grass mixture	5[b]	4[b]	NR

[a] For moderately resistant soils reduce reported velocities by 1 ft/s. For easily eroded soils reduce velocities by 2 ft/s.
[b] For easily eroded soil reduce velocities by 1 ft/s.
[c] Not recommended.

nel (cross-sectional area of the channel, ft^2, divided by P, ft, which is the wetted perimeter of the channel cross-section). This equation was developed by an Irish engineer, Robert Manning, between 1891–1895. If SI or metric units are used, the value of 1.5 in the equation is replaced by 1.0. Geometric relationships which are useful in determining the hydraulic radius and cross-sectional area are presented in Figure 7.5.

The bed slope of a natural stream or river can be found by dividing the change in elevation along a reach by the length of the reach. Manning's "n" can be estimated from tabular information for different bed and side slope materials (Table 7.4).

Equations 7.9 and 7.11 are often used to determine the depth of flow, d, the discharge, q, and the average flow velocity, v, in a river. A common method for determining discharge in a channel is to measure the flow depth and then calculate the discharge using equations that best describe the type of flow which occurred. This is the simplest scenario to evaluate because the depth of flow and channel geometry are known. The hydraulic radius and cross-sectional area can be determined by using Figure 7.5, Equation 7.11 can be used to determine the average flow velocity, and Equation 7.9 can then be used to determine the discharge. Estimations of this nature might be required to determine if bed and side slope scour occurred or to determine discharge associated with an observed storm event.

Example 7.2: Determine the flow velocity and discharge in the mountain stream shown in Figure 7.6. The flow depth is 10 ft, the bed slope is 0.005, the width is 25 ft, and the side slopes are 3:1 (horizontal:vertical). The stream has a clean gravel bottom with a few boulders.

Solution: From Table 7.4 it is determined that Manning's "n" might vary between 0.03 and 0.05. Use the normal value of 0.04.

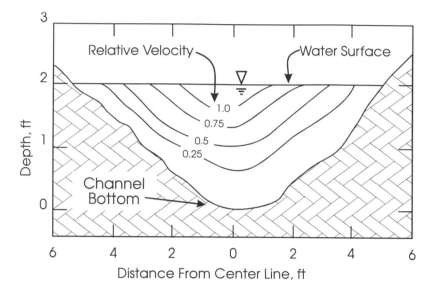

Figure 7.4. Velocity distribution in a stream channel.

From Figure 7.5:

$$R = \frac{bd + zd^2}{b + 2d \sqrt{(z^2 + 1)}}$$

therefore,

$$R = \frac{(10)(25) + 3(10)^2}{25 + 20 \sqrt{(9 + 1)}} = \frac{550}{88.25} = 6.23 \text{ feet.}$$

For uniform flow in a stream the average velocity, v, can be estimated by Manning's equation:

$$v = \frac{1.5}{n} \times R^{2/3} \times S^{1/2}$$

therefore,

$$v = \frac{1.5}{0.04} \times 6.23^{2/3} \times 0.005^{1/2} = 9 \text{ ft/s}$$

The discharge of a stream is:

$$q = v \, a$$

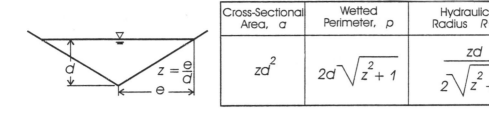

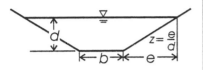

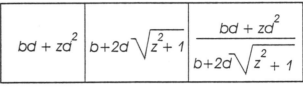

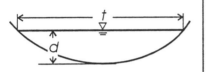

	Cross-Sectional Area, a	Wetted Perimeter, p	Hydraulic Radius $R = \dfrac{a}{p}$
	zd^2	$2d\sqrt{z^2 + 1}$	$\dfrac{zd}{2\sqrt{z^2 + 1}}$
	$bd + zd^2$	$b + 2d\sqrt{z^2 + 1}$	$\dfrac{bd + zd^2}{b + 2d\sqrt{z^2 + 1}}$
	$\dfrac{2}{3}\,td$	$t + \dfrac{8d^2}{3t}$	$\dfrac{t^2 d}{1.5\,t^2 + 4d^2}$

Figure 7.5. Channel cross section, wetted perimeter, and hydraulic radius equations.

From Figure 7.5:

$$a = bd + zd^2$$

therefore,

$$a = (10)(25) + 3(10)^2 = 550 \text{ ft}^2$$

and,

$$q = v\,a = 9 \times 550 = 4950 \text{ ft}^3/\text{s}$$

Answer: The flow velocity is 9 ft/s and the flow rate is 4950 ft³/s.

A more difficult but common scenario is to have a known estimate of the discharge (perhaps by using a predictive method described in Chapter 5) and be interested in evaluating the velocity and flow depth in an existing channel or river. An example of this scenario would be when an evaluation is being made of an upstream land use modification. Typically, an estimate would be made of the peak discharge associated with the land use modification, then an analysis would be made to see if existing channels are able to convey the estimated peak discharge.

Table 7.4.   Values of Manning's "n" roughness coefficient.(From Chow's *Open-Channel Hydraulics*, 1959.   McGraw-Hill.   Summarized with permission of McGraw Hill.)

Type of Channel and Description	Minimum	Normal	Maximum
Closed Conduits Flowing, Partly Full			
Steel	0.010	0.014	0.017
Cast Iron	0.010	0.014	0.016
Wrought iron	0.012	0.015	0.017
Corrugated Metal			
• Subdrain	0.017	0.019	0.021
• Storm drain	0.021	0.024	0.030
Concrete			
• Culvert	0.010	0.012	0.014
• Sewer with manholes, inlet, etc., straight	0.013	0.015	0.017
Clay			
• Vitrified sewer on drainage tile	0.011	0.014	0.017
• Vitrified subdrain with open joints, manholes	0.014	0.016	0.018
Brickwork			
• Sanitary sewers with bends and connections	0.012	0.013	0.016
• Paved invert, sewer, smooth bottom	0.016	0.019	0.020
Constructed Channels			
Smooth steel surface	0.011	0.013	0.017
• Corrugated steel surface	0.021	0.025	0.030
Concrete			
• Trowel finish or float finish	0.011	0.014	0.016
• Finished, with gravel on bottom	0.015	0.017	0.020
• Gunite, good section	0.016	0.019	0.023
• Gunite, wavy section	0.018	0.022	0.025
• On excavated rock	0.017	0.022	0.030
Concrete bottom float finished with sides of			
• Dressed stone in mortar	0.015	0.017	0.020
• Random stone in mortar	0.017	0.020	0.024
• Cement rubble masonry, plastered	0.016	0.020	0.024
• Cement rubble or dry rubble or riprap	0.020	0.027	0.035
Gravel bottom with sides of			
• Formed concrete	0.017	0.020	0.025
• Random stone in mortar	0.020	0.023	0.026
• Dry rubble or riprap	0.023	0.033	0.036
Asphalt	0.013	0.014	0.016
Earth, straight and uniform			
• Clean	0.016	0.020	0.025
• With short grass, few weeds	0.022	0.027	0.033

Table 7.4.   Continued.

Type of Channel and Description	Minimum	Normal	Maximum
Constructed Channels			
Earth, winding and sluggish			
• No vegetation	0.023	0.025	0.030
• Dense weeds or plants in deep channels	0.030	0.035	0.040
• Earth bottom and rubble sides	0.028	0.030	0.035
• Cobble bottom and clean sides	0.030	0.040	0.050
Rock cuts			
• Smooth and uniform	0.025	0.035	0.040
• Jagged and irregular	0.035	0.040	0.050
Channels not maintained, weeds and brush			
• Dense weeds	0.050	0.080	0.120
• Same, highest stage of flow	0.045	0.070	0.110
• Dense brush, high stage	0.080	0.100	0.140
Streams			
Streams on plain			
• Clean, straight, full stage, no rifts or deep pools	0.025	0.030	0.033
• Clean, winding, some pools, shoals, weeds & stones	0.033	0.045	0.050
• Same as above, lower stages and more stones	0.045	0.050	0.060
• Sluggish reaches, weedy, deep pools	0.050	0.070	0.080
• Very weedy reaches, deep pools, or floodways with heavy stand of timber and underbrush	0.075	0.100	0.150
Mountain streams, no vegetation in channel, banks usually steep, trees and brush along banks submerged at high stages			
• Bottom: gravels, cobbles, and few boulders	0.030	0.040	0.050
• Bottom: cobbles with large boulders	0.040	0.050	0.070
Pasture, no brush			
• Short Grass	0.025	0.030	0.035
• High grass	0.030	0.035	0.050
Cultivated areas			
• No crop	0.020	0.030	0.040
• Mature row crops	0.025	0.035	0.045
• Mature field crops	0.030	0.040	0.050
Brush			
• Scattered brush, heavy weeds	0.035	0.050	0.070
• Light brush and trees	0.035	0.060	0.080
• Medium to dense brush, in winter	0.045	0.070	0.110
• Medium to dense brush, in summer	0.070	0.100	0.160

Table 7.4.   Continued.

Streams (Continued)			
Trees			
• Dense willows, summer, straight	0.110	0.150	0.200
• Cleared land with tree stumps, no sprouts	0.030	0.040	0.050
• Same as above, but with heavy growth of sprouts	0.050	0.060	0.080
• Heavy stand of timber, a few down trees, little undergrowth, flood stage below branches	0.080	0.100	0.120
• Same as above, but with flood stage reaching branches	0.100	0.120	0.160
Major streams (top width at flood stage > 100 ft)			
• Regular section with no boulders or brush	0.025	...	0.060
• Irregular and rough sections	0.035	...	0.100

There are two approaches that could be used to evaluate this scenario. Equations 7.9 and 7.11 could be combined as follows in order to eliminate the flow velocity:

$$q = \frac{1.5}{n} \, a \, R^{2/3} \, S^{1/2} \qquad (7.12)$$

The hydraulic radius and the cross-sectional area can then be written in terms of the depth of flow, d, which is now the only unknown. Unfortunately, Equation 7.12 is not easily solved as it contains complicated functions of depth.

A practical approach which is commonly adopted is to use trial and error requiring a guess of flow depth. Equation 7.11 is then solved to determine the flow velocity at the guessed depth. This value is then substituted into Equation 7.9 together with the cross-sectional area associated with the guessed depth and an estimate of the discharge is obtained. The flow rate based on the guessed depth is then compared with the known discharge. If the discharge based on the guessed depth is too high, a shallower depth is tried and the exercise is repeated. If the discharge is too low, then a deeper depth is tried. With practice a solution within 10% of the correct answer can be obtained in three or four tries. Trying to obtain a more accurate estimate has little practical meaning because of uncertainties in determining the known discharge.

Example 7.3: Determine the flow depth and flow velocity in the mountain stream shown in Figure 7.6. The discharge rate is 2000 ft³/s, the bed slope is 0.005, the width is 25 ft, the side slopes are 3:1 (horizontal:vertical), and Manning's "n" is 0.04.

Solution: From Figure 7.5:

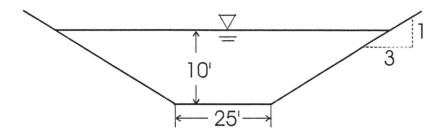

Figure 7.6. Flow geometry for Examples 7.2 and 7.3.

$$R = \frac{bd + zd^2}{b + 2d\sqrt{(z^2 + 1)}}$$

and,

$$a = bd + zd^2$$

Guess d = 5 ft, then:

$$R = \frac{(5)(25) + (3)(5)^2}{25 + 2(5)\sqrt{(3^2 + 1)}} = 3.53 \text{ ft}$$

and,

$$a = (5)(25) + (3)(5)^2 = 200 \text{ sq ft}$$

Manning's equation:

$$v = \frac{1.5}{n} R^{2/3} S^{1/2}$$

therefore,

$$v = \frac{1.5}{0.04}(3.53)^{2/3} (0.005)^{1/2} = 6.15 \text{ ft/s}$$

and,

$$q = v \, a = 6.15 \times 200 = 1230 \text{ cfs}$$

This is lower than 2000 cfs, so try a deeper depth. Guess d = 7 ft, then:

$$R = \frac{(7)(25) + (3)(7)^2}{25 + 2(7)\sqrt{(3^2 + 1)}} = 4.65 \text{ ft}$$

and,

$$a = (7)(25) + (3)(7)^2 = 322 \text{ sq ft}$$

therefore,

$$v = \frac{1.50}{0.04}(4.65)^{2/3}(0.005)^{1/2} = 7.39 \text{ ft/s}$$

Substituting in the continuity equation:

$$q = v \, a = 7.39 \times 322 = 2379 \text{ cfs}$$

This is higher than 2000 cfs, so try a shallower depth between 6 and 7 feet.

Answer: The depth of flow is about 6.4 feet and the approximate velocity is 7.0 ft/s.

The most difficult scenario to evaluate is when there is a known discharge and a need to determine a channel geometry which is capable of conveying this flow. This type of evaluation is commonly made by agricultural and civil engineers who are designing a ditch or channel. However, there are also many situations where a scientist might conduct a conceptual analysis for use in a cost-benefit analysis. The scenario is evaluated by first selecting a channel geometry based on technical and practical considerations. Once a geometry has been selected the approach is similar to the previously discussed scenario. However, at the end of the analysis it might be determined that the geometry is greatly over- or undersized, in which case it will be necessary to try a new geometry. Reference should be made to a hydraulic engineering text or manual for solutions of this type of scenario (Chow, 1959; Barfield et al, 1981; Hwang and Hita, 1987).

When there is severe flooding, the flow might overtop the banks of the main river channel and flow along the adjacent flood plain. In parts of the United States, earth and concrete levees have been constructed to contain the flow within the main channel and prevent flooding. However, as was evidenced in the 1993 flooding in the heartland of America (Nebraska, Iowa, Minnesota, Wisconsin, Kansas and Illinois), there will eventually be an event of sufficient magnitude to overtop or breach any levee. To estimate flow in a complex cross-section which consists of a main channel and flood plain, it is possible to use Manning's equation provided there is uniform flow. In this case, the cross-section can be divided into several subsections (Figure 7.7). Typically each subsection will exhibit different roughness coefficients and mean velocities. The total flow rate will be the sum of flow rates in all the subsections (see Chapter 12, Exercise 7).

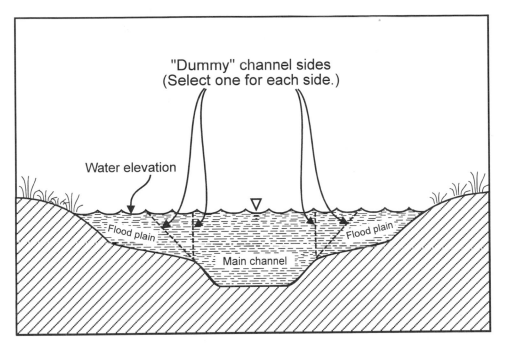

Figure 7.7.   Complex channel cross-section.

7.6. MODIFYING CHANNEL CAPACITY

Flood reduction can be achieved by: (a) constructing floodwater detention structures; (b) modifying the stream channel carrying capacity; or (c) using a combination of (a) and (b). For example, a detention structure might be designed to prevent flood discharges from exceeding the carrying capacity of a downstream channel. However, if this channel could be modified to carry a higher flow, the cost of the detention structure might be less. Of course in the combination system, if channel modifications are not maintained, then flood damages will not be prevented.

Channel modification might be accomplished by removing obstructions and decreasing the roughness coefficient (n) or resistance to water flow velocity; or increasing the hydraulic gradient (s) by reducing meanders through channel straightening. If stream channel meanders are cut off and the channel straightened and cleaned, the roughness would be decreased and the slope increased. If, prior to channel straightening, the fall in 1,000 ft of a stream reach was 25.0 ft and this same fall occurred in 500 ft of a straightened channel, the gradient would increase from 0.025 to 0.05. Based on Manning's equation this would have the effect of increasing the flow velocity by a factor of 1.44 (the ratio of the square roots of 0.025 and 0.05). There is some danger in channel straightening in areas of highly erodible soil. When the flow velocity increases its erosive force also increases. The erosive process could upset the channel stability and accelerate gully erosion as is evident in the Missouri Valley deep loess region.

It is often difficult to quantify the influence of making channel modifications if Manning's "n" is obtained from Table 7.4. However, Chow (1959) presents information based on the work of Cowan (1956) which can be used for this purpose (Table 7.5). The description of the following procedure is obtained from Chow's *Open-Channel Hydraulics*, 1959, published by McGraw-Hill and summarized with permission of McGraw-Hill. The value of Manning's "n" is:

$$n = (n_0 + n_1 + n_2 + n_3 + n_4)\, m_5 \tag{7.13}$$

where n_0 is the value of n for a straight, uniform, smooth channel in natural materials, n_1 is a value added to n_0 to correct for the effect of surface irregularities, n_2 is a value for variations in shape and size of the channel cross-section, n_3 is a value for obstructions, n_4 is a value for vegetation and flow conditions, and m_5 is a correction factor for meandering of the channel.

The values given in Table 7.5 were developed from a study of small and moderate channels. Therefore, the method is questionable when applied to large channels whose hydraulic radii exceed, say 15 ft. The method applies only to unlined natural streams, floodways, and drainage channels with a minimum *n* value of 0.02.

In selecting the value of n_1, the degree of irregularity is considered *smooth* for surfaces comparable to the best attainable for the materials involved; *minor* for good dredged channels, slightly eroded or scoured side slopes of canals or drainage channels; *moderate* for fair to poor dredged channels, moderately sloughed or eroded side slopes of canals or drainage channels; and *severe* for badly sloughed banks of natural streams, badly eroded or sloughed sides of canals or drainage channels, and unshaped, jagged, and irregular surfaces of channels excavated in rock.

In selecting the value of n_2, the character of variations in size and shape of cross section is considered *gradual* when the change in size or shape occurs gradually; *alternating occasionally* when large and small sections alternate occasionally or when shape changes cause occasional shifting of main flow from side to side; and *alternating frequently* when large and small sections alternate frequently or when shape changes cause frequent shifting of flow from side to side.

The selection of the value n_3 is based on the presence and characteristics of obstructions such as debris deposits, stumps, exposed roots, boulders, and fallen and lodged logs. In judging the relative effect of obstructions, consider the extent to which the obstructions occupy or reduce the average water area; the character of obstructions (sharp-edged or angular objects induce greater turbulence than curved, smooth-surfaced objects); and the position and spacing of obstructions transversely and longitudinally in the reach under consideration.

In selecting the value of n_4, the degree of effect of vegetation is considered:

1. *Low* for conditions comparable to the following: (a) dense growths of flexible turf grasses or weeds (Bermuda and blue grasses are examples) where the average depth of flow is 2 to 3 times the vegetation height; and (b) supple seedling tree switches, such as cottonwood, willow, or salt cedar, where the average depth of flow is 3 to 4 times the vegetation height.
2. *Medium* for conditions such as (a) turf grasses where the average depth of flow is 1 to 2 times the height of vegetation; (b) stemmy grasses, weeds, or tree seedlings with moderate cover where the average depth of flow is 2 to 3 times the height of vegetation; and (c) brushy growths, moderately dense, similar to willows 1 to 2 years old, dormant season, along side slopes of a

Table 7.5.   Values for the computation of the roughness coefficient by Equation 7.13 (Chow, 1959).

Channel Conditions			Values
Material Involved	Earth	n_o	0.020
	Rock Cut		0.025
	Fine Gravel		0.024
	Coarse Gravel		0.028
Degree of Irregularity	Smooth	n_1	0.000
	Minor		0.005
	Moderate		0.010
	Severe		0.020
Variations of Channel Cross Section	Gradual	n_2	0.000
	Alternating Occasionally		0.005
	Alternating Frequently		0.010–0.015
Relative Effect of Obstructions	Negligible	n_3	0.000
	Minor		0.010–0.015
	Appreciable		0.020–0.030
	Severe		0.040–0.060
Vegetation	Low	n_4	0.005–0.010
	Medium		0.010–0.025
	High		0.025–0.050
	Very High		0.050–0.100
Degree of Meandering	Minor	m_5	1.000
	Appreciable		1.150
	Severe		1.300

channel with no significant vegetation along the channel bottom, where the hydraulic radius is greater than 2 ft.

3. *High* for conditions comparable to the following: (a) turf grasses where the average depth of flow is about equal to the height of vegetation; (b) dormant season willow or cottonwood trees 8 to 10 years old, intergrown with some weeds and brush none of the vegetation in foliage, where the hydraulic radius is greater than 2 ft; and (c) growing season, busy willows about 1 year old intergrown with some weeds in full foliage along side slopes, no significant vegetation along channel bottom, where hydraulic radius is greater than 2 ft.

4. *Very high* for conditions comparable to the following: (a) turf grasses where the average depth of flow is less than one-half the height of vegetation, (b) growing season—busy willows about 1 year old, intergrown with weeds in full foliage along side slopes, or dense growth of cattails along channel bottom with any value of hydraulic radius up to 10 or 15 ft, and (c) growing season—trees intergrown with weeds and brush, all in full foliage, with any value of hydraulic radius up to 10 or 15 ft.

In selecting the value of m_5, the degree of meandering depends on the sinuosity of the channel. Meandering is considered *minor* for a sinuosity of 1.0 to 1.2, *appreciable* for sinuosities of 1.2 to 1.5, and *severe* for sinuosities of 1.5 and greater.

Example 7.4: It is planned to straighten and improve a channel which has been poorly maintained. Determine the change in Manning's "n" and flow velocity if the channel: is formed in earth; has moderately eroded side slopes; gradual changes in cross-section; has appreciable lodged logs and deposits of debris; cottonwood trees intergrown with weeds and brush (high to very high vegetation); and appreciable meandering. Following the channel improvements it is anticipated that the channel will have the side slopes repaired; the debris will be removed and lodged logs will be cleared; all weeds and brush will be removed; and there will be minor meandering.

Solution: From Table 7.5 and the descriptions presented in the text it is determined that, prior to straightening and improving the channel: $n_0 = 0.02$; $n_1 = 0.01$; $n_2 = 0.00$; $n_3 = 0.02$–0.03; $n_4 = 0.025$–0.050; and $m_5 = 1.15$. Use the average n_3 value of 0.025, and the upper end of the range for n_4 of 0.05 because the vegetation is described as high to very high, then:

$$n = (0.02 + 0.01 + 0.0 + 0.025 + 0.05)1.15 = 0.12$$

Following the straightening and improvements: $n_0 = 0.02$; $n_1 = 0.005$; $n_2 = 0.00$; $n_3 = 0.01$–0.015; $n_4 = 0.01$–0.025; and $m_5 = 1.0$. Let's assume $n_3 = 0.01$ and $n_4 = 0.025$, then:

$$n = (0.02 + 0.005 + 0.0 + 0.01 + 0.025)1.0 = 0.06$$

Answer: Straightening and improving the channel will reduce Manning's "n" from 0.12 to 0.06. This will result in an approximate doubling of the flow velocity. However, it is noted that a range of n_3 and n_4 values could have been used and the answer is approximate.

7.7. MEASURING STREAMFLOW

Streamflow measurements on major streams are made by the U.S. Geological Survey (USGS) of the Interior Department. Sites on stream systems have been selected for gaging flow to meet specific informational needs for city water supply, hydroelectric power, navigation, flood control, recreation, and the preservation of natural wildlife areas. Annual publications report daily, monthly, and annual runoff data along with values of maximum rates of streamflow. Selected sites have been equipped to obtain water quality information which is published in separate annual reports for each state. Streamflow gaging by the USGS is usually made in cooperation with individual state water resources and environmental protection agencies.

   Streamflow or runoff data that meet stringent reliability standards are very expensive and, in some locations, very difficult to obtain. For these reasons some hydrologists have bypassed the direct measurement of runoff for watersheds of 100 mi^2 and larger by equating annual-runoff values to the difference between total watershed precipitation and ET, after allowance is made for storage changes in soil water content and groundwater. Water years starting October 1 (dry season) or May 1 (wet season) are often selected to estimate annual runoff from rainfall because net soil water storage is minimized and yearly variability is reduced. The results are only approximate and subject to the ability of the hydrologist to estimate ET accurately. The effect of an error in estimating ET on the value of runoff is illustrated in Example 7.5.

Example 7.5: Determine the magnitude of the error in computed annual runoff depth, Q, from errors of 10% and 20% in estimates of evapotranspiration, ET. The annual precipitation, P, is 37.0 inches and estimated ET is 24.0 in.

Solution:

$$Q = P - ET = 37.0 - 24.0 = 13.0 \text{ inches}$$

For a 10% overestimate of ET:

$$ET = 1.1 \times 24.0 = 26.4 \text{ inches}$$

and,

$$Q = P - ET = 37.0 - 26.4 = 10.6 \text{ inches}$$

Therefore, the percent error in the annual runoff value = 100 (13.0 - 10.6) / 13.0 = 18%

For a 20% overestimate of ET:

$$ET = 1.2 \times 24.0 = 28.8 \text{ inches}$$

and,

$$Q = P - ET = 37.0 - 28.8 = 8.2 \text{ inches}$$

Therefore, the error in the annual runoff value = 100 (13.0 - 8.2) / 13.0 = 37%

Answer: A 10% overestimate of ET resulted in an 18% error in the runoff value. A 20% overestimate of ET resulted in a 37% error in the runoff value.

Runoff may be identified and measured at a number of locations on the land surface according to the specific need for runoff data. Small watersheds of 1 to 3 acres (0.4 to 1.2 ha) mostly have surface runoff with little or no interflow. Watersheds of several hundred acres usually include surface flow, interflow, and groundwater flow. Gage locations of different watershed areas are shown on a typical cross-section of a hill of North Appalachia (Figure 7.8). Runoff from small watersheds (not more than a few acres) is usually gaged fairly close to the ridge of larger watersheds. These gages do not include interflow return to the land surface. Annual surface runoff in this region is usually not greater than 10% of the average annual precipitation depth of 37 in. (Figure 7.8). Runoff ceases soon after rainfall stops.

Gage locations for watersheds of larger size are downstream and more interflow comes to the surface and passes through these gages as shown in Figure 7.8. Annual runoff for the 100-acre watershed of about 12 in. is comprised of surface and interflow as well as some groundwater flow. Finally, at about 1,000 acres, annual flow reaches 15 in. on the average. Downstream from this last gage site, the stream gradient is relatively flat and there is practically no increase in return flow or groundwater flow as the watershed areas increase.

A stream where discharge values per unit of area increase, as shown in Figure 7.8, is termed a gaining stream. It is possible to find places where a gaining stream turns into a losing stream as the flow passes through areas of gravel where water escapes into the bed and banks of the stream. In humid areas of the eastern U.S., gaining streams for watersheds of 1 to 17,500 acres are common. In the semiarid and arid West, losing streams are more common. In Arizona, storm runoff is often completely lost in a few miles of a dry, gravely stream channel.

Another runoff characteristic of note is the difference in seasonal flow for l-acre and 17,500-acre watersheds (Figure 7.9) in the wet winter-spring period and June. In June, all areas of the l-acre watershed contribute to runoff. However, June storms are of local nature and cover only part of the 17,500-acre area, and consequently only parts of the larger watershed contribute runoff resulting in a smaller depth per unit of area. In the wet period, January through April, storms are generally of large area coverage, the soil is wet and has lower infiltration rates, and although rainfall rates are low, large amounts of runoff occur (2 in./month and greater). These runoff process concepts were observed under North Appalachian conditions. Hydrologic characteristics of other regions can be developed in like manner.

Essentially there are three related operations in measuring discharge at a gaging station. The first involves the determination of the gage height or stage (H, ft or m) of the water surface flowing through the gaging section (Figure 7.10). The second involves determining the mean velocity (v, ft/sec or m/sec) of flow through the gaging section. The third operation is derivation of the relationship between water stage and discharge (q, ft^3/sec or m^3/sec) termed "stage-discharge relationship."

The measurement of water surface stage may be made periodically by observing the level of water on a staff gage graduated in ft, 0.1 ft, 0.01 ft, or in m, cm, and mm. This method is acceptable for very large watersheds where the stage changes slowly. Otherwise, it is necessary to use devices that make continuous records of stream water surface level. Data loggers or recorders are usually installed in a shelter set over a well that is connected by open pipes to the stream. Water-stage fluctuations in a stilling well will coincide with those in the stream.

Determining the mean velocity of a stream cross-section is the most difficult step in the field calibration of stage-discharge relationships. Estimates of velocity can be

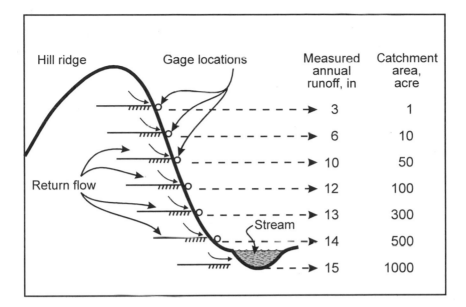

Figure 7.8. Annual runoff and catchment area at gage location below the ridge.

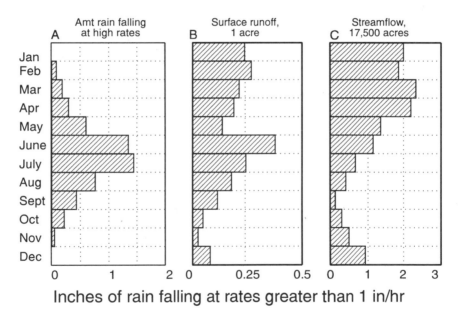

Inches of rain falling at rates greater than 1 in/hr

Figure 7.9. Seasonal pattern of high rates of rainfall and runoff from 1 acre and 17,500 acre catchments, Coschocton, Ohio.

made by observing the speed of a floating object at several points across the width of the water surface when it is unaffected by wind currents. The speed of the surface float is approximately 1.2 times the mean velocity of flow through the measuring section. The variation in velocity with depth is logarithmic and the mean velocity is approximately equal to the average of the velocities at 0.2 and 0.8 of the depth (refer back to Figure 7.4). Stream velocity measurements are usually made with a more accurate method such as a current-meter (see Chapter 12, Exercise 7).

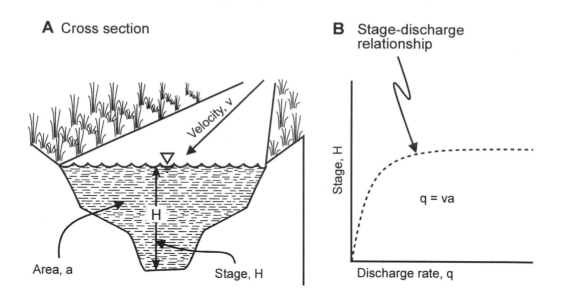

A Cross section

B Stage-discharge relationship

Velocity, v

H

Area, a

Stage, H

Stage, H

Discharge rate, q

q = va

Figure 7.10.   Essentials of A) runoff measuring section, and B) stage-discharge relationship.

One practical approach which is used by the U.S. Geological Survey is known as the midsection method and uses a current-meter to measure stream velocities at several points along a stream cross-section (Buchanan and Somers, 1984).

For calibrated weirs and flumes, no measurement of velocity is required. The calibration formula of the device relates discharge to stage. When these devices are installed and operated properly, runoff records are more accurate than those from a natural stream channel that is field calibrated. A discussion on weirs and flumes is presented in Section 7.8.

In some cases tracers are injected into a channel or river and water samples are taken at a downstream point which is a known distance from the injection location. Common tracers include chemicals, dyes, and radioactive materials. If complete mixing has occurred, the steady flow rate, q, is:

$$q = q_t \frac{(C_1 - C_2)}{(C_2 - C_0)} \qquad (7.14)$$

where q_t is the tracer application rate, C_1 the applied concentration of the tracer, C_2 the measured concentration of the tracer in the dosed flow at the downstream point, and C_0 is the initial concentration of tracer at the downstream point prior to dosing.

Example 7.6: A tracer test is conducted to determine the discharge in a small stream. The tracer application rate is 0.2 cfs and the concentration of the tracer at application is 2000 ppm (parts per million). The measured concentration of the tracer in the dosed flow at a downstream location is 100 ppm. The initial concentration of the tracer at this downstream location was 0 ppm.

Solution: The following information has been given: $C_1 = 2000$ ppm, $C_2 = 100$ ppm, $C_0 = 0$ ppm, and $q_t = 0.2$ cfs. Substituting this information into Equation 7.14:

$$q = q_t \frac{(C_1 - C_2)}{(C_2 - C_0)}$$

and,

$$q = 0.2 \frac{(2000 - 100)}{(100 - 0)} = 3.8 \text{ cfs}$$

Answer: The discharge of the small stream is 3.8 cfs.

*7.8. WEIRS AND FLUMES

Weirs and flumes are rigid structures whose cross-sectional areas are considerably less than the channel in which they are installed and the dimensions are closely defined and stable. A weir is a control structure which raises the bed of the channel. A flume narrows the width of the channel. If both the width and bed height are raised, then the control structure is normally classified as a flume. The constriction in the stream channel forces the flow to form a pool at the entrance to the structure. This still water then plunges freely over the weir or flume outlet.

The discharge over a weir, q in cfs, can be estimated by using the equation:

$$q = C L H^{3/2} \tag{7.15}$$

where C is a weir coefficient, L is the weir length in feet, and H is the height of the flow above the riser crest. For a pipe, a value of 3.1 may be used for C, and L will be the circumference of the pipe riser.

A sharp-crested weir with a 90° "V" notch outlet shown in Figure 7.11 has the following discharge and head relationship:

$$q = 2.5 \, h^{5/2} \tag{7.16}$$

where q is the discharge rate, ft^3/sec or cfs, and h is the head, or depth of water above the point of zero flow, ft. Calibration formulas for weirs with broad crests and different shapes are given by Brater and King (1976) or Bos et al. (1991).

7.9. RECORDING STREAM DISCHARGE DATA

Mean daily discharge values are computed from mean daily gage heights and published by the U.S. Geological Survey. Gage heights at some measuring stations are obtained from periodic manual observations and converted to discharge values by the use of calibration curves or tables of stage-discharge relationships. At most stations, gage heights are recorded continuously and mean daily values are converted to mean

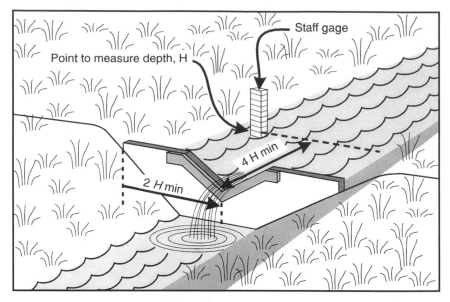

Figure 7.11.   A 90 degree "V" notch, sharp-crested weir.

Table 7.6.   **Daily runoff from Alum Creek watershed (189 mi^2) at Columbus, Ohio, April 1972.**

Date	Mean Daily Discharge (cfs)
April 1	100
omitted	omitted
6	70
7[1]	2,200
8	1,800
omitted	omitted
20	3,300
omitted	omitted
30	80
April Mean	798
Maximum	3,300
Minimum	70

1. The maximum discharge on April 7 was 3,730 cfs.

daily discharges.   A partial tabulation of mean daily discharge values of the type published by the USGS (1973) appears in Table 7.6.

Mean daily discharge tabulations are adequate for problems dealing with runoff from watersheds of 1,000 mi^2 and larger.   However, for smaller watersheds where runoff rates vary widely in a day, more detailed data are needed.   This variation is apparent in Table 7.6 which shows that on April 7th, the average daily discharge rate was 2,200 cfs, whereas the maximum rate on that day was 3,730 cfs. This indicates a sizeable fluctuation of flow rates within a 24 hour period.

Example 7.7: Use the data presented in Table 7.6 to determine the depth of runoff during April, 1972 for the Alum Creek watershed of 189 mi².

Solution: The mean daily discharge rate is 798 cfs. Therefore the total volume of runoff, Q, is:

$$Q \text{ (acre–ft)} = 798 \text{ ft}^3/s \times \frac{60s}{1 \text{ min}} \times \frac{60 \text{ min}}{hr} \times \frac{24 \text{ hr}}{day} \times \frac{30 \text{ days}}{April} \times \frac{Acre}{43,560 \text{ ft}^2}$$

therefore, Q is 47,484 acre-ft and,

$$Q = 47484 \text{ acre ft} \times \frac{1}{189 \text{ mi}^2} \times \frac{1 \text{ mi}^2}{640 \text{ mi}} = 0.3926 \text{ ft}$$

and,

$$Q = 0.3926 \text{ ft} \times \frac{12 \text{ inches}}{ft} = 4.7 \text{ inches}$$

Answer: The depth of runoff on the watershed is 4.67 inches for the month of April.

Annual runoff estimates may be made from precipitation records using relationships between gaged precipitation and runoff (Figure 7.12). A relationship like the one shown in Figure 7.12 is useful in extending the runoff record for years when precipitation data are available but runoff data are not. Probable average interval of expected recurrence (return period) of flood peak values is estimated from annual series of recorded maximum flood peak rates at this site and plotted on log-probability graph paper (see Chapter 2).

7.10. ROUTING FLOW THROUGH RESERVOIRS

Flood routing through reservoirs and stream channels is a practical means for evaluating the effect of storage on hydrograph shapes. In the flood-routing procedure for reservoirs, the factors to be considered are: (1) inflow hydrograph; (2) relationship between reservoir spillway water depth and detention storage volume; (3) relationship between spillway water depth and outflow discharge rate, and (4) outflow hydrograph. These factors pertain to a reservoir in which detention storage modifies the inflow hydrograph to produce an outflow hydrograph having a lower peak discharge but essentially the same flood volume (see Figure 7.13).

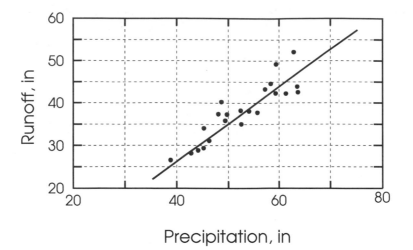

Figure 7.12.   Example of a relationship between annual precipitation and runoff.

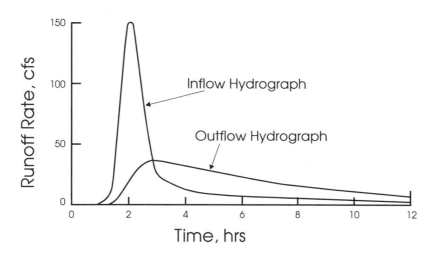

Figure 7.13.   Typical inflow and outflow hydrographs.

Features of reservoirs involved in flood routing are as follows (see Figure 7.14):

1. The principal spillway is designed to carry all the frequent discharges.
2. The flood spillway is usually an open channel and is designed to operate for a short time during which flood flows exceed the capacity of the other spillways. It is a safety factor and is designed to prevent overtopping or breaching of the embankment.
3. Freeboard, another factor of safety, is always provided to prevent waves or any other water from overtopping the dam.

The "permanent" pool of water which is formed up to the crest of the principal spillway is usually sized based on sediment inflows during the life of the impoundments, recreational needs, and water supply requirements. Water will be lost from

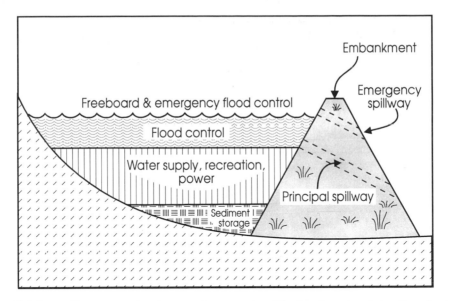

Figure 7.14.   Elements of a dam and reservoir with flood-water detention features.

this pool due to evaporation, seepage, pumped withdrawals, and storage lost due to sediment inflows.

The outflow from most reservoirs is controlled by a principal spillway and the outflow rate is proportional to the height of water above the spillway inlet for a passive gravity flow system. At the beginning of a flood event there may be little or no flow over the spillway. In fact evapotranspiration or pumped withdrawals might have caused the water elevation to be lower than the spillway invert. As inflow continues the water elevation and outflow rate increase. This process will continue until the inflow and outflow rates are equal. When this occurs the highest water elevation (stage) in the impoundment and the peak outflow occur. The peak outflow occurs after the peak inflow and, by continuity, the peak outflow rate intersects the receding limb of the inflow hydrograph.

The change in reservoir water storage can be determined from the continuity equation as follows:

$$\delta S = \frac{(I_2 + I_1)}{2} \, \delta t - \frac{(O_2 + O_1)}{2} \, \delta t \qquad (7.17)$$

where δS is the change in water storage, δt is the change in time between time t_1 and t_2, I_1 and I_2 are the inflow rates at times t_1 and t_2, and O_1 and O_2 are the outflow rates at times t_1 and t_2. Equation 7.17 is best solved by using a computer program or a spreadsheet.

The Soil Conservation Service has related inflow and outflow rates, runoff volume, and storage volume as follows:

$$\frac{S}{V} = 1 - 2 \left(\frac{q_{po}}{q_{p_i}}\right) + 1.8 \left(\frac{q_{po}}{q_{p_i}}\right)^2 - 0.8 \left(\frac{q_{po}}{q_{p_i}}\right)^3 \qquad (7.18)$$

where V is the runoff volume (area under the inflow hydrograph), S is the flood storage volume in the impoundment, q_{pi} is the peak inflow, and q_{po} is the peak outflow. This relationship should only be used for watershed areas of less than 250 acres. The relationship between the storage volume ratio and the flow rate ratio has been plotted in Figure 7.15.

Example 7.8: Estimate the temporary storage volume necessary to provide a peak outflow rate of 100 cfs if the peak inflow rate is 250 cfs and the inflow volume is 80 acre-ft.

Solution: Substituting in Equation 7.18 the values of V, q_{pi}, and q_{po}:

$$\frac{S}{80} = 1 - 2 \left(\frac{100}{250}\right) + 1.8 \left(\frac{100}{250}\right)^2 - 0.8 \left(\frac{100}{250}\right)^3$$

therefore,

$$S = 80(1 - 0.8 + 0.288 - 0.051) = 35 \text{ acre-ft}$$

Answer: Approximately 35 acre-ft of temporary flood storage will be required to provide a peak discharge of 100 cfs for the specified inflow hydrograph.

Ward et al. (1979) developed a simple procedure which gives a conservative estimate of the required reduction in peak runoff rate and the storage volume:

$$\frac{S}{V} = 1 - \frac{q_{p_o}}{q_{p_i}} \tag{7.19}$$

This relationship has also been plotted on Figure 7.15 (curve A). For a given ratio between the peak inflow and outflow rates, the required temporary storage will normally be between the values determined by the two methods.

Example 7.9: Determine the temporary storage volume needed for the same problem as in Example 7.8 if the double triangle procedure of Ward et al. (1979) is used.

Solution: Substitute into Equation 7.19 the peak outflow, q_{po}, of 100 cfs; the peak inflow, q_{pi}, of 250 cfs; and the inflow volume of 80 acre-ft. Then:

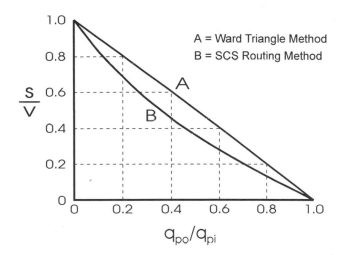

Figure 7.15.   Determination of temporary flood storage in a reservoir.

$$\frac{S}{80} = 1 - \frac{100}{250}$$

therefore,

$$S = 0.6 \times 80 = 48 \text{ acre-ft}$$

Answer: The temporary storage volume is 48 acre-ft. This could have been estimated by using Figure 7.15. It can be seen that the answer is conservative and requires a storage volume which is 13 acre-ft more than the SCS method (see Example 7.8).

*7.11. RESERVOIR DETENTION TIME (also see Chapter 12, Exercise 8)

Small impoundments called sediment ponds are often constructed downstream from a disturbed area. The purpose of these ponds is to trap soil eroded from the disturbed areas. The ponds slow down the flow, and soil particles fall from suspension and are deposited. The time it takes a particle to fall from suspension will depend on the particle size and the physics of the process is described by Stokes' Law (Equation 3.3). The time flow is stored in a pond is called the detention time and can be determined by the procedures developed by Ward et al. (1979) which are outlined below.

If the inflow and outflow hydrographs are approximated by triangular hydrographs, relationships can be established between the peak inflow and outflow rates, the base time of the inflow hydrograph, the volume of inflow, and the volume of flood storage in the impoundment.

From the geometry of a triangle, the base time of an inflow hydrograph, t_{bi}, is related to the inflow volume, V, and the peak inflow, q_{pi}, as follows:

$$V = 0.5 q_{pi} \, t_{bi} \tag{7.20}$$

If V is in acre-ft, q_{pi} is in cfs, and t_{bi} is in hrs, Equation 7.20 becomes:

$$V = 0.0413 \, q_{pi} \, t_{bi} \tag{7.21}$$

A similar relationship can be established for the temporary storage:

$$S = 0.5 \, t_{bi} \, (q_{pi} - q_{po}) \tag{7.22}$$

where S is the temporary storage volume, t_{bi} is the base time of the inflow hydrograph, q_{pi} is the peak inflow rate and q_{po} is the peak outflow rate (Figure 7.16). Equation 7.19 which was presented earlier is obtained by dividing Equation 7.22 by Equation 7.20.

If the storage volume is expressed in acre-feet, time in hours, and flow rates in cfs, then:

$$S = 0.0413 \, t_{bi} \, (q_{pi} - q_{po}) \tag{7.23}$$

S may also be approximated by the equation:

$$S = 0.124 \, t_d \, q_{po} \tag{7.24}$$

where t_d is the detention time in hours between the centroids of the hydrographs and q_{po} is the peak discharge rate. When Equations 7.23 and 7.24 are combined, q_{po} can be estimated from the relationship:

$$q_{p_o} = \frac{t_{bi} \, q_{pi}}{(3t_d + t_{bi})} \tag{7.25}$$

Graphical solutions to these equations are given in Figures 7.17 and 7.18.

7.11.1. Determination of Impoundment Geometry

Once a potential impoundment site has been located, a survey should be conducted and a topographic map prepared. For small structures with a capacity of less than 20 acre-ft storage, the contour interval should be 2–5 feet and the scale between 1:600 to 1:2400. An example map is shown in Figure 7.19. From the map, a stage-area and stage-capacity curve should be calculated. The incremental volume between each stage elevation is determined from the equation:

$$V = \frac{(A_1 + A_2)(Z_2 - Z_1)}{2} \tag{7.26}$$

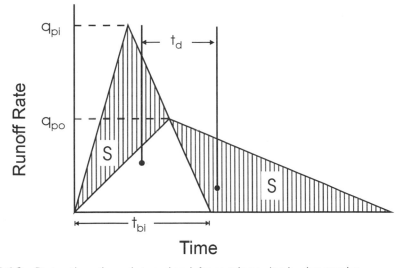

Figure 7.16. Detention time determined from triangular hydrographs.

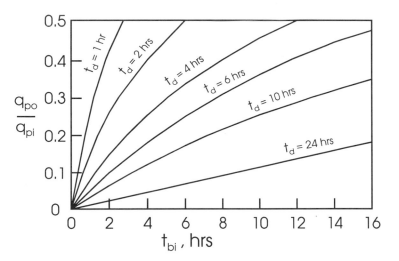

Figure 7.17. Determination of peak flow rate reduction.

where A_1 and A_2 are the areas at elevations Z_1 and Z_2, and V is the volume of storage between elevations Z_1 and Z_2.

As soon as a stage capacity relationship has been established, preliminary design calculations to estimate the riser configuration and pond embankment size can be made. Sizing of the pond embankment and the spillway systems will depend on the magnitude of the design runoff event and the stage-capacity relationship for the pond.

Example 7.10: It is required to size a sediment and flood control pond for a 10-year, 24-hour postmining event on a 120-acre watershed (Figure 7.20). Flow should be retained in the structure for an average time of 24 hours. The watershed is located on the Kentucky-West Virginia border. One seam of coal

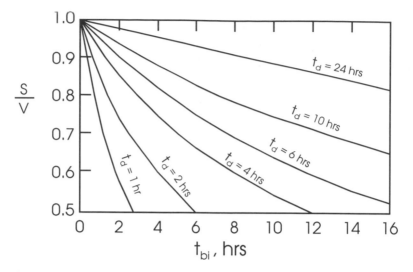

Figure 7.18. Temporary storage to satisfy detention time requirements.

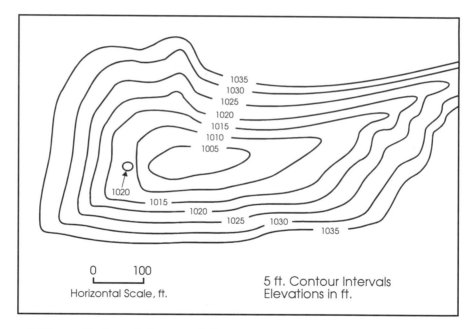

Figure 7.19.   Sedimentation pond site map.

is to be surface mined using a contour mine operation. The disturbed area will total about 24 acres. The postmining runoff hydrograph has a peak of 140 cfs and a volume of 11.87 acre-ft. Average land slopes vary between 20–60%, with an average of 45%. The maximum flow length is 3200 feet with an average grade of 7%. The soils are predominantly Muse-Shelocta association and land uses have been marked on the topographic map. The required sediment storage has been assumed to be 0.1 acre-ft for each disturbed acre. The pond is to be located at the site shown in Figure 7.19. A stage-storage-discharge relationship for the pond has been developed and is

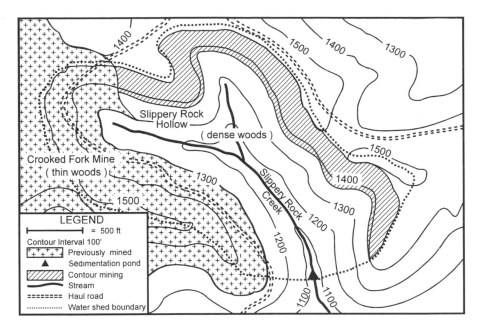

Figure 7.20. Topographic map for Example 7.10.

Table 7.7.   Stage-capacity relationship at the proposed pond location.

Elevation (feet)	Stage (feet)	Area (acres)	V (acre-ft)	Storage (acre-ft)
1005	0.0	0.046	0.0	0.00
1010	5.0	0.368	1.04	1.04
1015	10.0	0.735	2.78	3.79
1020	15.0	1.194	4.82	8.62
1025	20.0	1.560	6.89	15.50
1030	25.0	2.112	9.18	24.68
1035	30.0	2.755	12.17	36.85

presented in Table 7.7. Size a sediment and flood control pond for the 10-year, 24-hour post-mining runoff hydrograph.

Solution: The sediment storage is 0.1 times 24 or 2.4 acre-ft.

The base time for the inflow hydrograph may be estimated by using Equation 7.21:

$$V = 0.0413\ t_{bi}\ q_{pi}$$

$V = 11.87$ acre-ft and $q_{pi} = 140$ cfs, therefore $t_{bi} = 24.2\ (11.87)/140$ hrs or $t_{bi} = 2.06$ hours. Use a value of two hours. From Figure 7.17 the peak outflow rate should be about 0.03 times the peak inflow rate for a detention time of 24 hours. Therefore, the peak discharge will be 4.2 cfs. Using

Figure 7.18, the required temporary storage is 11.5 acre-ft. (97% of the inflow volume).

From Table 7.7 it is determined that to provide 2.4 acre-ft of sediment storage, the lowest elevation the riser crest can be located is 1013.0 feet. The crest of the emergency spillway should be located at a minimum elevation of 1024.0 feet to provide an additional 11.5 acre-ft of temporary storage. A procedure has been described for determining the lowest permissible elevation of a dewatering device and also for estimating the lowest elevation of the emergency spillway crest. It will now be necessary to design a spillway system which will provide the desired detention time and discharge rating curve.

Answer: Total storage volume = 11.5 + 2.4 = 13.9 acre-ft.

Most sedimentation ponds have a drop inlet riser or a hood inlet riser as the principal spillway. Diagrams of typical drop inlet and hood inlet riser systems are shown in Figure 7.21a and 7.21b. For a drop inlet riser the change in discharge with head is controlled by three different phases of flow (weir, orifice, and pipe). The discharge at any stage is taken as the minimum flow resulting from weir, orifice, or pipe flow at that stage. Generally, the conduit riser is six inches smaller in diameter than the vertical drop inlet riser.

Orifice flow occurs when flow is restricted by the size of the opening and is determined as:

$$Q = Ca \ (2gh)^{1/2} \tag{7.27}$$

where C is a coefficient depending on the orifice geometry, a is the cross-sectional area of the pipe, g is the gravitational constant, and h is the head in feet. For a sharp-edged orifice, a value of 0.6 may be used for C.

As the head continues to increase, the outlet begins to flow full and the flow is controlled by the outlet pipe. Pipe flow is determined by using the equation:

$$Q = \frac{a \ (2gH)^{1/2}}{(1+K_e+K_b+K_cL)^{1/2}} \tag{7.28}$$

where K_e is an entrance loss coefficient, K_b is a correction factor for energy losses in bends and K_c is a friction factor. H is the head or difference in water elevations between the flow in the pond and at the outlet.

Estimates of the pipe size that should be used for a particular flow geometry can be determined by using Table 7.8.

Example 7.11: For the pond that was sized in Example 7.10, determine the size for the drop inlet riser and pipe through the embankment to convey the peak discharge of 4.2 cfs. Assume that the crest of the riser is at the elevation

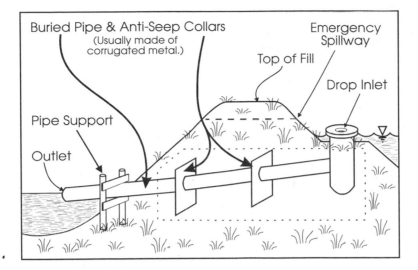

Figure 7.21a. Drop inlet riser.

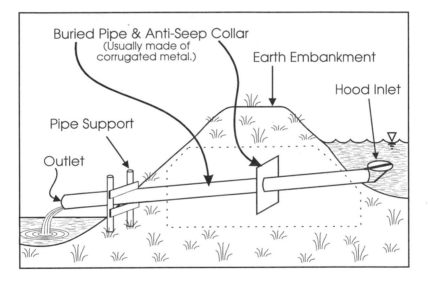

Figure 7.21b. Hood inlet riser.

of the top of the sediment storage volume. A corrugated metal pipe will be used for the riser.

Solution: The peak discharge should be 4.2 cfs and a head, at the inlet of 12.0 feet, occurs at the peak rate. An illustration of the embankment and spillway geometry is presented in Figure 7.22. The difference in elevation between the inlet and outlet, H, is 22 feet.

From Table 7.8 it is determined that a 6-12 inch pipe will be required. A 12 inch pipe discharges 9.3 cfs for a head of 22 ft and a pipe length of 70 ft. A correction factor of about 0.7 needs to be used because the pipe length is

170 feet. Therefore, the discharge through a 12 inch pipe would be 6.5 cfs (9.3 × 0.7 cfs). Generally, the vertical riser would be six inches large than the pipe through the embankment.

Answer: Use an 18 inch drop inlet riser and a 12 inch pipe through the embankment.

*7.12. ROUTING FLOWS THROUGH CHANNELS AND RIVERS

The influence of a channel or river on runoff hydrographs is not very different than that of a reservoir. Water storage in a river will usually attenuate and delay flows. For an impoundment it was illustrated in Figure 7.13 that the peak outflow rate intersected the receding limb of the inflow hydrograph. This will not always be the case in a river because storage is a function of both inflows and outflows for a reach. Local inflows to the reach might include flow from tributaries, overland runoff, groundwater flow, and rainfall. In some cases such as karst areas, there might be substantial losses of water due to seepage. At many points along rivers there will also be withdrawals to satisfy irrigation, industrial, and public water supply needs.

Procedures for routing flows through impoundments will generally not provide good estimates of flow attenuation and storage because the flow depth along a reach will vary as the inflow and outflow rates are different. Two commonly used channel routing procedures are the kinematic method and the Muskingum Method. A brief discussion of the use of kinematic procedures as they relate to describing surface runoff was presented in Chapter 5. For more comprehensive discussions the reader should refer to Barfield et al. (1981) or Bedient and Huber (1988). Both books contain an excellent level of detail for the practicing engineer.

In the Muskingum method, channel storage is the following linear function of inflow and outflow rates:

$$S = KO + KX (I - O) \tag{7.29}$$

where S is the storage in the reach, K and X are constants, and I and O are the simultaneous inflow and outflow, respectively.

For flood routing applications, the Muskingum method is combined with the continuity equation presented in Equation 7.17 to give:

$$O_2 = O_1 + C_1 (I_1 - O_1) + C_2 (I_2 - I_1) \tag{7.30}$$

where the subscripts indicate the beginning and end of the time period, t. The coefficients C_1 and C_2 are defined as follows:

$$C_1 = \frac{2\delta t}{2K(1 - X) + \delta t} \tag{7.31}$$

Table 7.8. Pipe flow design flow rates (Source: SCS, 1975b). Pipe flow chart n = 0.025 for corrugated metal pipe inlet $K_m = K_e + K_b = 1.0$ and 70 feet of corrugated metal pipe conduit (full flow assumed). Note correction factors for pipe lengths other than 70 feet of pipe.

H (feet)	Pipe Diameter (inches) and Discharge (cfs)										
	6"	12"	18"	24"	30"	36"	48"	60"	72"	84"	96"
1	0.33	1.98	5.47	11.0	18.8	28.8	55.7	91.8	137	191	255
2	0.47	2.80	7.74	15.6	26.6	40.8	78.8	130	194	271	360
3	0.58	3.43	9.48	19.1	32.6	49.9	96.5	159	237	331	441
4	0.67	3.97	10.9	22.1	37.6	57.7	111	184	274	383	510
5	0.74	4.43	12.2	24.7	·2.1	64.5	125	205	306	428	570
6	0.82	4.86	13.4	27.0	46.1	70.6	136	225	336	469	624
8	0.94	5.61	15.5	31.2	53.2	81.5	158	260	388	541	721
10	1.05	6.27	17.3	34.9	59.5	91.2	176	290	433	605	806
12	1.15	6.87	19.0	38.2	65.2	99.9	193	318	475	663	883
15	1.29	7.68	21.2	42.8	72.8	112	26	355	531	741	987
20	1.49	8.87	24.5	49.4	84.1	129	249	410	613	856	1139
25	1.66	9.92	27.4	55.2	94.0	144	279	556	685	957	1274
30	1.82	10.9	30.0	60.5	103	158	305	620	750	1048	1396

Table 7.8.   Continued.

H (feet) Pipe Length	Pipe Diameter (inches) and Discharge (cfs)										
	6"	12"	18"	24"	30"	36"	48"	60"	72"	84"	96"
	Correction Factors for Other Pipe Lengths										
20	1.69	1.53	1.42	1.34	1.28	1.24	1.18	1.14	1.11	1.10	1.08
30	1.44	1.36	1.29	1.24	1.21	1.18	1.13	1.11	1.09	1.07	1.06
40	1.28	1.23	1.20	1.17	1.14	1.12	1.10	1.08	1.06	1.05	1.05
50	1.16	1.14	1.12	1.10	1.09	1.08	1.06	1.05	1.04	1.04	1.03
60	1.07	1.06	1.05	1.05	1.04	1.04	1.03	1.02	1.02	1.02	1.02
70	1.00	1.00	1.00	1.00	1.00	1.00	1.00	1.00	1.00	1.00	1.00
80	.94	.95	.95	.96	.96	.97	.97	.98	.98	.98	.99
90	.89	.90	.91	.92	.93	.94	.95	.96	.96	.97	.97
100	.85	.86	.88	.89	.90	.91	.93	.94	.95	.95	.96
120	.78	.80	.82	.83	.85	.86	.89	.90	.89	.93	.94
140	.72	.75	.77	.79	.81	.82	.85	.87	.86	.90	.91
160	.68	.70	.73	.75	.77	.79	.82	.84	.92	.88	.89

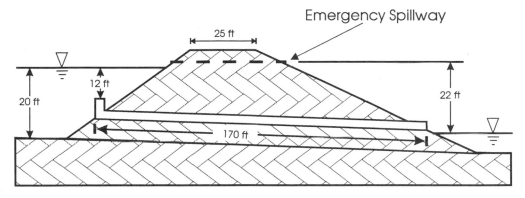

Figure 7.22.   Embankment and spillway geometry for Example 7.11.

and,

$$C_2 = \frac{\delta t - 2KX}{2K(1 - X) + \delta t}$$ (7.32)

The coefficient K is called the storage constant and is approximated as the travel time of flow in the reach if streamflow records are not available for I and O. If X is zero, then the procedure describes flow routing in an impoundment. If X is 0.5, the storage is a function of the average flow rate in the reach. A good account of the method is presented by Cudworth (1989).

7.13. FLOOD FORECASTS

Each new decade shows vast improvements in the accuracy of forecasts of time and crest stage of floods at critical points in river systems. Such forecasts have enabled local authorities to plan and carry out flood plain evacuation or to provide temporary increases in height of levees. In large basins, flood levels increase slowly day after day, and refinements can be made almost daily in the early forecast of crest stages as the crest approaches. The ability to accurately forecast peak discharges was illustrated during the devastating flooding of the Mississippi and Missouri Rivers that occurred in 1993. In most cases river crests were correctly forecast to within a few tenths of one foot, and the time of the crest was predicted to within a few hours. This was truly remarkable considering the widespread extent of flooding and extended periods (many weeks) during which abnormal rainfall and flooding persevered.

For many years the National Oceanic and Atmospheric Administration (NOAA) has used flood forecast procedures on main streams of large river basins. There is at least one forecasting office in each major river basin of the country. Flood forecasting depends on extensive knowledge of land uses, topography, conveyance systems, and real-time climatic information. Presentation of the methodologies used in making flood forecasts is beyond the scope of this text. Reference should be made to texts such as Kraijenhoff and Moll (1986).

Flood forecasts on a similar basis for small watersheds are not yet possible. The small-area flood often results from short-duration, local storms with high rainfall rates. The storm may not have been recorded in the widely-spaced Weather Bureau gage network. Also, a flood may crest in a few hours or less and there is no time to warn occupants of the flood plain. In such situations, NOAA is aware of climatic conditions in which severe storms may occur, and issue warnings of the possibility of flash floods. However, the forecast cannot provide information on the time and height of the flood crest.

For a headwater watershed, the peak flood discharge for a particular return period can be predicted from a probability plot of gaged data. The statistical approach is identical to that described for precipitation in Chapter 2. Methods for predicting peak rates and amounts of flood runoff on ungaged watersheds were presented in Chapter 5. The smaller the watershed, the more accurately the peak discharge, runoff volume, and hydrograph shape can be predicted because (1) the rainfall-runoff relationship for a small area is simpler; and (2) channel storage influences on hydrologic relationships increase with increases in watershed size.

As the flood from a small headwater watershed proceeds downstream, its hydrograph shape and peak discharge are modified greatly by channel storage. It may be better to develop hydrographs for headwater tributary watersheds and route them downstream through the river channel storage system to the point of interest rather than developing a single rainfall-runoff relationship for the entire watershed. This routing requires more effort than the direct estimate of flood volume and peak flow rate.

7.14. FLOOD ZONES

Information on floodwater inundation depths and frequencies is valuable in land-use planning or zoning for flood plains. For decades the land in many flood plains has been in use by residential, industrial, utility, transportation, and other highly developed business areas. The hydrologist obtains information on the flood hazards in the area needing protection, the engineer works out plans for structures required to provide protection and their cost, and the economist prepares reports on the dollar value of benefits expected from the protection planned. If the benefits exceed the cost, the flood reduction plan is approved by state and federal agencies and the project is funded. It is highly important that the community be informed and involved in projects of this type from the start and that public hearings are held before plans are submitted for final approval and activation. Areas identified as high-hazard from frequent flooding could be reserved for recreational use where little damage results from floods, instead of trying to justify the construction of costly flood-reduction systems. But in some cases where the land has a very high value for uses other than recreation, the flood reduction system may be the best solution. Generally, it is more expensive or not possible to obtain insurance for structures built on flood plains even when the return period for any potential flooding is 100 years.

The hydrologist develops data for a flood-hazard evaluation for an area to show the stage of the flood, the flood zone boundary, area inundated, flood discharge volume, and frequency of floods with different probability of recurrence (Figure 7.23).

Flood Zoning, Zip City

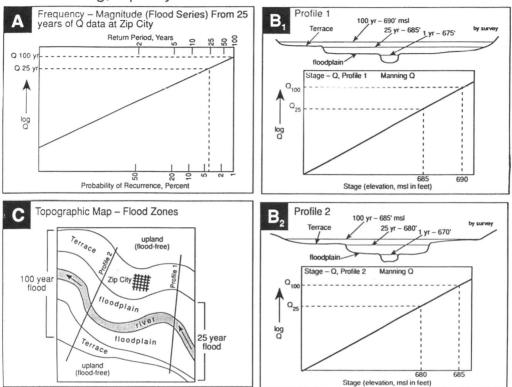

Figure 7.23.   Illustration of flood zoning for Zip City, U.S. (Courtesy of S.W. Trimble, UCLA.)

Field surveys supply basic information which takes the form of stage vs. area inundated for various depths of flooding; flood peak discharge vs. area inundated; and frequency vs. area inundated. Field surveys are made in the area of flood concern to determine from local residents' testimony or recorded data on the elevation of high-water marks for as many annual floods as possible.

7.15. FLOOD-FLOW REDUCTION METHODS

Reduction of flood flow may be accomplished in varying degrees by watershed land treatment in which the storage of storm water is increased on the surface and in the soil profile, using flood-water detention structures, and increased groundwater storage. Groundwater storage increases are accomplished primarily in arid regions where storm runoff, diverted from the stream and spread over a considerable area, percolates to groundwater aquifers. Measures that retard or reduce runoff tend to conserve water resources, providing more water for crops and groundwater storage, the latter causing more uniform seasonal distribution of streamflow.

Table 7.9.   Example of downstream benefits of headwater flood-control reservoirs.

Drainage Area (mi^2)	Number of Reservoirs	Peak flow Reduction at Subwatershed outlet (%)
10	1	90
100	6	40
500	25	30

Land-treatment measures include increased vegetation with more soil surface protection, along with practices such as terraces and waterway development and stabilization.   The choice of practices depends on climate, hydrology, soils, and economics.   For example, we would not expect much flood-flow reduction from increased vegetation where the soil depth is very thin, or from heavy clay soil, because the storage opportunity is small and not readily available.   Likewise in areas of heavy and shallow soil, graded terraces with outlets will not reduce floods.   But on deep, well-drained soil, level terraces have a marked influence on flood flows as well as on water conservation.   In the deep loess region of western Iowa, where there are level terraces with a retention capacity of 1.5 in. of runoff and an infiltration capacity of 0.3 in./hr, it has been estimated that a flood peak from a 882-mi^2 watershed could be reduced from 66,000 cfs to 45,000 cfs by constructing level terraces on 17% of the area (Schwab et al., 1966).

There are two main types of flood control reservoirs, flood storage and flood detention.   Flood water detention reservoirs are mainly equipped with automatic outlets with fixed openings.   Flood storage reservoirs are operated with adjustable outlet gates, but many of these reservoirs are used to detain flood waters just long enough to prevent flood damage downstream before returning the reservoir pool level to its normal elevation as quickly as possible.

Headwater flood-control reservoirs are designed for protection of a designated reach of channel downstream from the reservoir.   For protection further downstream, other flood-control reservoirs are needed.   Both headwater and downstream reservoirs are needed for complete protection.   This is illustrated in Table 7.9 for a 500-mi^2 watershed in which flood-reduction evaluations are made for the effect of one reservoir on a 10-mi^2 subwatershed.   Six reservoirs are located in the headwater areas of a 100-mi^2 subwatershed which is a tributary to the 500-mi^2 watershed.

In regions of nearly level topography, flood water detention reservoirs on the streams are not practical.   Channel modification in these areas would be possible if it were not for the expense involved and public concern over the effect of such works on ecology.   Nolte and Schwab (1971) reported that flood water damage reduction in these regions may be possible by pumping excess storm runoff from streams into detention reservoirs located off-stream and later releasing the stored water back to the stream in dry weather.

During flood periods, water pumped from the stream would reduce the flood discharge downstream.   The potential for this method of flood reduction was computed for a flood series on a 346 mi^2 watershed using a 173,000 ac-ft storage reservoir and a 3,460 cfs pumping rate.   It was determined that the flow downstream was reduced to 450 cfs compared to the natural streamflow of 3,460 cfs.   A downstream channel of 450 cfs carrying capacity with a reservoir would provide the same degree of drainage as a 3,460 cfs capacity channel carrying natural flow.

REFERENCES

Barfield, B.J., R.C. Warner, and C.T. Haan. 1981. Applied Hydrology and Sedimentology for Disturbed Areas. Oklahoma Technical Press, Stillwater, OK.

Bedient, P.B. and W.C. Huber. 1988. Hydrology and Floodplain Analysis. Addison-Wesley Publishing Company, Reading, MA.

Bos, M.G., J.A. Replogle, and A.J. Clemmens. 1991. Flow Measuring Flumes for Open Channel Systems. American Society of Agricultural Engineers, St. Joseph, MI.

Brater, E.F., and H.W. King. 1976. Handbook of Hydraulics (6th Edition). McGraw-Hill Book Company, New York.

Buchanan, T.J., and W.P. Somers. 1984. Discharge measurements at gaging stations. Techniques of Water-Resources Investigations of the United States Geological Survey, Chapter A8, Book 3, Applications of Hydraulics. U.S. Geological Survey, Alexandria, VA.

Chow, V.T. 1959. Open-channel Hydraulics. McGraw-Hill Book Company, New York.

Cowan, W.L. 1956. Estimating hydraulic roughness coefficients. *Agricultural Engineering* 37(7):473–475.

Cudworth, A.G. 1989. Flood hydrology manual. A Water Resources Technical Publication. United States Department of the Interior Bureau of Reclamation, Denver, CO.

Hack, J.T. 1957. Studies of longitudinal stream profiles in Virginia and Maryland. Professional Paper 294B, United States Geological Survey, 97 pp.

Horton, R.E. 1945. Erosional development of streams and their drainage basins: hydrophysical approach to quantitative morphology. *Bulletin of the Geological Society of America* 56:275–370.

Hwang, N.H.C., and C.E. Hita. 1987. Fundamentals of Hydraulic Engineering Systems, 2nd Edition. Prentice-Hall, Inc., Englewood Cliffs, NJ.

Ikeda, S., and G. Parker. 1989. River meandering. Water Resources Monograph 12, American Geophysical Union, Washington, D.C.

Jarvis, R.S., and M.J. Woldenberg (Eds). 1984. River Networks. Benchmark Papers in geology. vol. 80, Hutchinson Ross Publishing Co., Stroudsburg, PA.

Knighton, D. 1984. Fluvial Forms and Processes. Edward Arnold Ltd., London, England.

Kraijenhoff, D.A., and J.R. Moll. 1986. River Flow Modelling and Forecasting. D. Reidel Publishing Company, Dordrecht, Holland.

Leopold, L.B., and Wolman, M.G. 1957. River channel patterns–braided, meandering and straight. Professional Paper 282B, United States Geological Survey.

Leopold, L.B., Wolman, M.G., and J.P. Miller. 1964. Fluvial Processes in Geomorphology. W.H. Freeman Publishers, San Francisco.

Morisawa, M. 1968. Streams: their dynamics and morphology. McGraw-Hill Book Company, NY.

Nolte, B.H. and G.G. Schwab. 1971. An alternate to channelization. *Ohio Agr. Res. and Devel. Center, Ohio Report* 56(5):70–71.

Richards, K. 1982. Rivers, form and process in alluvial channels. Methuen & Co., New York City, NY.

Schwab, G.O., D.D. Fangmeier, W.J. Elliot, and R.K. Frevert. 1993. Soil and Water Conservation Engineering, 4th ed. John Wiley & Sons, Inc., NY.

Schwab, G.O., R.K. Frevert, T.W. Edminster, and K.K. Barnes (1966). *Soil and Water Conservation Engineering,* 2nd ed. John Wiley and Sons, New York City, NY.

Schumm, S.A. 1956. The evolution of drainage systems and slopes in badlands at Perth Amboy, New Jersey. *Bulletin of the Geological Society of America,* 67:597–646.

Schumm, S.A. 1977. The Fluvial System. Wiley-Interscience, New York, NY.

Shreve, R.L. 1967. Infinite Topologically Random Channel Network. *Journal of Geology* 75:178–186.

Soil Conservation Service. 1984. Engineering Field Manual (including Ohio Supplement). U.S. Department of Agriculture, Washington, D.C.

Strahler, A.N. 1952. Hypsometric (area-altitude) analysis of erosional topography. *Bulletin of the Geological Society of America* 63:1117–42.

Ward, A.D., C.T. Haan, and J. Tapp. 1979. The DEPOSITS Sedimentation Pond Design Manual. Institute for Mining and Minerals Research, University of Kentucky, Lexington, KY.

Problems

7.1 Use the drainage network laws to determine the bifurcation ratio, stream length ratio, and drainage area ratios for the watershed data in the following table.

Stream Order (n)	Number of Streams (N_n)	Average Stream Length (L_n, ft)	Average Drainage Area (A_n, acres)
1	64	300	7
2	20	800	23
3	6	2,000	72
4	1	4,000	230

7.2 The sinuosities of a stream in 1950, 1970, and 1990 were found to be 1.0, 1.4, and 1.6. Describe how the stream is probably changing with time and the channel pattern in each of the three years.

7.3 A stream with a bed slope of 0.005 has a bankfull discharge (flow rate) of 1,000 cfs. If the stream is not straight, is it more likely to be braided or meandering?

7.4 A grassed waterway with resistant soils has a bed slope of 6% and is vegetated with Buffalo grass. What is the maximum permissible velocity of flow in this waterway?

7.5 Determine the discharge in a stream if the average flow velocity is 5 ft/s and the cross-sectional area of the flow is 25 ft^2.

7.6 Determine Manning's roughness coefficient, n, for a channel constructed in earth, with minor irregularities, occasional changes in cross-section geometry, severe obstructions (use an average value for this condition), trees intergrown with weeds and brush during the growing season (use an average value for this condition), and a sinuosity of 1.7.

7.7 Use Manning's equation to determine the average velocity of a stream if the roughness coefficient, n, is increased by weed growth from 0.03 to 0.06. Assume S and R do not change and the average velocity with a roughness coefficient of 0.03 is 4 ft/s.

7.8 Use Manning's equation to determine the velocity of a mountain stream if the slope of a stream was increased from 0.04 to 0.09 and R and n did not change. At the 0.04 slope the velocity was 3.0 ft/s.

7.9 Compute the discharge for the channel in Figure 7.6 if the flow depth is 6 ft, the channel is concrete-lined, and the slope of the channel bed is 0.01 ft/ft.

7.10 Determine the flow depth and flow velocity in a channel when the discharge is 500 cfs. The bottom width is 12 ft and the channel has side slopes of 1:1 (1 ft horizontal to every 1 ft vertical)? The channel is cut in smooth uniform rock (use the normal Manning's "n" for this condition).

7.11 Use Manning's equation to determine the flow velocity (m/s) and discharge (m³/s) in a stream on a plain which has sluggish reaches and weedy deep pools. The main stream channel is trapezoidal and has a bottom width of 2 m, a bed slope of 4%, and 3:1 side slopes. The depth of flow is 1 m (use the normal Manning's "n" for this condition).

7.12 Develop a conceptual design for a trapezoidal concrete channel which is sized to convey a design discharge of 40 m³/s. Evaluate the feasibility of constructing a channel with a bed slope of 0.005, a bed width of 2 meters and side slopes of 2 (horizontal): 1 (vertical). Determine the design flow depth. If the flow depth exceeds 2 m the design will not be feasible. Comment on the feasibility of the design and, if necessary, discuss ways the channel might be modified to comply with the maximum flow depth of 2 m.

7.13 The average cross-sectional geometry of a channel is: bed width of 1.0 m; side slopes of 2:1; maximum depth of 2 m; bed slope of 0.5%; Kentucky bluegrass; erosion resistant soils; minor irregularity, gradual variations in cross section, negligible obstructions, medium height vegetation, and minor meandering. Will the channel be able to convey the 25-year, 24-hour peak discharge of 12 m³/s? Will channel erosion be a problem?

7.14 Determine the potential error in the annual runoff depth if it is determined as the difference between annual precipitation and annual ET. The annual precipitation is 45 inches and the accuracy of the calculated ET of 28 inches is plus or minus 25%.

7.15 Determine the discharge in a stream based on the results of a tracer study where the tracer application rate was 0.4 cfs and the applied concentration of the tracer was 4,000 ppm. At a downstream location the measured concentration of the tracer initially was 50 ppm and 150 ppm in the dosed flow.

7.16 Estimate the temporary storage volume necessary to provide a peak outflow rate of 50 cfs if the peak inflow rate is 250 cfs and the inflow volume is 100 acre-ft. What is the base time of the inflow hydrograph and the average detention time of the flow?

CHAPTER 8

Forests and Wetlands

Charles H. Luce

8.1. INTRODUCTION

Much of the discussion thus far in this book has focused on agricultural, urban, and rangeland hydrology. Two other land uses, forests and wetlands, require further discussion because hydrologic processes in those settings behave differently than in agricultural and urban land uses.

Forests cover a large percentage of the land area worldwide. The United Nations Food and Agriculture Organization (FAO, 1993) estimates that forests constitute 30 percent of the land area with a total area of over four billion hectares. Thirty percent of the continental United States is dedicated to forests. Such an extensive coverage makes the unique aspects of water movement in forests an important aspect of global and regional hydrologic cycles.

Globally, forest cover is disappearing at an alarming rate (FAO, 1993). Rates are estimated at 11–20 million hectares of land converted from forests to other land uses each year. Much of the conversion of forests is occurring in developing countries. A particularly important aspect of the disappearance is the role that some of these forests play in the global climate and hydrologic cycle. The Amazon Rain Forest in South America has been the subject of much study on evaluating the role of the large tropical rain forests in the global climate (Balek, 1983).

An additional important aspect of forest loss over the world is the increased erosion that often accompanies forest removal. The erosion problem tends to be more regional in scope than climate feedback problems, but it is repeated in so many regions of the world that it is truly a global problem for water quality and soil productivity. A good understanding of the hydrologic cycle as affected by forests at a local and regional scale is important in solving these problems.

Within the United States, a great deal of debate over how forestlands should be managed has focused on water and the hydrologic cycle at regional and local scales. Water-related issues have been a focus of forest management in this country since the beginning of forestry as a science. The Organic Act of the Forest Reserves, enacted in 1891, stated that one of the purposes of the forest reserves was to "obtain favorable conditions of flow." Since then many debates have taken place over the effects of forest management on floods, water supply, water quality, and fisheries production. The issues discussed in the past continue to be discussed today in scientific journals and the courts.

One important issue is the simple expectation that water coming from forests be clean. Once it was believed that forests purified water, and the belief continues today. The water supply for several major metropolitan areas in the country is surface water from forested watersheds. Among these are Seattle and Tacoma, Washington; Portland, Oregon; and San Francisco and Los Angeles, California. As our populations increase with an increasing demand for water for consumption and irrigation, people are concerned about the amount of water available as well as its quality. Once it was believed that forests produce water; now our understanding is better and we realize that trees consume water like any plant. In this case, water quality and water quantity can become conflicting goals in watershed management.

Less specifically related to human needs, but no less important as far as quality of life is concerned, is protection of endangered species. Many fish, amphibian, plant, and insect species are dependent on forest streams for habitat. Some spend only a small part of their life cycle in forest or mountain streams, but the quality of the habitat is important nonetheless. One important example is the debate over how the forests of the Columbia River Basin must be managed to preserve salmon habitat for the endangered Salmon River Sockeye and Chinook runs. These fish use forest and mountain streams only for spawning and the first months of rearing. The remainder of their life cycle includes travel through the Snake and Columbia River hydroelectric reservoir and dam system and several years in the Pacific Ocean. Their success in the fragile birth and rearing stages is critical to their survival as a species.

Careful consideration of the hydrologic and erosion processes of forests is important in resolving these debates. Understanding how the forest affects hillslope hydrology at the scale of a stand of trees during a single storm event is as important as understanding how forest management affects distributions of fish spawning and rearing habitat in a watershed over two or three hundred years.

8.2. HOW ARE FORESTS DIFFERENT?

The hydrologic cycle and processes discussed thus far in the book apply to forests as they do to any other land use, but the differences in rates and combinations of important processes unique to forests and the large areas covered by forests mandate a chapter devoted to forest hydrology. The differences between forest hydrology and cropland, urban, and rangeland hydrology stem from the presence of large amounts of vegetation.

There is a tremendous diversity in forest ecosystems around the world. Trees have evolved to fill niches as diverse as bayou swamplands and nearly barren mountaintops on the North American continent alone. In some cases the soil and physical topography alone can describe the important aspects of the hydrologic cycle, and the influence of the forest is less significant. This chapter will focus on forests in the temperate zones in typical hilly or mountainous environments, making only occasional points about unique forest ecosystems.

The biomass per unit area in forests is many times the biomass per unit area of cropland or rangeland. With so much plant matter, the soil properties become less important than the vegetation in determining how the system behaves. The tree canopies modify the cycling of water through the system significantly through their effects on precipitation and evapotranspiration. The large canopy system above ground is mirrored by an equally large root system beneath the ground. These roots gather water from a large area and can reach deeper than roots in croplands. Tree roots are, furthermore, typically much larger in diameter, creating a network of large pores through which water travels much faster than the soil matrix itself.

A mass of fine roots and decaying leaves form a mat of organic matter at the soil surface. This layer is not present in croplands and is much less developed in most rangelands. In forests, the thickness of this layer depends mostly on the decay rate of leaves and time since the last time the layer was burnt. Typical depths in western Oregon and Washington forests range from 5 to 10 cm. Interior forests with warmer temperatures and faster decay and the occasional ground fire typically have an organic horizon from 1 to 5 cm thick. Coastal Alaskan forests can have organic horizons in excess of 1 m deep.

With a large leaf area per unit ground area and soil development incorporating a thick organic mat and extensive root systems, the rates of some hydrologic processes in forests differ from those in croplands and grasslands. The large canopies increase interception of precipitation, a process often considered negligible in croplands and grasslands. The large leaf area and root access to deeper water allows for much greater transpiration rates. The root system and high organic content of the soil create a highly conductive soil. Figure 8.1 shows some typical hydraulic conductivity values for the various soil layers in a forest. The fibrous root and litter layer is also resistant to movement from raindrop impact and protects the underlying mineral particles from moving; thus, erosion rates in forests are typically much smaller than on croplands.

Another difference between forestlands and croplands is that forests frequently occur where the land is not suitable or profitable for other uses. Forest uses bring in less money per acre than agricultural and urban uses (Bradley, 1984). Consequently, if land is suitable for other uses it often has been converted to those uses. Not all steep land is unsuitable for other purposes, but a good portion of steep land in mountainous and hilly areas is forested. Shallow, young, and coarse-grained soils are less useful for agricultural purposes, so if not put into urban uses, these lands are often forested. Not all wetlands are forested, and a great deal of wetland habitat has been converted to both urban and agricultural uses, but where suitable conditions and species exist to have forested wetlands, such as in Arkansas and Mississippi, forests commonly occupy wetlands.

Forests grow only where the natural climate and soil water conditions are suitable. Large sections of the continental United States have not historically maintained forests. Figure 8.2 shows a map of the historical extent of forests in the continental United States. Precipitation in comparison to evapotranspiration is one of the primary controls on the extent of forest. In areas with low precipitation, forests grow only near streams or similar concentration points for water. Temperature and long term snowpack are other controls important in high elevations and latitudes. Irrigation, which allows crops and pasture to thrive in arid environments, is seldom used to maintain forests.

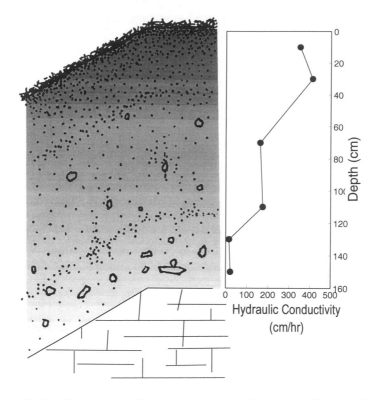

Figure 8.1. Hydraulic conductivity changes with depth in a forest soil. (Data from Harr, 1977.)

Human modification, management practices, and disturbances in forests differ from agricultural and urban uses. The primary human actions in forestlands are road building, timber harvest, and site preparation for replanting. Forest roads are more sparsely distributed than agricultural and urban roads and are usually smaller. Timber harvest is done by cutting the tree with a chainsaw and removing the log. Ground-based removal systems, such as skidders and tractors, are common where much of the landscape is level enough, but in areas of the country where there are more mountains and steeper slopes, cable-suspended yarding is much more common. Site preparation usually involves burning the residue left after harvest. However, on relatively flat and productive land, such as the converted cotton fields in the southeastern United States, site preparation may be more complex, involving machine scarification and piling of residue and, occasionally, tillage. Forests are disturbed much less often than croplands. The harvest and site preparation procedures might be carried out once every 30 years in some regions; 50 to 90 years is a more common rotation in other regions. Sometimes rangeland and forestlands overlap, so grazing practices common to grasslands are used in these forests. Forestlands used for grazing are a small subset of all forests, however.

In the following sections the specific processes in forests will be discussed. Discussion will start with precipitation and the energy balance, move through interception and evapotranspiration in the canopy, snowmelt and infiltration on the hillslope below the canopy, shallow subsurface flow (sometimes termed interflow) and over-

Figure 8.2. Historical extent of forest in the United States (shaded area).

land flow on the hillslope scale, and finally, watershed scale hydrology. Further discussion will focus on erosion processes in forestlands.

8.3. FOREST CLIMATES—RAIN AND SNOW

Many benefits have been attributed to forests with regard to water. For much of history, people have considered forests the fonts of fresh water. The Ancient Greeks believed that forests brought salt water up through their roots and purified the water, which was released through springs. Later civilizations, although perceiving the connection between precipitation and streamflow, still attributed the creation of clouds to forests. Even as recently as the end of World War II there was some debate on the effects of forests on precipitation (Kittredge, 1948), but the rain gage observations fueling the debate may well have been in error because gages in forest clearings have improved catches due to reduced wind velocities. Today, it is generally considered that forests in North America are a result of precipitation and not the cause of it. There is some recognition of the role that forests have in recycling rainfall in the Amazon (Balek, 1983). Desertification following deforestation in the Middle East and North Africa was influenced by the reduced infiltration of grazed lands and the global climate shift that has occurred over the last 3000 years.

One relationship between forests and climate that is relatively certain is that the presence of forests indicates more precipitation than the presence of shrub and grassland communities. The ecotone—a zone of transition between two ecosystems—between juniper shrublands and ponderosa pine forestlands in eastern Oregon and Washington occurs at about 32 cm of precipitation annually (Franklin and Dyrness, 1973). Areas with more precipitation have forests, and areas with less have shrub and grass communities.

The complex associations of the amount of water available, its seasonal distribution, and the seasonal temperatures create the amazing diversity of forest ecosystems. The upper treeline in the Cascades of Washington State is defined by too much precipitation in the form of snow. While there is certainly as much water on the Pacific coastal rain forest, the snow falling at high elevations in the Cascades prevents trees from having a long enough growing season to get started. Too much moisture is seldom a problem otherwise. Other trees, such as the bald cypress, have adapted to swampy conditions in the southeast. The black spruce has adapted to swampy conditions, cold temperatures and snow loads in order to survive in the sub-boreal taiga regions of North America.

8.4. INTERCEPTION—RAIN, SNOW, AND FOG

You may have observed dry areas underneath trees near the beginning of a storm, and perhaps taken shelter from storms under trees. The process of trees capturing some of the falling rain and snow is called interception.

8.4.1. Rain Interception

Figure 8.3 shows the disposition of rain or snow coming into a forest canopy. Rain falling on a forest canopy will adhere to the leaves and branches in the canopy. As the surfaces of the leaves and branches become wetted, any additional rainfall will cause some of the water to fall off and drip onto lower leaves and branches. Eventually, all of the branches and leaves are wetted, and some water will fall through to the ground, while other water will flow down the trunks and stems of the forest and shrub cover. The amount of water stored on leaves and stems is called interception, the amount flowing down the stem is called stemflow, and the amount falling from the canopy to the ground is called throughfall. Because the intercepted water frequently evaporates, it is often called interception loss.

Some hydrologists consider interception storage within the leaf litter lying on the ground. This layer can be dealt with as a soil layer or as interception. This chapter will follow the convention of calling the litter layer a soil layer and consider only interception in the canopies of trees, shrubs, and standing plants.

Interception is measured as a comparison of rain gages placed above the forest canopy, or in a nearby open area, to gages placed under the canopy, and gages collecting stemflow. While one or two standard precipitation gages are sufficient to measure the gross precipitation above the canopy or in a nearby opening, many randomly placed gages are necessary to estimate throughfall. Stemflow is caught on collars placed around tree trunks and directed to collection cans. It is necessary to have several trees in a plot to reliably estimate throughfall and stemflow. The volume of stemflow is divided by the total area of the plot to estimate the depth of precipitation abstracted by stemflow. Interception, I, is then calculated as the difference between gross precipitation, P, and the sum of stemflow and throughfall, S + T.

$$I = P - (S + T) \tag{8.1}$$

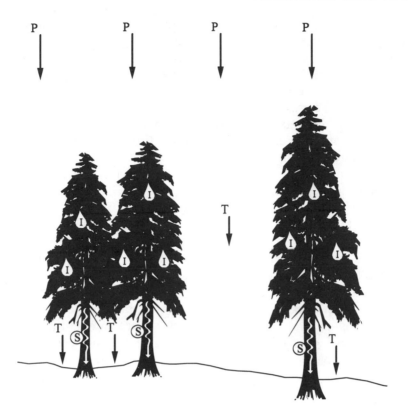

Figure 8.3.   Interception processes in forests, P is total precipitation, T is throughfall, S is stemflow, and I is interception.

Note that when using Equation 8.1 in a water balance, all intercepted water is considered to evaporate after or during a storm.

Example 8.1: During a 3.51 cm storm, throughfall and stemflow were measured on a 1/10th hectare plot. The throughfall was collected in rain gages scattered over the plot and averaged 2.15 cm. Tanks collected 11,200 liters of stemflow. Calculate the depth and percentage of rainfall intercepted.

Solution: First convert the stemflow volume to depth by dividing by the plot size.

11,200 liters $= 11.2$ m^3
0.1 ha $= 1,000$ m^2
11.2 m^3 / 1,000 m^2 $= 0.0112$ m
$S = 1.12$ cm

$S + T = 1.12$ cm $+ 2.15$ cm $= 3.27$ cm

$I = P - (S+T) = 3.51$ cm $- 3.27$ cm $= 0.24$ cm

$I/P = 0.24$ cm/3.51 cm $= 0.068$

Answer: 0.24 cm of the storm or 6.8% was intercepted.

Interception and stemflow rates are controlled by a variety of factors. Perhaps the most important of these is the type of tree. This determines the type of leaf, leaf size and shape, leaf density and arrangement, whether it is deciduous or evergreen, the branching form, and bark roughness. The average interception over a forest stand is clearly dependent on the density of the stand, and in particular on the density of water holding surfaces as might be represented by the leaf area index (LAI) of the stand. The characteristics of the storm event can be important also. Rain intensity and wind speed control the efficiency with which leaves hold water.

Interception rates for different types of vegetation are presented in a variety of forms. Some researchers have presented the proportion of annual rainfall that is lost as evaporation of intercepted water (Kittredge, 1948). Others present the proportion lost in a storm (Hewlett, 1982). Because the water holding capacity of a tree's surface is limited, this type of information is difficult to consider. Very little rain becomes throughfall until the interception capacity of a tree is nearly full, after which all of the rain passes through as either stemflow or throughfall. Therefore, the percentage captured by any storm is a function of whether or not the interception capacity of the tree was exceeded. These sorts of values are more dependent on the number and characteristics of storms in a year or in the sample than on the type of vegetation intercepting the water. It is becoming more common just to express the interception capacity of particular species. Table 8.1 presents the interception capacities of fully developed canopies for a variety of species.

8.4.2. Snow Interception

Snowfall is intercepted too. Losses of intercepted snow to sublimation (evaporation from the ice state of water) are generally smaller than evaporation losses of intercepted rainfall. Colder temperatures and the much greater energy required to sublimate snow lessen this impact. Most of the intercepted snow either blows out of trees, melts, or falls as clumps of wet snow.

The process of snow interception differs from that of rainfall interception in that the rainfall interception capacity is defined more by the amount of surface available to hold water droplets where snowfall interception capacity is more a function of branch strength and canopy shape. Snow falling on needles initially bridges small gaps, forming a platform for further accumulation. Eventually the snow builds up high enough on branches that either the snow sloughs off, being unstable itself, or the branches bend. Snow falling at temperatures near freezing is quite cohesive and generally does not slough off unless a branch bends. After the storm, the snow metamorphoses into less cohesive forms and falls to the ground or it melts.

Few reliable measurements of snow interception exist, and most work that has been done relates to coniferous species because deciduous trees have few leaves in the winter and there are few evergreen hardwoods in snowy climates. The magnitude of the snowfall interception capacity is generally greater than the rainfall interception capacity. Western white pine saplings (approximately 4 m in height) sampled at the Priest River Experimental Forest in Idaho intercepted 0.25 cm of snow water equivalent in one storm and Douglas-fir saplings (4 m high) intercepted 0.37 cm in the same storm (Haupt and Jeffers, 1967). Compare these values to the values of rainfall interception for a fully developed canopy in Table 8.1.

Table 8.1.   Interception capacity of fully developed canopies of several eastern species (after Helvey, 1971).

Vegetation	Interception Capacity (in)	Interception Capacity (cm)
Red Pine	0.15	0.38
Loblolly Pine	0.15	0.38
Shortleaf Pine	0.14	0.36
Ponderosa Pine	0.13	0.33
Eastern White Pine	0.14	0.36
Pine (avg)	0.14	0.36
Spruce-fir-hemlock	0.26	0.66
Hardwoods, leafed out	0.10	0.25
Hardwoods, bare	0.05	0.13

8.4.3. Fog Drip

Fog drip is a form of interception that works somewhat differently than the other forms and is hydrologically important only in a few locations. Fog drip occurs when trees intercept horizontally moving water droplets in clouds and fog that would not otherwise fall to the ground. This effect occurs mostly in coastal forests and in mountain ranges not far from coasts. Some foresters say that without fog interception, the Sequoia trees of California's coast could not survive. Harr (1982) noted the contribution of fog drip to the hydrologic cycle in the Cascades near Portland, Oregon. Deposition of rime ice on trees in the northern Appalachians is a similar process.

8.5. ENERGY BALANCE IN FORESTS

In order to understand the evapotranspiration and snowmelt processes in forests, it is important to understand the energy balance in forests. The forest canopy absorbs and emits radiation, changes wind flows, and releases water vapor, all of which change the energy balance in the forest.

There are four primary forms of energy considered in an energy balance: short-wave radiation, long-wave radiation, latent heat, and sensible heat. Short-wave radiation is usually considered to be the wavelengths of light emitted by the sun and is sometimes called solar radiation. This includes light in the ultraviolet, visible, and near-infrared ranges. Long-wave radiation is light in the far-infrared range (greater than 4 microns) and is sometimes called terrestrial radiation, because the earth's surface radiates in this wavelength. Latent heat refers to the latent heats of vaporization and fusion for water, which refer to the energy necessary to turn water from a liquid to a vapor or to turn water from a solid to a liquid. Approximately 2500 kiloJoules per kilogram (590 cal/g) of water are necessary to change liquid into vapor, and 335 kJ/kg (80 cal/g) are required to melt ice. The reverse is also true: 2500 kJ of heat are released per kg of water condensed, and 335 kJ/kg are released when zero-degree water freezes. When water vapor in the air over the snowpack condenses and freezes onto the snowpack, 2835 kJ/kg (670 cal/g) are released. Sensible heat refers to the heat energy due to the temperature of some substance (air, rain, snow, or soil for example), and the specific heat of the substance refers to how much energy is required to change the temperature of that substance.

Example 8.2: The branches of a small tree hold 3 kg of snow. Calculate the energy required to melt the snow and the energy required to sublimate (change directly from a frozen to gaseous state) the snow.

Solution:
Melt: 3 kg × 335 kJ/kg = 1,005 kJ

Sublimation: 3 kg × 2835 kJ/kg = 8,505 kJ

Answer: The three kg of snow in the branch will require 1,005 kJ to melt and 8,505 kJ to sublimate. Is it more likely that the snow will melt from the branch or sublimate? Note that there are conditions when sublimation will occur when melt cannot, like windy, dry days when the temperature is below freezing.

An energy balance considers the fluxes of all of these forms of energy to find the net energy flux (Figure 8.4). The equation of conservation of energy is

$$Q_{sw} + Q_{lw} + Q_{le} + Q_h - \frac{\Delta S}{\Delta t} = 0 \qquad (8.2)$$

where, Q_{sw} is the short-wave energy flux, Q_{lw} is the long-wave energy flux, Q_{le} is the latent heat flux, Q_h is the sensible heat flux (all flux units are energy per unit time, e.g., watts), and $\Delta S/\Delta t$ is the change in stored energy per unit time. This equation must hold for any defined volume. When calculating evapotranspiration, it is easiest to consider the primary surfaces at the top of the canopy and under the soil, and when trying to calculate snow melt, it is easiest to consider the primary surfaces at the top of the snowpack and the bottom of the soil.

Short-wave radiation input can be calculated based on time of year, aspect, slope, and cloudiness. The reflectiveness of the location for which the balance is being calculated is called the albedo, and the reflected light is subtracted from the short-wave balance. The long-wave balance can be calculated based on air temperature and cloudiness. Latent heat transfer is considered in regard to transpiration, which requires an energy input, and in condensation and freezing of water vapor to a snowpack, where energy is released. Rates of latent heat transfer are regulated primarily by wind, which carries moist air away in the case of transpiration, and carries moist air to the snowpack in the case of condensation. Sensible heat is transferred primarily by air convection; however, a sensible heat flux from the earth is transferred by conduction. An increase in stored energy raises the temperature of the material of which the volume is composed.

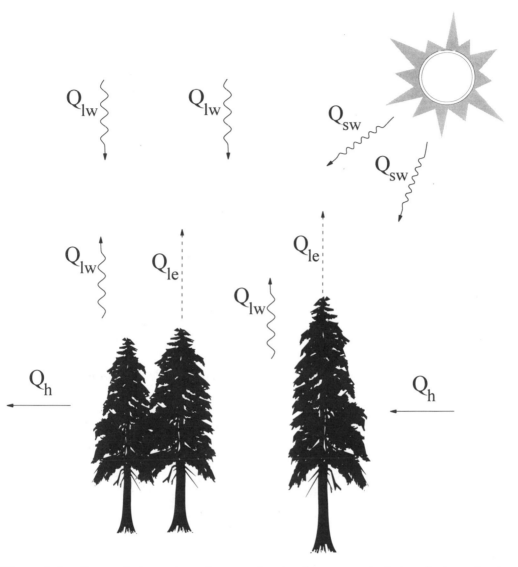

Figure 8.4. Energy balance in a forest. Q_{sw} is short-wave (solar) radiation, Q_{lw} is long-wave (terrestrial) radiation, Q_{le} is the latent heat of evaporation, and Q_h is sensible heat.

A basic point to make about the energy balance is that heating caused by positive short-wave radiation, long-wave radiation, and sensible heat fluxes, such as found on a warm summer day, can result either in evapotranspiration (a negative latent heat flux) or an increase in the stored heat energy, or temperature. In a wheat field in late August, after all available soil moisture has been used, the wheat field will become quite warm, whereas in a more deeply rooted forest, transpiration will keep the temperature lower. This transpiration effect, more than shading, is what keeps forests relatively cool compared to adjacent open areas.

8.6. EVAPOTRANSPIRATION

With their large vegetation masses and deep root systems, forests are capable of transpiring a great deal more water than dryland crops. Evapotranspiration in forests is usually considered only in terms of its overall effect on water supply, although academic interest is being shown with regard to its importance in modeling tree growth, microclimate modification, and shallow groundwater flow. Most studies of evapotranspiration have been done at a watershed level using a simple water balance for the stream by ignoring deep seepage of groundwater. This section will discuss the process, and the basin-wide importance will be described later in the chapter.

Evapotranspiration has been covered in some detail in Chapter 4. In forests, the process is essentially the same, differing only in a few minor details. The most important differences are that forests transpire much more than they evaporate; forest have greater rooting depth, so have a deeper water supply to draw on; and forests are seldom irrigated.

There is generally little direct evaporation in forests as compared to transpiration. The soil surface is well protected by the deep canopy of living, transpiring vegetation and an insulating layer of litter. Evaporation occurs mainly from intercepted rainfall. Some evaporation occurs during storms and between rainfall events. Intercepted snow generally melts before much of it sublimates as can be deduced from the energy requirements for melting versus sublimation (335 kJ/kg to melt vs. 2835 kJ/kg to sublimate).

Transpiration from forests is best calculated using the Penman-Monteith equation because the stomatal resistance of trees is the primary control on transpiration rates. The Penman-Monteith method can be modified to ignore the boundary layer resistance for coniferous forests, because the canopy projects so high into the air, and the needles have such a high surface area (Waring and Schlesinger, 1985). The energy balance component of the Penman-Monteith method considers net radiation $(Q_{sw}+Q_{lw})$ as input, ignores Q_h, and uses the change in air temperature to find ΔS. Evapotranspiration is thus estimated from Q_{le} and the ability of the stomata to retain water. The stomatal resistance in turn is a function of the soil moisture and plant moisture stress.

One notable difference between forests and croplands is the rooting depth. Figure 8.5 shows the difference in soil water taken as evaporation and transpiration from plots with aspen, herbaceous species, and bare ground. The main factor causing the differences in Figure 8.5 is rooting depth.

Because forests are not irrigated, it is difficult to apply equations that assume a well watered crop. This difference also makes it difficult to compare gross amounts of water use by crops and trees. There are proposals to apply wastewater effluent as irrigation water to forest stands as a form of sewage treatment. Because of the dangers of nitrate leaching to deep groundwater, where it poses health threats to water supplies, it is important to be able to estimate how much water trees will transpire in a year if irrigated.

8.7. SNOWMELT

Another important application of the energy balance in forests is snowmelt. Snowmelt is important in rangeland and agricultural systems as well, but most re-

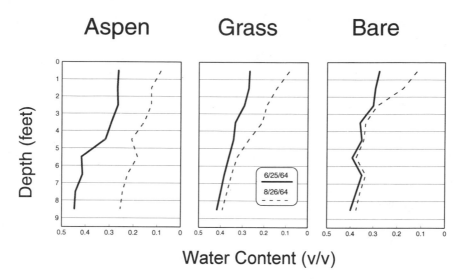

Figure 8.5. Evapotranspiration from aspen, grass, and bare plots (after Johnston, 1970).

search in snowmelt has focused on forest lands because forests can be manipulated to change timing and yield of water from snowpacks. Many forested areas are important as reservoirs of snow for spring and summer runoff used to irrigate lower elevation rangelands. Consequently, much effort has gone into trying to determine how forests can be manipulated to augment flows or delay melt later into the summer. About 70% of the precipitation in the western United States falls as snow, making an understanding of snowmelt very important in managing western water resources. A quite thorough work on the subject of snow accumulation and melt (Hathaway et al., 1956) focuses on forecasting flow rates of major stream systems for hydroelectric reservoir regulation. More recently, the effects of forest harvest on peak flow of stream systems that rely on snowmelt has focused attention on modeling rapid melt rates that occur over short time scales.

While many of the studies focusing on water supply concluded that the primary effect of forests on snowmelt was due to redistribution of snow, a general consideration of the energy balance was shown to be important in determining watershed or basinwide melt rates. Studies on rapid melt rates have focused on the energy balance (Marks and Dozier, 1992).

All of the terms in Equation 8.2 are important in determining melt from a snowpack. During the spring, the components of the energy balance for a snowpack are: incoming solar radiation during the day, outgoing long-wave radiation during the night, latent heat of sublimation added to the snowpack as vapor condenses on the snowpack under certain weather conditions, and added sensible heat from the air, rainwater, and ground heat flux. These components all act to change the energy content of the snowpack. Any energy added above that necessary to bring the pack to 0°C in the frozen state will melt the snow.

Solar radiation during daylight hours is dramatically affected by trees. Shading from trees causes a reduction in short-wave energy reaching the snowpack and is a function of tree density and crown height and whether the tree is coniferous or deciduous.

Trees, like snow, are near perfect black bodies with regard to long-wave radiation, that is they absorb almost all long-wave radiation and emit strongly in the long-wave band. Heating of the forest canopy by solar radiation, therefore, increases the long-wave radiation output by the trees. Under a forest canopy the long-wave radiation into and out of the snowpack is well balanced, or even positive, whereas in an open area, the long-wave energy balance is negative with a net outflow of energy.

Sensible heat is an important factor, but it is actually a relatively small part of the energy budget until late spring, when heat conduction from bare earth heated by radiation speeds late season melt. Air temperatures are typically below freezing until spring, at which time there is some warming. Sensible heat is transferred most effectively by wind. When air is still, a thin layer of cold air will develop directly over the snow, insulating it from the warmer air above. Turbulence created by wind breaks up insulating layers.

In spite of sensible heat's relatively minor contribution to the snowmelt energy budget, the temperature index snowmelt model has been the most popular for most basin modeling approaches. These models are typically of a simple form:

$$M = K(T_a - T_b) \tag{8.3}$$

where M is melt for the day (in units of depth, usually inches), K is the degree-day factor (units of depth divided by temperature), T_a is the average air temperature, taken as the average of the minimum and maximum temperature for the day, and T_b is the base temperature, usually taken as the melting temperature of snow, 0°C or 32°F. The idea is that there is no melt for temperatures below freezing, and that melt is directly proportional to the number of degrees above freezing. There are a variety of values for K and T_b that must be determined empirically for each general area (Hathaway et al., 1956).

Example 8.3: At the Central Sierra Snow Laboratory, the degree-day factor, K, for April was determined to be 0.036 inches/degree day above base temperature, and the base temperature, T_b, was found to be 26°F. The Weather Service forecasts an average temperature of 40°F on April 10. Find the average depth of snowmelt expected over a 5,000-acre basin.

Solution:
$M = K(T_a-T_b)$
$K = 0.036$ in/°F
$T_b = 26°F$
$T_a = 40°F$

$M = 0.036$ in./°F $(40°F-26°F) = 0.50$ in.

Answer: On April 10, 0.50 inches of melt are expected.

The primary reason that temperature index snowmelt models are popular is because temperature is often one of the only reliable and consistently available weather variables measured at weather stations. Radiation measurements are still uncommon. Also, in spite of the fact that they ignore an explicit representation of the variety of processes affecting snowmelt, a surprising number of variables correlate to air temperature. Days of melt with high solar input are often warm, and warmer air holds more moisture for condensation than cold air. For these reasons, these models have worked reasonably well for springtime melt estimates over large basins.

The latent heat of sublimation can provide a strong positive input of energy during periods of snowmelt. The effect has been somewhat misnamed as "rain-on-snow" because during rain events, the moisture content of the air is the highest, and the transfer of latent heat to the snowpack is the highest. The wind often associated with rainstorms helps carry warm moisture-laden air to the snowpack. As the air flows over the snowpack, the moisture condenses onto the snow and releases heat to the snow. This heat is conducted into the rest of the snowpack through ice, meltwater, and rainwater. Sensible heat carried by rainwater is relatively small, because rainfall temperature is usually close to freezing during these events. The volume of simultaneously melting snow and falling rain can be much greater than when either event happens alone, however, making such events hydrologically important. Under trees, the wind speed is lower and there is less turbulence, causing less movement of moist air onto the snowpack and, consequently, a lower snowmelt rate.

8.8. INFILTRATION

Free water available at the soil surface, due to rainfall or snowmelt, infiltrates into the ground. The same basic principles apply in forests as in agricultural and rangeland conditions. Forest soils are often highly conductive. Table 8.2 shows typical forest and agricultural soils with the respective range of hydraulic conductivity values.

The reasons for higher hydraulic conductivity relate primarily to the presence of vegetation. Leaves, needles, and roots form a mat of litter on the forest floor. This layer of organic matter is often highly conductive itself, and contributes to the conductivity of lower layers as it decomposes and is incorporated into the mineral soil by earthworms and small burrowing animals. The roots create macropores through these shallow layers and into deeper layers. The litter, roots, and forest canopy all protect the soil from damaging raindrop splash and prevent the formation of the surface crusts that can be important in the hydrologic behavior of agricultural soils.

Infiltration capacities of the native forest soil can be dramatically altered by forest management practices. Compaction by machinery is one of the more obvious ways in which forest soils are altered; road construction is an extreme example of this. The removal of litter layers by fire and machinery also works to retard movement of water into the soil. Fire can sometimes lead to hydrophobicity, a form of water repellency, in soil that can prevent infiltration altogether for some period of time (DeBano, 1981). Collectively, most of these areas of reduced infiltration capacity are a small proportion of most watersheds, and so create little concern for increased peak flows on a large scale. Many examples exist of decreased infiltration in a small area due to road construction leading to erosion, landslides, and gullying, showing us that reduced infiltration is important at a local scale.

Table 8.2.   Hydraulic conductivity values for a few surface treatments of forest soils on two soils and hydraulic conductivity values for several agricultural soils.

Soil Name or Treatment	Texture	Hydraulic Conductivity (mm/hr)	Hydraulic Conductivity (in./hr)
Forest Soils			
High Severity Burn	Sandy Loam	76.90	3.03
Low Severity Burn	Sandy Loam	80.90	3.19
Root Mat Removed	Sandy Loam	77.83	3.06
Light Use Skid Trail	Sandy Loam	81.73	3.22
Heavy Use Skid Trail	Sandy Loam	58.41	2.30
Moderate Severity Burn	Silt Loam	35.96	1.42
Light Use Skid Trail	Silt Loam	50.88	2.00
Forest Road	Any	<4	<0.2
Agricultural Soils			
Hersh	Sandy Loam	58.27	2.29
Sverdrup	Sandy Loam	29.56	1.16
Whitney	Sandy Loam	21.87	0.86
Williams	Loam	18.12	0.71
Academy	Loam	14.01	0.55
Barnes	Loam	10.84	0.43
Zahl	Loam	9.52	0.37
Barnes	Loam	9.15	0.36
Woodward	Silt Loam	29.55	1.16
Keith	Silt Loam	28.67	1.13
Portneuf	Silt Loam	2.55	0.10
Nansene	Silt Loam	1.73	0.07
Palouse	Silt Loam	1.19	0.05

8.9. SUBSURFACE FLOW (INTERFLOW)

Because hydraulic conductivities of forest soils are so high, subsurface flow is the primary form of water movement in forested areas. Water in forest soils flows in one of two modes: saturated and unsaturated. Unsaturated subsurface flow occurs in damp soils where pore spaces are only partially filled and the water is often not continuous between pores. Saturated flow occurs through soils that are fully saturated. Saturated subsurface flow occurs where a layer with low hydraulic conductivity, such as bedrock, underlies a more conductive region, such as a forest soil. Saturated subsurface flow is usually much slower than overland flow, so it is still not an important contributor to the storm hydrograph except when it prevents infiltration in particular areas as will be discussed in the next section.

In order to discuss lateral subsurface flow in forests it is easier to use the basic equation relating to water movement in soil, Darcy's Law,

$$V = -K \frac{\partial h}{\partial L} \tag{8.4}$$

where V is the average velocity over the cross section of soil considered, K is the hydraulic conductivity, and $\partial h/\partial L$ is the change in head per unit distance in the direction of flow. The negative sign shows that water flows from places with high head to places with low head. Multiplying V by the depth of the soil gives the flow per unit width of hillslope being considered. Saturated flow responds mostly to slopes in the water table, whereas unsaturated flow can respond to differences in head caused by the soil being drier in one area than another. For shallow saturated subsurface flow, as may be found on steep mountain slopes, $\partial h/\partial L$ is sometimes taken as the slope of the ground.

In forest soils, hydraulic conductivity is not uniform in direction. Leaves fall with their long axes parallel to the ground, and more area is covered in roots spreading radially from trees, again, parallel to the ground, so the hydraulic conductivity in a direction parallel to the ground surface is greater than in the direction perpendicular to the ground. This is a condition called anisotropy.

Forests often occur on steeper slopes than agricultural land. The steeper slopes lead to higher gradients and, as indicated by Darcy's Law, faster flow. In mountainous areas, these steeper slopes often have young, shallow soils. Because they are young, their clay content is typically low, and the saturated hydraulic conductivity is high. The shallowness creates a system of temporary perched water tables during the rainy season, with no connection to a deeper regional groundwater system. By the late dry season, many of these water tables no longer exist, with only unsaturated soils remaining. This effect is due more to drainage than to evapotranspiration.

In steep areas with shallow soil, the greatest groundwater depths are found in swales, sometimes called bedrock hollows. Groundwater will flow perpendicular to contours and will be concentrated in these topographic depressions. These areas are the most vulnerable to landsliding because of the generally higher pore water pressures that can exist there.

Forest management can affect interflow substantially. The most important effect is that forest road construction often intersects temporary groundwater tables. In these cases, the road system can act like an agricultural drainage system in a field by providing faster paths for interflow to reach streams. Roads can also add significantly to groundwater depths when culverts concentrate water from a large area into a small area. Forest roads commonly have ditches to carry water from the cutslope seepage and road runoff. The ditches have ditch relief culverts spaced periodically down the road for drainage and often discharge to hillslopes and not to streams. Very often, all of the water infiltrates and the groundwater level is raised locally by this concentrating effect. The increased groundwater level increases the risk of landsliding.

Forests have deep root systems and can withdraw a great deal of water from groundwater. In areas where water supply is short, the trees compete with other potential uses of the water. In areas with excess water, such as in many parts of the Mississippi Delta area, when trees are removed, groundwater levels rise and the harvested area becomes a swamp. Some claim that similar benefits are important in areas with potential for landsliding. In general this is seldom the case. Areas that are prone to landsliding are steep, and water moves out by saturated and unsaturated flow much faster than it is evapotranspirated. Also, the depth of flow necessary for landsliding occurs over a very short time period during rainstorms, and evapotranspiration will not affect this peak.

One exception to this rule is on geologic formations called earthflows. They are most common in volcanic mountain ranges with a high annual precipitation, such as the Cascades, and they also occur in the Siskiyou Mountains of Northern California and Southern Oregon on ancient sea floor sediments. Earthflows are made up of fine grained material that flows slowly down mountains like a glacier made of mud. At the toes, they commonly fail in large landslides into streams. Flow rates on earthflows have been tied to groundwater levels (Swanston et al., 1988).

8.10. SURFACE RUNOFF

Surface runoff is rare on most forest hillslopes. Particular conditions must exist for surface runoff to occur, and it is best to discuss the overland flow occurring by the two primary processes separately. The first process, often called Horton overland flow, occurs when the rainfall intensity exceeds the infiltration capacity of the soil. The second process, called saturation overland flow, occurs when the soil is so full of water that no additional water can infiltrate.

Circumstances leading to Horton overland flow include high rainfall intensities and compacted or otherwise low-infiltration soils. During a storm the initial infiltration capacity is quite high and decreases to a steady state infiltration rate over time, as described in Chapter 3. If the rainfall intensity at any time exceeds the infiltration capacity, there will be rainfall excess. If enough rainfall excess occurs to fill surface depressions, runoff will occur. Summer thunderstorms are the most common source of high intensity rainfall for Horton overland flow. Almost all areas of the United States experience thunderstorms, but they are less frequent on the West Coast than in other areas. High mountainous areas and the southeastern United States are particularly subject to these events (See Chapter 2).

Most forest soils have high infiltration capacities and do not produce Horton overland flow unless they have been disturbed, even during these extreme events. Road building will decrease infiltration capacities to less than 4 mm/hr for most soils (Luce and Cundy, 1994). Skid trails and skid roads similarly reduce hydraulic conductivities, although not to as great a degree. In unusual circumstances, burning causes hydrophobicity, a water repellent condition, in the soil (DeBano, 1981).

Saturation overland flow is more common in forests, particularly where there are shallow soils. Saturation overland flow is linked to a concept called variable source area (Hewlett and Hibbert, 1967; Hewlett, 1982), which notes that the pieces of the watershed that are contributing to the stream hydrograph are spread throughout the watershed near the stream network. The stream network and contributing areas of surface runoff expand as the storm continues in response to increasing depth of interflow in these source areas (Figure 8.6). Dunne and Black (1970) expanded on Hewlett and Hibbert's point by describing the process involved in the creation of these source areas and how they contribute to the stormflow.

When looking at a hillslope in profile at a particular point during a storm, the depth of saturated soil is greater at the bottom of the slope than at the top (Figure 8.7). This follows from the fact that the "watershed" near the top of the slope is much smaller than the "watershed" for the bottom of the slope, and for the same reasons that a river gets bigger as one goes downstream, the volume of water flow-

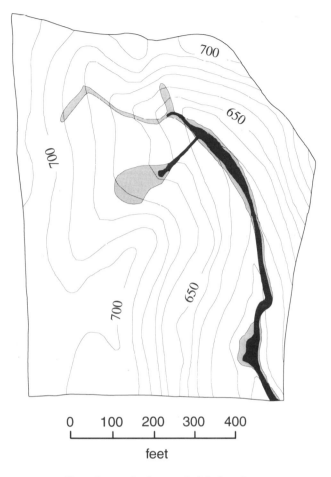

700

650

700

650

700

0   100   200   300   400

feet

Contour Interval 10 feet

Figure 8.6.   Expansion of saturated area in a basin during a single storm.   The dark
shaded area shows the saturated area at the beginning of the storm, and
the lighter shading is the saturated area at the end of the storm.   All
precipitation falling in saturated areas is converted to runoff (after Dunne
and Leopold, 1978).

ing through the hillslope must increase as one goes downslope.   Now consider the
discussion of Darcy's law earlier.   Flow velocity in groundwater is a function of
slope and soil properties only, and the flow does not get faster as it gets deeper.   So,
the velocities of the water at the top and bottom of the slope in Figure 8.7 are the
same, and therefore to pass the greater volume, the groundwater must be thicker at
the bottom of the slope.   When it rains hard enough and long enough, there will be
more volume coming from above than can pass laterally through the soil, and some
must seep onto the surface to flow overland.   In addition, no additional water can
enter, so that both the water seeping out and the rainfall will flow overland.

Several situations promote surface runoff by the saturation overland flow process.
Concave slopes are the most susceptible, as water concentrates in the lowest part.

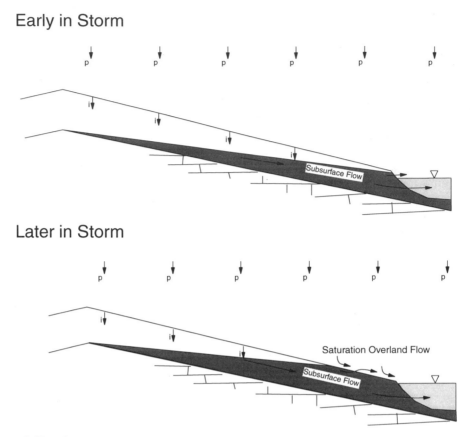

Figure 8.7. Changes in depth of subsurface flow leading to saturation overland flow. "p" refers to precipitation and "i" to infiltration.

Road cuts are also susceptible to saturation overland flow. Road cuts cause an abrupt shallowing of the soil, and the excess interflow that cannot pass under the road prism must seep out on the cutslope. Removing a duff layer can change an area from one that never has overland flow to an area that will have saturation overland flow. Removal of the duff layer seems like an innocuous shallowing of the soil but considering that hydraulic conductivities of duff layers can be 100 times those of the soil below, removal of the duff layer can represent a significant loss of interflow capacity for the soil.

8.11. STREAMFLOW AND WATERSHED HYDROLOGY

Many questions raised in forest hydrology focus on the watershed scale as opposed to the hillslope scale described above. However, understanding how water behaves at a point and on hillslopes makes it easier to understand how the watershed acts as a whole. The two main areas of concern usually discussed with regard to watershed hydrology in forests are floods and water supply.

The controversy of how well forests protect against floods has existed for the better part of a century. The Organic Act of the National Forest Service states that part of the reason for the forests is to "secure favorable conditions of flow." In general it is agreed that forest cover has little effect on flood size for large stream systems, but is very important for small watersheds (Lull and Reinhart, 1972). The debate continues, however.

Originally, it was believed that the forest canopy and litter layer prevented quick surface runoff, thus water soaked into the sponge-like forest only to seep out slowly later. As understanding increased about the role of interflow and saturation overland flow, it was found that storms that caused floods on major rivers tend to have long durations and low intensities. While interception and soil storage manage to delay runoff somewhat, the amount of water intercepted and the amount of water detained are small in comparison to total storm volume (Lull and Reinhart, 1972). Even though the only runoff-producing process in low intensity storms such as this is interflow discharging to streams, the long time span allows for many parts of a watershed to contribute. When the Ohio River floods, the Mississippi downstream can handle the increased flow if no other streams are flooding, but if all of the tributaries are having minor floods, the mainstem of the Mississippi may experience a major flood.

On small watersheds, forests can reduce flooding. Thunderstorms with high intensities, covering small areas and lasting a short time, can create severe flooding in streams with other land uses. Storms with high intensities can create Horton overland flow on lands used for agriculture and grazing, but interception and infiltration capacities of forests are great enough that overland flow will not occur even in a thunderstorm. Pathways for stormflow in forests are slower than in agricultural, range, and urban lands, and streams respond only to long duration events.

In some parts of the United States, removal of forest canopy can increase the potential for flooding at larger scales. Snowmelt by latent heat transfer during long, low-intensity rain events in the Pacific Northwest can augment a currently occurring rainstorm, creating a situation with sustained high water input rates. Early evidence showed that small watersheds (100 acres) and even medium sized (50,000 acres) watersheds showed an increase in peak flows following removal of trees from a portion of the watershed. Berris and Harr (1987) showed that the removal of the forest canopy decreased interception of snow and increased turbulent transfer of latent heat to the snowpack. There is still some question about the magnitude of the effect at large scales.

The effect of forests on water yield is much less controversial. Trees consume water. Research done to examine how to augment snowpacks and reduce canopy interception and evaporation of snow showed that changed forest-cutting patterns serve primarily to influence how snow is redistributed by the wind and little to change how much evaporates. Researchers were also able to influence the timing of the melt to some extent, but timing still depended much more on weather than on harvest practices. The only effective means to increase water yield from a forest is to prevent transpiration, and some water supply organizations have done so.` The Los Angeles Department of Water and Power, for example, has removed all riparian forests from the Owens River near Bishop, California in an effort to increase the water yield to their aqueduct. Such practices are rare because water quality is often as important as the amount available, and forest removal promotes erosion.

8.12. EROSION—SEDIMENT BUDGET

Erosion in forestlands is one of the key concerns of forest managers. The discussion on hydrology points out that, on a watershed scale, the forestland manager can cause only slight changes in water yield. It is, however, easy to increase sediment yields many hundred-fold with careless management. The sediment budget of a forest stream is the combined sources of sediment less the change in sediment storage along the channel, which must equal the amount of sediment transported out of the watershed. When estimating the sediment budget of a forest, three distinct upland sediment sources must be considered: sheet and rill erosion, gully erosion, and mass wasting. In addition, many hydrologists consider channel erosion as a part of the sediment budget. Channel erosion is more properly considered a change in channel storage when considering basin-wide sediment budgets.

8.12.1. Sheet and Rill Erosion

Sheet and rill erosion is commonly seen on agricultural, urban, and range lands. This form of erosion shows signs of minor soil displacement and rilling on the surface of the soil as sediment is removed. Sheet and rill erosion occurs only under conditions of surface runoff characterized by shallow overland flow as opposed to the channelized flow of gully and channel erosion. Surface runoff in the form necessary for sheet and rill erosion is rare in forest environments. It is confined primarily to road surfaces but may occasionally be seen on harvest units that have had severe burns, extensive tractor skidding, or mechanical site preparation. Dissmeyer and Foster (1981) extended the C factor of the Universal Soil Loss Equation for forest environments. They suggest changes to the ground cover, canopy, soil consolidation, high organic content, fine root, residual binding effect, depression storage, step, and contour subfactors to calculate the C factor.

8.12.2. Gully Erosion

Gully erosion is dominated by concentrated flow, and contributions of sediment from interrill areas are generally insignificant compared to the amount of material removed from the gully headwall and banks. Gullies are often associated with roads and skid trails. These disturbances create and concentrate overland flow. When the concentrated flow is directed to an area susceptible to erosion, a gully may be started that will propagate both up- and downstream. Where gullies occur, they can provide an order of magnitude more sediment than sheet and rill erosion. There are no practical methods to predict gully erosion volumes because most gullies occur unexpectedly. Sheet and rill erosion are often accepted as the "price of doing business," but managers work hard to prevent gullies because they have a high potential of damaging stream ecosystems and constructed facilities, such as roads.

Gully erosion is common around forest roads with poorly designed or maintained drainage systems. Normally operating road ditches and culverts concentrate flow only to the degree that ditch materials can safely handle the water, and the water is placed downslope of the road only in areas that will not form gullies. Rip rap, and other stabilizing materials are commonly placed at culvert discharges to prevent gully formation, and culverts are extended down long hillslope embankments to prevent gully erosion in the highly erodible road fill material. When maintenance fails to keep culverts open and ditches clean, excessive flows can build up and be directed to areas where the concentrated flow will be detrimental. Also, all culverts have a limited lifetime and will eventually fail, either by rusting through the bottom, plugging with branches and sediment, or by being over-topped by a large storm. When culverts fail, large gullies can be formed.

8.12.3. Channel Erosion

Channel erosion is another important process in forestlands. Channel erosion refers to any erosion in channels formed by water. It is a step up in scale from rill and gully erosion but the underlying process, removal of sediment by concentrated flow, is similar. Channel erosion can be considered in two distinct forms. Channel erosion in steep stream reaches, where most of the geologic processes are erosional, is very similar to gully and rill erosion, perhaps differing only in scale. Channel erosion in depositional reaches of streams, typically where the gradient is lower and the valley is wider, is somewhat of a misnomer, and it is better described as a change in the amount of sediment in storage. Sediment transport takes place in all channels containing non-consolidated materials, like gravel, sand, or mud, and erosion of river bank materials is usually a re-entrainment of material temporarily stored by the river. So why is this form of channel erosion called "erosion"? Probably because valuable land lies along the sides of larger streams, and any removal of stored soil is viewed as detrimental. In addition, it visibly scars streambanks, and the impacts to water quality from so-called channel erosion may be as great as from any other form of erosion.

In steep stream reaches, where most detached soil is removed from the site, channel erosion shows up as either an enlargement of the stream channel or as the creation of a new channel. This process is caused by a change in the local hydrology that increases the peak flow. This change can come about by natural events or management. Wildfires and unusually intense storms are typical natural causes, and road construction and prescribed burning are common management causes. Earlier sections describe how the hydrology is changed by these events and practices. This type of erosion is common in areas with high road densities.

In depositional reaches, channel erosion is often caused by regular meanders of the stream. Stream meanders cutting into stream banks is a normal and natural process. The erosion is often balanced by the deposition of a point bar on the bank opposite the one being eroded. It is important to note, however, that not all meanders actively erode. Some meander systems formed under different climates than now exist, and the meander pattern and flows under the current climate are such that no migration of the meanders is expected.

Channel erosion is also caused by changes in the amount of sediment in transport from upstream reaches. Decreases in the sediment load from upstream mean that more material can be removed from the stream reach in question to fill the transport capacity of the stream. This leads to downcutting of the channel and slumping of the banks as the old floodplains are abandoned. Increases in sediment load from above can lead to channel widening as the stream changes to accommodate the increased load (Sullivan et al., 1987). Changes in sediment load caused by forest management practices can have impacts on stream systems far downstream with this effect.

In forest lands, channel erosion is sometimes seen near where a tree has fallen into the water. This observation has been used as a reason to remove all trees from the streambanks in some places. Such a strategy often backfires, as the importance of the tree roots in holding the banks in place is far greater than the effect of trees redirecting flow. Trees and brush along streams also act to slow water, reduce channel erosion, and provide fish habitat. This same slowing of the water increases depths of floodwaters, however, and can be a nuisance to people dwelling in the floodplain.

8.12.4. Mass Wasting

Mass wasting is the form of erosion where soil is transported in a mass, such as a landslide. Mass wasting is the primary form of sediment delivery in many mountainous forest lands in the western United States. In addition to being a major contributor to the sediment budget of forested basins, mass wasting events can deliver larger sized materials than are typically delivered with sheet and rill or gully erosion.

Mass wasting occurs in many forms. Some typical names for mass wasting phenomena include landslide, debris slide, debris flow, slump, and earth flow. These names represent specific types of movement in specific materials as described by Varnes (1978) (Table 8.3). Of greatest concern in forest environments are the debris slide/debris avalanche/landslide movement type, the debris flow movement type, and the earth flow movement type. Drawings and descriptions for these three common mass movement types are in Figure 8.8. Debris slides are translational failures of soil common to steep slopes, and are typically shallow, one to a few meters thick. Debris flows occur when sufficient water is incorporated in the failed mass to liquefy the mass, as might happen when a debris slide moves into a steep headwater stream channel. Debris flows travel further and faster because of the increased liquid content. Earth flows occur as a slow glacier-like deformation of the soil, typically several meters thick. Earth flows are common in soils with high clay contents with a great deal of water available, such as the Franciscan Melange geology of the Siskiyou Mountains in California and Oregon and some types of volcanically derived soils that weather rapidly. Earth flows occur on lower angle slopes than do debris slides.

Water is an important contributor to all types of slope instability. It is important in terms of increasing the pore water pressure, decreasing particle contact within the soil, and adding to the weight that must be supported. Landslides typically occur in steep hollows. The shape of these swales helps to concentrate the water making the saturated soil depth greater than would exist on a plane or convex slope. Swanston et al. (1988) studied the effects of interflow fluctuations on earth flow movement and found it to be an important predictor in the rate of movement of the mass. Force balance equations used by engineers for risk analysis of potential slide areas show the importance of water in increasing instability.

Land managers can strongly affect mass erosion within forested watersheds. Redistribution of water by forest roads is one of the greatest problems. Placement

Table 8.3. Classification of slope movement types (after Varnes, 1978).

Type of Movement			Type of Material		
			Bedrock	Engineering Soils	
				Predominantly coarse	Predominantly fine
Falls			Rock fall	Debris fall	Earth fall
Topples			Rock topple	Debris topple	Earth topple
Slide	Rotational	Few Units	Rock slump	Debris slump	Earth slump
			Rock block slide	Debris block slide	Earth block slide
	Translational	Many Units	Rock slide	Debris slide	Earth slide
Lateral Spread			Rock spread	Debris spread	Earth spread
Flow			Rock flow (deep creep)	Debris flow (soil creep)	Earth flow
Complex			Combination of two or more principal types of movement		

of ditch relief culverts can inadvertently concentrate water into an area likely to fail. Consultation by geotechnical engineers is therefore an important part of any road design project, so that unstable areas can be identified and avoided. Additional problems can be created on sidecast roads because of oversteepening. Proper compaction of the road fill mitigates this problem. Removal of trees decreases the amount of cohesion given to the soil by fine roots (Burroughs and Thomas, 1977). The intertwining of root systems between trees forms a fibrous network that will hold unstable patches of soil in place. Roots begin to decay as soon as a tree is cut and the reinforcement to soil strength offered by the roots is lost quickly. Root systems of the new stand may take several years to become as effective in holding the soil.

8.13. WETLANDS

8.13.1. Definitions and Importance

The term "wetlands" frequently brings to mind scenes of swamps and marshes and ducks taking wing. This is one aspect of wetlands but by no means a complete picture. In its broadest context, the term "wetlands" also includes coastal beaches and estuaries, lakes, rivers, and poorly drained farmlands. In its use in ecological and regulatory literature today, the term "wetland" refers to any site whose soil development, biotic community, or hydrologic behavior is dominated by the fact that it is at least periodically saturated with water. Most definitions exclude deepwater habitats, where water, not soil or air, provides the primary medium for biota.

Because saturation with water can be a nuisance for many land uses, a great deal of effort has been spent in draining and filling wetlands, and they have been retained

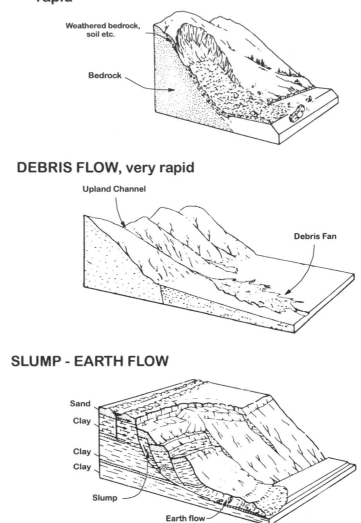

DEBRIS AVALANCHE, very rapid to extremely rapid

Weathered bedrock, soil etc.

Bedrock

DEBRIS FLOW, very rapid

Upland Channel

Debris Fan

SLUMP - EARTH FLOW

Sand

Clay

Clay

Clay

Slump

Earth flow

Figure 8.8. Descriptive drawings of three common forms of mass failure in forest-lands (after Varnes, 1978).

only where it is most convenient or too expensive to change. As wetlands have become rarer, their importance is being realized and more effort is being spent to understand and protect them.

Wetlands provide an ecologically important role in the landscape. They are the only sites in which a great diversity of hydrophytic (water loving) plants will grow. Without wetlands we would lose this diversity of plant life. They are well known for their role in supporting nesting and migration of waterfowl, who depend on these ecosystems for food and cover.

Wetlands are showing increasing value for improving water quality. Wetlands promote deposition of soil eroded in upland areas for example. Recent research has shown that they can effectively treat industrial and municipal wastewater as well.

Wetlands are less well recognized in their many hydrologic roles. They are sometimes the sites of groundwater recharge. Surface runoff collected from surrounding areas flows into wetlands during the rainy season and ponds temporarily until soaking into a local or regional aquifer system. They are important areas in flood control. Wetlands will act as detention areas and slow the flow of runoff. Floodplains are very important in this process; if flood-waters are not allowed to spread into floodplains, laws of hydraulics make it clear that flooding rivers will flow deeper and faster (see Chapter 7).

Today, the legal and regulatory aspects of wetlands are also an important issue. All levels of government—federal, state, county, and municipal—have laws regarding use and development of wetlands. Most of the regulation is in county and municipal zoning and building codes. Federal and state laws primarily provide guidance in objectives and methods. Any person considering development or modification of wetland areas needs to be familiar with often rapidly changing federal and local regulations and must be aware of the effects of their development on wetlands.

8.13.2. Identification of Wetlands

Identifying and delineating wetlands constitute the bulk of the legal and technical work associated with wetlands. Many methods are available to people examining their lands for the presence and extent of wetlands (Lyon, 1993). For small sites, methods include detailed soil analyses from individual pits and ecological evaluation of plant communities. In examining larger areas, aerial photography and remote sensing become valuable tools.

To discuss identification of wetlands it is important to understand their causes. The basic cause is an abundance of water. The abundance may be due to a large amount of water available from surface water, as in the case of a river; from groundwater, as in the case of hillside seeps; or as a combination of the two, a common case including lake shores, rivers, and most swamps. Technically, most wetlands could be said to be areas where the groundwater table intersects the surface, regardless of the direction that the water is flowing. Hillside springs, for example, are purely an expression of the groundwater supplying water to the surface. Many wetlands have been created by irrigation water, which is supplied from the surface, and wetlands are formed in pockets as it flows to a shallow groundwater table.

Most groundwater/surface-water systems in wetlands are more complex. Near lake shores, for example, one cannot separate the effect of the lake and its tributary streams on the groundwater level from the effect of the groundwater level on the lake. The groundwater level and lake level are expressions of the same hydrologic system. When rivers cover their floodplains during the spring, this water soaks into the local shallow aquifer at relatively large distances from the river. This water returns to the stream gradually through the summer to provide water during low flows. Some rivers and streams are surface expressions of regional or local groundwater tables seeping out through the channel banks. Many rivers in the basin and range province of the western United States, however, transport water from the mountains into colluvial basins where water from the rivers soaks into the ground to deep water tables. The potentially complex ways that groundwater and surface water can interact is what makes identification of wetlands so challenging.

Given that wetlands can generally be connected to groundwater, the first step in identifying and classifying a wetland is to look at the groundwater system to which it is connected. Many wetlands, particularly small wetlands, are connected to small perched water tables. This type of condition is common in Alaska and other boreal climates, where permafrost provides a shallow impermeable layer turning any depression in the ground into a place that water will collect. Compacted glacial till underlying many soils once covered by continental glaciation in the northern United States provides a shallow impermeable layer that leads to formation of seasonal wetlands in many small pockets. Larger groundwater systems with surface expression, such as the Everglades in Florida, have proven to be immense challenges to analyze.

Recognizing the role of groundwater makes it easier to identify wetlands. The primary clues are shallow water, hydric soils (soils showing signs of flooding), and hydrophytic vegetation. If there is shallow water at the site, either flowing or standing, it is usually quite clearly a wetland. Without water at the site, clues from the soil and vegetation may be needed. Soil that is waterlogged usually shows a grey or mottled grey and rust coloration. A rust coloration typically comes about when soils are periodically waterlogged and drained. Gravel and rock streambeds have no soils, but are clearly wetlands. Hydrophytic vegetation is vegetation that requires or is tolerant of large amounts of water and is an important clue in identifying wetlands. The list of potential hydrophytic plants in wetlands is far too long for this textbook, and is best learned regionally. One clue, however, is if there is an abrupt change in vegetation in an area that is topographically probable as a wetland. A good example of this is a mountain meadow. Assuming the area is not farmed, a natural reason must exist for the grass to outcompete the trees, and a shallow groundwater table is often the cause.

Identified wetlands are classified by the type of water system to which they are connected. An overview of the classification system used by the U.S. Fish and Wildlife Service (Cowardin et al., 1979) shows the variety of places wetlands can be found. The first level of classification separates wetlands into marine, estuarine, riverine, lacustrine, and palustrine. Marine wetlands are connected with tidal and intertidal environments of high-salt, high-energy coastal systems. Estuarine systems are generally lower energy and lower salinity but can be temporarily higher in salinity from evaporation. They frequently have some degree of freshwater mixing from streams or groundwater. Riverine systems are closely associated with stream beds and banks. They can be intermittent in nature as with the washes of the southwest desert. Lacustrine systems are associated with shores and shallows of large freshwater bodies. Palustrine systems encompass any other non-tidal system and are meant to describe most upland wetland sites. Palustrine systems include the types of systems most often thought of when discussing wetlands, including bogs, fens, marshes, swamps, ponds, and prairie potholes.

The above is a brief introduction to some important aspects of wetlands. Because of the complex scope of wetlands, any person who will be working with the legal aspects of delineation and classification of wetlands is strongly encouraged to seek specialized training.

REFERENCES

Balek, J. 1983. Hydrology and Water Resources in Tropical Regions. (Developments in Water Science, Vol. 18). Elsevier. New York. 271 pp.

Berris, S.N. and R.D. Harr. 1987. Comparative snow accumulation and melt during rainfall in forested and clear-cut plots in the western Cascades of Oregon. Water Resources Research, 23(1):135–142.

Bradley, G.E., Editor. 1984. Land Use and Forest Resources in a Changing Environment. The Urban/Forest Interface. University of Washington Press. Seattle, WA. 222 pp.

Burroughs, E.R., Jr. and B.R. Thomas. 1977. Declining root strength in Douglas-fir after felling as a factor in slope stability. Research Paper INT-190. USDA Forest Service, Intermountain Forest and Range Experiment Station. Ogden, UT. 27 pp.

Cowardin, L., V. Carter, F. Golet, and E. La Roe. 1979. Classification of wetlands and deepwater habitats of the United States. U.S. Fish and Wildlife Service, Report Number FWS/OBS-79/31, Washington, D.C. 103 pp.

DeBano, L.F. 1981. Water repellant soils: a state-of-the-art. USDA Forest Service General Technical Report PSW-46. Berkeley, CA.

Dissmeyer, G.E. and G.R. Foster. 1981. Estimating the over-management factor (C) in the universal soil loss equation for forest conditions. Journal of Soil and Water Conservation, 36(4):235–240.

Dunne, T. and R.D. Black. 1970. Partial area contributions to storm runoff in a small New England watershed. Water Resources Research, 6(5):1296–1311.

Food and Agriculture Organization of the United Nations. 1993. Forest Resources Assessment 1990: Tropical Countries. FAO Forestry Paper 112. Rome, Italy.

Franklin, J.F. and C.T. Dyrness. 1973. The natural vegetation of Oregon and Washington, USDA Forest Service, General Technical Report PNW-8. Portland, OR.

Harr, R.D. 1977. Water flux in soil and subsoil on a steep forested slope. Journal of Hydrology, 33:37–58.

Harr, R.D. 1982. Fog drip in the Bull Run Municipal Watershed, Oregon. Water Resources Bulletin, 18(5):785–789.

Hathaway, G.A., et al. 1956. Snow Hydrology. North Pacific Division Corps of Engineers. U.S. Army. Portland, OR. 437 pp.

Haupt, H.F. and B.L. Jeffers. 1967. A system for automatically recording weight changes in sapling trees. USDA Forest Service, Intermountain Forest and Range Experiment Station. Research Note INT-71. Ogden, UT. 4 pp.

Helvey, J.D. 1971. A summary of rainfall interception by certain conifers of North America. In: Proceedings of the Third International Seminar for Hydrology Professors: Biological Effects in the Hydrological Cycle. E.J. Monke, Editor. pp 103–113. Purdue University, West Lafayette, IN.

Hewlett, J.D. 1982. Principles of Forest Hydrology. University of Georgia Press, Athens, GA. 183 pp.

Hewlett, J.D. and A.R. Hibbert. 1967. Factors affecting the response of small watersheds to precipitation in humid areas, In: Forest Hydrology. Sopper, W.E. and H.W. Lull, Editors. Pergamon Press, Oxford. pp. 275–290.

Johnston, R.S. 1970. Evapotranspiration from bare, herbaceous, and aspen plots: a check on a former study. Water Resources Research, 6(1)324–327.

Kittredge, J. 1948. Forest Influences. McGraw-Hill, New York. 360 pp.

Luce, C.H. and T.W. Cundy. 1994. Parameter identification for a runoff model for forest roads. Water Resources Research, 30(4):1057–1069.

Lull, H.W. and K.G. Reinhart. 1972. Forests and Floods in the Eastern United States. USDA Forest Service Research Paper NE-226. Upper Darby, PA.

Lyon, J.G. 1993. Practical Handbook for Wetland Identification and Delineation. Lewis Publishers, Boca Raton, FL. 157 pp.

Marks, D. and J. Dozier. 1992. Climate and energy exchange at the snow surface in the alpine region of the Sierra Nevada. Water Resources Research, 28(11): 3043–3054.

Sullivan, K., T.e. Lisle, C.A. Dolloff, G.E. Grant, and L.M. Reid. 1987. Stream Channels: The link between forests and fishes. In: Streamside Management: Forestry and Fishery Interactions. Salo, E.O. and T.W. Cundy, Editors. University of Washington Institute of Forest Resources, Seattle, WA. pp 39–97.

Swanston, D.N., G.W. Lienkamper, R.C. Mersereau, and A.B. Levno. 1988. Timber harvest and progressive deformation of slopes in southwestern Oregon. Bulletin of the Association of Engineering Geologists, 25(3):372–381.

Varnes, D.J. 1978. Slope movement types and processes, In: Landslides: Analysis and Control, Special Report 176. Transport. Res. Board, National Academy of Science. Natl. Res. Counc., Washington, D.C. pp 11–33.

Waring, R.H. and W.H. Schlesinger. 1985. Forest Ecosystems, Concepts and Management. Academic Press. Orlando, FL. 340 pp.

Problems

8.1 Your manager asks for an estimate of the annual evapotranspiration for a forest in southern Missouri. You find 12 years of records from a weather station at a stream gage that collects daily maximum and minimum temperatures, wind speed and direction, daily precipitation, and daily flow. Describe the data and equation(s) you would use to estimate annual evapotranspiration.

8.2 The commercial landowner in Example 5.5 expects to have a thick canopy of maples covering 80% of the proposed 500-acre development within 5 years because of his aggressive landscaping using planters. How much will the peak flow from the watershed be reduced for the storm of example 5.5 by the forest canopy? The property currently has a good forest cover; what is the approximate peak flow in its current state? Is the aggressive maple landscaping a reasonable mitigation for the hydrologic impacts of the proposed commercial development?

8.3 A friend suggests that if the forest industry left 300-foot buffers of undisturbed forest along every perennial stream, substantial flood control benefits would result. Do you agree? Explain your position.

8.4 On 2 consecutive days in September 1983, two greatly different melt rates were observed on a glacier. The first day was sunny and warm, and 2 cm of melt were recorded. On the second day, a storm straight off the Pacific Ocean with 60 mile per hour winds blew across the glacier and 15 cm of melt were recorded. Calculate the energy transferred to the glacier's snow on each of these two days. The snow surface of the glacier is at 0°C and has a density of 760 kg/m^3.

8.5 A flume was installed at the bottom of a 10-acre watershed in a forest in central Idaho that has a silt loam soil. The watershed was clearcut and burned and several heavily used skid trails were installed cutting across the watershed. After two summers and one winter of data collection, no flow was collected during the brief but intense (up to 3 in./hr) summer thunderstorms. Flows around 30 gpm were collected during the two week melt season. Given that snowmelt rates seldom exceed 0.5 in./hr, and the information from Table 8.2, explain these observations.

8.6 A forester in Southeast Alaska has been told by a local fisherman that if she were to use partial cutting (removal of 50%) of the forest canopy instead of clearcutting, she would dramatically reduce the hydrologic impacts and erosion from forest harvest. She calls you to ask your opinion. Describe factually the relative impacts of the two types of harvest. (Southeast Alaska is wet all year long with generally low to moderate rainfall intensities. The winter is snowy and cold. The ground is steep and strongly affected by glaciation. Cable yarding is the primary yarding method used.)

8.7   Many states require stream buffers as a part of their forest practice regulations because of their benefits to water quality. Using your knowledge of runoff-producing processes in forests and the variety of erosional processes, describe the relative effectiveness of 10, 50, and 300-foot buffers for 1) a steep forested site in the Colorado Rockies, 2) a gently sloped tree farm in the piedmont soils of South Carolina, and 3) a moderately sloped stand in the basalt-derived soils of coastal southwest Washington's Willapa hills.

CHAPTER 9

Hydrogeology

E. Scott Bair

9.1. INTRODUCTION

The need to understand the fundamental concepts describing groundwater flow is becoming increasingly important as our groundwater resources become more and more threatened by contamination and overuse. This understanding involves an appreciation of several fields of science and engineering. Without it, we are destined to continue to believe that the movement of subsurface waters is secret and occult, as stated in an 1856 ruling by the Ohio Supreme Court, which was not changed until 1984, and to continue to mismanage one of our most valuable, renewable natural resources.

9.2. CHARACTERIZATION OF GROUNDWATER FLOW

Groundwater flow through porous material occurs under the influence of energy, flowing from regions of high energy to regions of low energy. As such, it is similar to the flow of heat and the flow of electricity. The amount of energy a particle of water possesses at any position within a flow system is the sum of its three forms of potential energy—elevation energy, pressure energy, and velocity energy. A particle of water has elevation energy by virtue of its position in the flow system relative to some standard measurement plane, sea level, for example. The pressure energy possessed by the particle of water is analogous to the energy of a compressed spring (Price, 1985). Under most conditions, except in conduit flow in limestone terranes and lava tubes, the energy derived from the movement of flowing groundwater is negligible and can be ignored.

Figure 9.1 shows three wells completed in a permeable layer such as sandstone. At well A, the elevation head (h_e) is less than the pressure head (h_p), whereas the opposite is true at well B. Flow, however, is governed by the total hydraulic head, which is the sum of h_e plus h_p. As a result, the total head (h) at well A is greater than the total head at well B and groundwater flow moves from left to right, up the dip of the sandstone layer. The total head at well C is less than that at wells A or

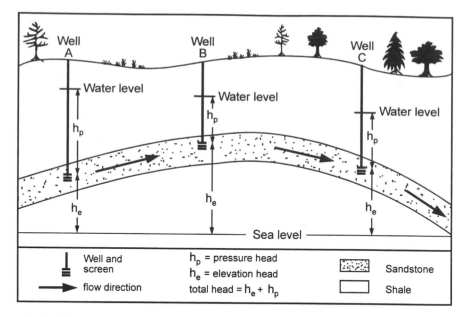

Figure 9.1. Components of total hydraulic head, elevation head (h_e) and pressure head (h_p), controlling flow in a sandstone layer.

B and flow moves down the dip of the sandstone layer between wells B and C. Thus, groundwater can flow laterally, up or down the dip of a layer depending only on the distribution of total hydraulic head.

Measurement of total hydraulic head—or head, as it is commonly known—usually is accomplished using a device that is lowered down a well to measure the depth of water in the well. This value then is subtracted from the elevation of the measuring point. If measurements of head are made in several wells completed in the same permeable layer, the values of head can be contoured to construct a potentiometric surface and used to determine directions of groundwater flow. Figure 9.2 shows a hypothetical potentiometric surface and its cross-sectional presentation based on contouring several measurements of head. The lines of equal head are called equipotential lines or potentiometric contours and represent lines of equal energy. Groundwater flows from areas of high energy (hydraulic head) to areas of low energy (hydraulic head). The path a particle of water makes through the flow field is called a flowline. In most flow systems, flowlines will be perpendicular to equipotential lines so that particles of water flow along the steepest gradient between pairs of equipotential lines, as shown on Figure 9.2.

The volumetric flow of groundwater is dependent on the gradient between equipotential lines, which is known as the hydraulic gradient, the cross-sectional area through which flow occurs, and the permeability of the material. Henry Darcy, an engineer employed by the town of Dijon, France, was the first to recognize and quantify this relation. While trying to filter spring water through layers of sand as a means of purifying the city's water supply, Darcy performed a series of experiments in 1855 and 1856 that led him to observe several relations between the quantity of water flowing through a cylindrical column filled with a particular grade of sand and the amount of energy (head) loss measured between an upper manometer,

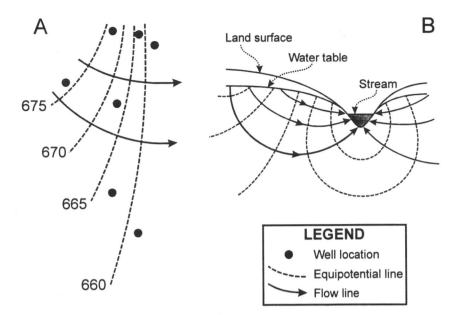

Figure 9.2. Relation of flowlines to equipotential lines in a small drainage basin. (A) View of upland area. (B) Cross-section view of basin. (All units in feet.)

X and a lower manometer, Y that are a distance, l apart (Figure 9.3). In the experiments, Darcy allowed a constant rate of water Q, to flow through the cylinder of known cross-sectional area, A (Figure 9.3).

From his experiments Darcy observed that if the flow rate, Q was doubled, the loss of head (dh = h_1 - h_2) between manometers X and Y also doubled. Thus, Q is directly proportional to head loss, dh and the hydraulic gradient, dh/l. Mathematically this is expressed as:

$$Q \approx \frac{h_1 - h_2}{l} = \frac{dh}{l} \qquad (9.1)$$

Based on this relation, it can be seen that it takes twice as much energy to drive the water through the sand at twice the flow rate. If a different cylinder is used, one in which the cross-sectional area of flow is twice as large as in the first cylinder, for a particular grade of sand, twice as much water will flow through the larger cylinder than through the smaller cylinder. Thus:

$$Q \approx A \qquad (9.2)$$

Combining these results, the following relation is obtained:

$$Q \approx A\frac{dh}{l} \qquad (9.3)$$

Darcy repeated his experiments using several different grades of sand and found that for a given grade of sand:

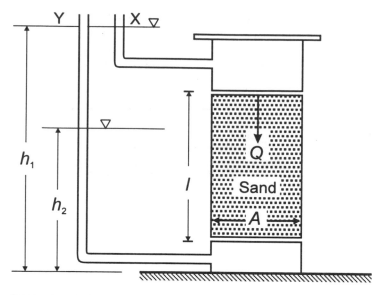

Figure 9.3. Darcy apparatus.

$$Q = (K)\ (A)\ \left(\frac{dh}{l}\right) \tag{9.4}$$

where K is a constant of proportionality for a given grade of sand. The experiments showed that the value of K was larger for a coarse sand than for a fine sand. Darcy deduced that the value of K was related to the ability of the sand to transmit fluid. Thus, K is referred to as hydraulic conductivity and represents the properties of the porous material and the properties of the fluid. (Intrinsic permeability is a term used to describe the ability of a porous material to transmit fluid, but refers solely to the properties of the porous material.) The units of K are length per time and are commonly expressed as feet per day or centimeters per second. Although Darcy did not perform any experiments using fluids other than water, it can be visualized that a more viscous fluid would flow more slowly through the column of sand than a less viscous fluid. The form of Darcy's Law written in Equation 9.4 can be expanded to separate the transmitting properties of the porous material from the properties of the fluid such that:

$$Q = (k)\ \left(\frac{\gamma}{\mu}\right)\ (A)\ \left(\frac{dh}{l}\right) \tag{9.5}$$

where k is the intrinsic permeability of the porous material, γ is the specific weight of the fluid, and μ is the dynamic viscosity of the fluid. Thus, the rate of fluid flow is directly proportional to the specific weight of the fluid and inversely proportional to its viscosity.

Laboratory and field experiments indicate that values of K for natural earth materials range over more than 11 orders of magnitude. Figure 9.4 shows that the most poorly permeable materials, massive clay deposits and unfractured igneous and

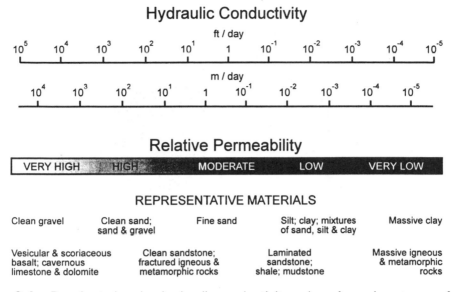

Figure 9.4. Bar chart showing hydraulic-conductivity values for various types of rock and sediment (modified from Bureau of Reclamation, 1977).

metamorphic rocks, have K values close to 10^{-5} ft/day, whereas the most highly permeable materials, clean gravel and cavernous limestone, have K values close to 10^5 ft/day. This figure also indicates that the range of K values for sand and gravel deposits alone extends over five orders of magnitude. This represents a tremendous range of uncertainty and potential error when attempting to compute the volumetric rate of groundwater flow in the absence of site-specific values of K obtained from laboratory or field experiments.

Example 9.1: Darcy's Law is used in various ways to make calculations of groundwater flow rates and travel times. For example, it may be necessary to determine the amount of groundwater flow occurring within a permeable material. Figure 9.5 shows two wells located 800 feet apart along a flowline. The difference in head values between the wells is 2.0 feet. Thus, the hydraulic gradient is 0.0025 ft/ft. The amount of groundwater flowing through a cross-sectional area of thickness, b times width, w is:

$$Q = (K)\ (b)\ (w)\ (\frac{dh}{l})$$

If it is determined from a field test that the hydraulic conductivity of the material is 125 ft/day and it is determined from well logs that the material is 25 feet thick and it is assumed that the cross section is 65 ft wide, then the quantity of groundwater flowing across the cross-sectional area is

$$Q = (125\ \text{ft/day})\ (25\ \text{ft})\ (65\ \text{ft})\ (0.0025\ \text{ft/ft}) = 508\ \text{ft}^3/\text{day}$$

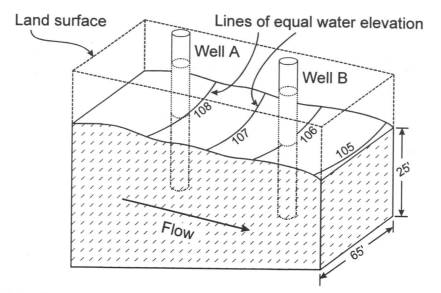

Figure 9.5. Block diagram for calculation of volumetric flow through a unit area perpendicular to the direction of flow. (All units in feet.)

Answer: The quantity of groundwater flowing across the specified cross-sectional area is 508 ft³/day.

Another way to express the ability of a porous material to transmit water is to account for its transmission across the entire saturated thickness of the permeable materials. Transmissivity is expressed in terms of length squared per time and is equal to:

$$T = Kb \tag{9.6}$$

Figure 9.6 shows a bar graph for estimating the capability of porous materials to transmit water to wells based on the measured transmissivity of the materials. This graph is especially useful in estimating whether a particular geologic material is capable of transmitting sufficient water to various types of wells.

Subsurface materials can be classified according to their ability to transmit groundwater. An aquifer is a body of rock, soil, or unconsolidated material that can transmit groundwater in sufficient quantities to supply wells or springs. An unconfined aquifer (Figure 9.7) is one in which infiltrating water moving through the unsaturated zone has pressures less than atmospheric until they reach the fully saturated pores and possess a pressure equal to atmospheric pressure at the water table. The water table is a surface that is conventionally determined by water-level measurements made in wells that penetrate only a short distance into the saturated portion of an unconfined aquifer.

A confined aquifer does not receive direct infiltration of precipitation because it is overlain by a layer of low hydraulic conductivity. As such it does not have a water table and the water is confined under sufficient pressure that water rises above

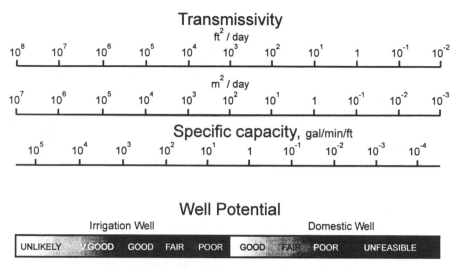

Well Potential

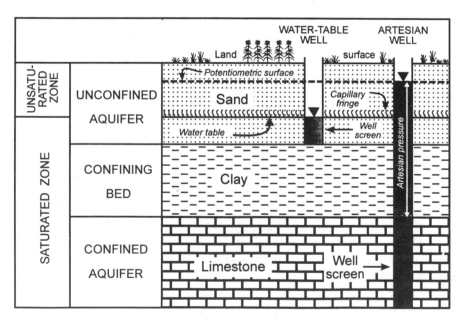

Specific-capacity values based on pumping period of
approximately 8 hours but are otherwise generalized.

Figure 9.6. Bar chart showing range of transmissivity values relative to the potential of various types of wells (modified from Bureau of Reclamation, 1977).

Figure 9.7. Diagram showing hydrogeologic conditions for an unconfined aquifer and a confined aquifer (modified from Heath, 1983).

the base of the overlying confining layer (Figure 9.7). If several wells are constructed in the same confined aquifer and a set of water-level measurements are made at the same time, the resulting contour map represents the potentiometric surface of the aquifer.

Fluctuations in water levels are normal in unconfined and confined aquifers and are related to various natural and anthropogenic causes. Seasonal fluctuations in unconfined aquifers reflect variations in precipitation, evaporation, and transpiration. Figure 9.8 shows a hydrograph from a well completed in unconfined sand and gravel deposits in southwest Ohio. The periods of rising water level indicate that recharge to this aquifer occurs predominantly in late winter and early spring, when precipitation is plentiful and evaporation and transpiration rates are relatively low. Analysis of over 60 annual hydrographs from eight wells in unconfined aquifers in Ohio indicates that the average duration of this recharge period is 23 weeks with a standard deviation of 9 weeks and the average height of water-level rise is 6.8 feet with a standard deviation of 4.3 feet (de Roche, 1993). This represents the period when recharge to the aquifer from precipitation exceeds discharge to springs and streams, which results in a prolonged water-level rise. The period of declining water levels represents the period when more water discharges to springs and streams than recharges the aquifer.

On a hydrograph, a drought would be indicated as a period of falling water levels that corresponds to a period of diminished precipitation. A prolonged drought occurred in Ohio during 1987 and 1988. The decline in water levels due to this drought is seen in Figure 9.8. Over-pumping wells in an unconfined aquifer causing groundwater levels to decline would be indicated by a period of several years in which water levels decline while precipitation remains near normal. In these cases, the volume of groundwater in storage is decreased as the pores in the unconfined aquifer drain. The volume of groundwater that will drain under the influence of gravity relative to the total volume of aquifer is called the specific yield (Sy). It is a measure of the storage characteristics of an unconfined aquifer.

Values of Sy commonly are expressed as a percentage. Sy values are less than the porosity of the aquifer materials and range from less than 1% to as much as 45%. The porosity (n), also commonly expressed as a percentage, of a soil or rock is defined as the volume of void space relative to the total volume. The difference between n and Sy is the volume of groundwater that does not drain under the influence of gravity, which is called the specific retention (Sr). Thus:

$$n = Sy + Sr \qquad (9.7)$$

Grain-size distribution and grain shape are the primary factors controlling the distribution of pore sizes and the rate of gravity drainage. Figure 9.9 shows the range of Sy, Sr, and n values measured from alluvial deposits in California.

Seasonal water-level fluctuations in confined aquifers also reflect changes in the amount of water in storage but not by the same mechanism as for unconfined aquifers. As potentiometric levels in a confined aquifer change, the pores remain saturated. The change in potentiometric level represents a change in the amount of pressure head in the aquifer. The amount of water-level change is controlled by the coefficient of storage (S), which is defined as:

$$S = \rho gb \, (\alpha + n\beta) \qquad (9.8)$$

where ρ is the density of the fluid, g is the gravitational constant, b is the saturated thickness of the aquifer, α is the compressibility of the aquifer skeleton, n is the porosity, and β is the compressibility of the fluid. Values of S are small, on the

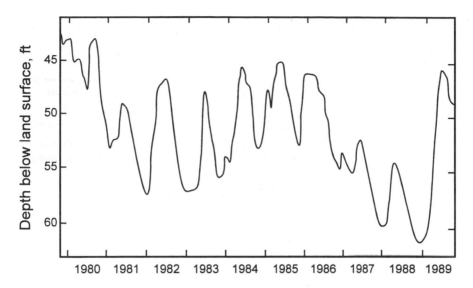

Figure 9.8. Hydrograph of a well completed in an unconfined aquifer in southwest Ohio showing seasonal and annual water-level fluctuations.

order of 0.0001 to 0.000001. Thus, it may require 1000 times the volume of soil/rock to store the same amount of groundwater in a confined aquifer as in an unconfined aquifer.

Short-term water-level fluctuations in unconfined and confined aquifers can be caused by the withdrawal of water from wells. Figure 9.10 shows the hydrographs from two wells completed in sand and gravel deposits at a municipal wellfield. The regularly spaced, small amplitude water-level rises are related to the decrease in pumping rates on weekends and are superposed on the cycle of seasonal water-level fluctuations.

Differences in the rate of infiltration of water through the unsaturated zone can cause differences in the rate at which recharge from precipitation reaches the water table, creating nonuniform rates of water-level rise. This results in changes in hydraulic gradient that are manifest as changes in the rate and direction of groundwater flow. The rate of groundwater flow is computed using Darcy's Law and accounts for the fact that flow occurs only through the pores. Thus, the average linear velocity (v) of groundwater flow is:

$$v = \frac{K\frac{dh}{l}}{n_e} \tag{9.9}$$

where n_e is the effective porosity, expressed as a decimal, and is defined as the volume of interconnected pore space relative to the total volume.

Example 9.2: Assume that the porosity of the aquifer shown on Figure 9.5 is 0.25. The rate of average linear velocity between wells A and B would be:

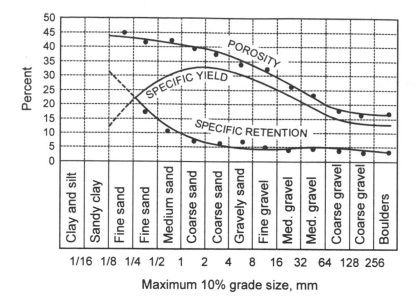

Figure 9.9.   Relation of specific yield and specific retention to total porosity for various sediment sizes (modified from Johnson, 1967).

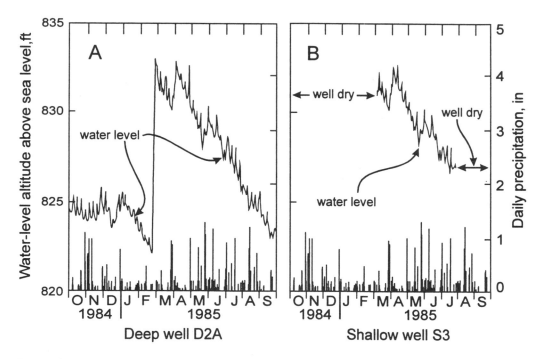

Figure 9.10.   Precipitation records and seasonal water-level fluctuations measured in a deeper well (A) and a shallower well (B) in a sand and gravel aquifer showing weekly fluctuations due to pumping stress (modified from Breen, 1988).

$$v = \frac{(125 \text{ ft/day})(0.0025 \text{ ft/ft})}{0.25} = 1.25 \text{ ft/day}$$

Answer: The average linear velocity would be 1.25 ft/day.

Figure 9.11 is a polar plot showing changes in the hydraulic gradient during a period of several months at a site in southern Ohio, based on water-level measurements made in three sets of three wells completed in an unconfined aquifer. Differences in the rate of infiltration cause the direction of the hydraulic gradient to change by 40 degrees and the magnitude of the hydraulic gradient and, therefore, the velocity of groundwater flow to change by a factor of 3. Natural, temporal changes in the rate and direction of groundwater flow in surficial aquifers are important in determining the direction of movement and the areal extent of anthropogenic contaminants.

If large changes in water levels or hydraulic gradients occur over long periods of time, the flow system is said to be transient. If only small changes occur, the flow system is said to be steady state because there is no net change in the amount of water in storage and, therefore, no changes in flow directions or flow rates.

9.3. LOCAL AND REGIONAL GROUNDWATER FLOW PATTERNS AND STREAM INTERACTION

Groundwater flow patterns are controlled by several factors including the elevation and location of recharge and discharge areas, the heterogeneity of the geologic materials, the thickness of materials, and the configuration of the water table. One of the simplest types of flow systems is shown on Figure 9.12 where recharge and discharge areas are adjacent. In Figure 9.12, both streambeds have the same elevation, the geologic materials are homogeneous, and groundwater divides correspond to surface-water divides. In this flow system, infiltration from precipitation enters the groundwater flow system in topographically high regions and discharges in topographically low regions to streams and rivers. During periods of drought, the water table will decline, particularly beneath the upland areas, in response to the lack of precipitation, but groundwater will continue to discharge to the streams until the water table drops below the bottom of the streambed. Thus, at low surface-water flow conditions the water in the streams is sustained by groundwater discharge.

This type of interaction between the groundwater flow system and the surface-water flow system is consistent with the flow patterns shown in Figure 9.12 along the potentiometric profile marked A-A'. Along potentiometric profile B-B' a different type of interaction occurs between the stream and the aquifer. In this case, the water in the stream has a greater level than the water in the aquifer and water in the stream recharges the groundwater flow system as the stream loses water along its course. Losing streams are common in the western part of the United States where streams begin in mountainous areas and flow across alluvial fans formed at the base of the mountains before flowing onto the relatively flat basins. Losing streams also can be found in similar but smaller settings in the glaciated parts of the Midwest and New England.

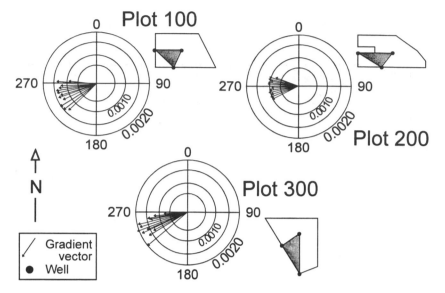

Figure 9.11.   Vectors of hydraulic gradient at three 10-hectare farm plots showing seasonal changes in direction, magnitude, and velocity of groundwater flow (Finton, 1994).

This type of interaction can be created artificially by pumping wells that are located sufficiently close to a stream to reverse the direction of local groundwater flow and induce water to flow out of the stream and toward the well (Figure 9.13). Many municipal, agricultural, and industrial water supplies rely on this type of interaction. The amount of induced infiltration from the stream can be approximated using analytical solutions (Jenkins, 1970) or measured by gaging streamflow upgradient and downgradient of the wells.

One of the most highly publicized cases involving induced infiltration of streamflow as a possible cause of contamination of water-supply wells occurred in Woburn, Massachusetts, in the 1980s. In 1982, a law suit was filed on behalf of eight families who alleged that industrial contamination of two water supply wells completed in an unconfined aquifer, which were located adjacent to the Aberjona River, led to the death of six children (GeoTrans, 1987). In the discovery phase of the law suit a test was performed to reproduce the pumping stress created by the two wells during their operation between 1964 and 1979. During a one-month period, the two wells were pumped at their normal discharge rates while routine measurements of groundwater levels and stream discharge rates were made. The water-level measurements indicated that the groundwater system became steady-state after approximately one month of pumping the wells at a combined rate of 1100 gallons per minute (gpm). Figure 9.14 is a water-level map of the unconfined aquifer at the end of the one month pumping stress when the system is steady-state. It shows that flow converges on the two wells from all sides of the valley and, therefore, the source of contamination could be in almost any direction.

The discharge of the Aberjona River was measured upstream and downstream of the wells. The difference between the two discharge rates indicates whether the stream is gaining water from groundwater discharge or losing water as induced infiltration from pumping the two wells. Figure 9.15 shows that prior to the start of

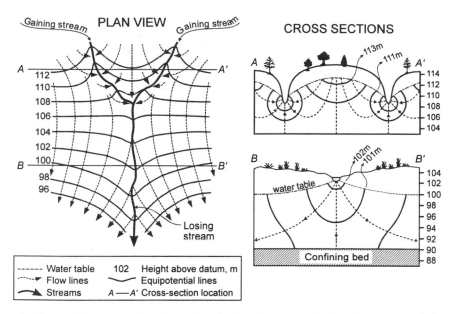

Figure 9.12.   Diagrams showing the relation between the configuration of the water table at a gaining and a losing stream (from Heath, 1983).

the pumping test the Aberjona River gained approximately 760 gpm between the upstream gaging station and the downstream gaging station, indicating groundwater discharge to the stream. As the pumping test proceeded, less and less flow was recorded at the downstream gaging station. When steady-state conditions were apparent, over 550 gpm less water was measured at the downgradient gaging station than at the upstream gaging station. This strongly suggests that about half the water pumped by the two municipal wells originated either as surface water in the Aberjona River or as intercepted groundwater flow that would have discharged to the Aberjona River. Thus, the quality of water in the Aberjona River as well as the quality of the water in the unconfined aquifer needed to be considered as possible sources of the groundwater contamination measured in the municipal wells.

In a simple, homogeneous aquifer, more complex regional flow patterns can develop, depending on the slope and shape of the water-table surface and the thickness of the saturated material. In the early 1960's, J.A. Toth performed a series of mathematical experiments using a solution to a steady-state flow equation that demonstrated these controls on groundwater flow patterns. Contrast the flow patterns in Figures 9.16a and 9.16b. The gently undulating configuration of the water table is the same in both figures. Figure 9.16b shows the flow patterns that develop in a deep basin, whereas Figure 9.16a shows the flow patterns that develop in a shallow basin.

In the deep basin local, intermediate, and regional flow systems develop, whereas in the shallow basin only local flow systems develop. In a local flow system recharge along a groundwater mound (surface-water divide) flows to adjacent discharge areas (streams). In intermediate and regional flow systems, recharge occurs only in selected areas, flow bypasses one or more adjacent discharge areas, and flow occurs to greater depths in the basin.

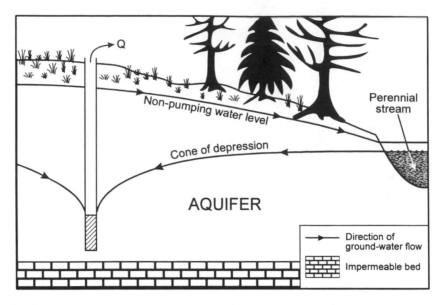

Figure 9.13.   Diagram showing the relation between induced infiltration from a stream and the cone of depression developed by a pumping well (modified from Peters, 1987).

Figures 9.17a and 9.17b contrast the flow patterns that develop in basins of the same thickness but differing water-table configurations. Figure 9.17a shows that both local and regional flow systems develop in basins with gentle topographic relief and gentle undulations to the water table, whereas Figure 9.17b shows that only local flow systems develop in basins with greater topographic relief and more pronounced undulations in the water table.

These concepts can be important in the consideration of alternative sites for the disposal of hazardous wastes and nuclear wastes. For example, many states allow the injection of hazardous liquid wastes into deep saline aquifers. From a purely hydrogeologic viewpoint of protecting shallow freshwater supplies, it would be prudent to inject these wastes into the downflow limb of a deep flow system, as is found in Figures 9.16b and 9.17a where the path of groundwater flow is deeper, longer, and slower, allowing more time for dilution and chemical and biological transformation of the wastes. The difficulty in applying these concepts lies in the assumption that the basin is homogeneous. In reality, most basins are geologically complex and have complex distributions of hydraulic conductivity that also affect groundwater flow patterns, as shown on Figures 9.18a and 9.18b (Freeze and Witherspoon, 1967). The task of characterizing flow patterns in deep basins is made more difficult because of the lack of wells. In most cases the geologic and hydraulic data from oil/gas exploration and production wells must be used along with the limited number of wells that can be drilled and tested as part of a characterization study.

9.4. FLOW TO WELLS

The response of a groundwater flow system to the stress created by pumping a well can be conceptualized using Darcy's Law. Figure 9.19 shows a cross-section

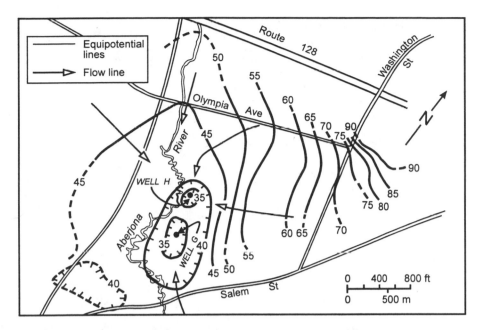

Figure 9.14. Measured steady-state potentiometric surface (January 1986) while pumping municipal wells G and H at a combined rate of 1100 gallons per minute (modified from GeoTrans, 1987).

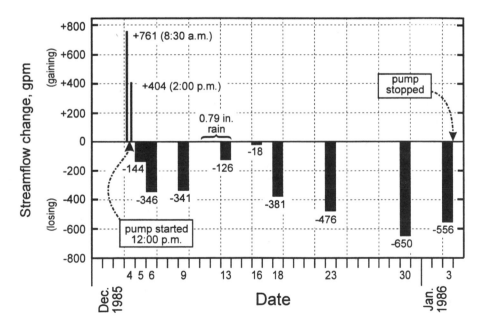

Figure 9.15. Bar graph showing streamflow depletion due to pumping wells G and H (modified from GeoTrans, 1987).

through a confined aquifer in which a pumping well is screened throughout the entire saturated thickness of the aquifer. Two concentric cylinders are drawn about the

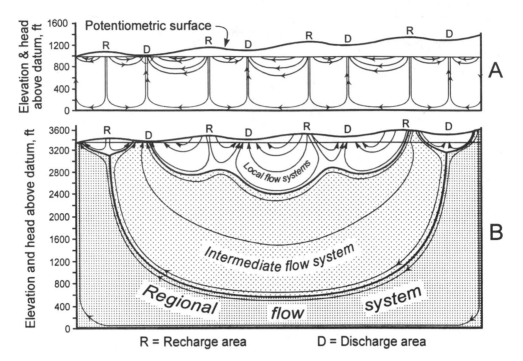

Figure 9.16.   The influence of basin thickness on the development of local flow patterns (A) versus the development of local, intermediate, and regional flow patterns (B) (modified from Toth, 1962).

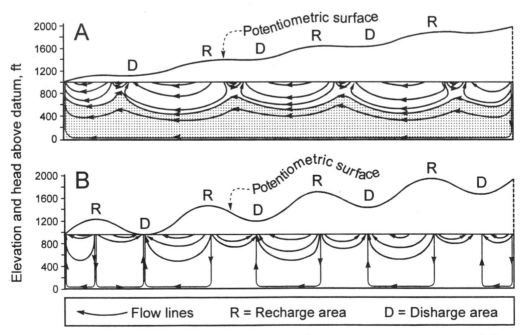

Figure 9.17.   The influence of water-table gradients on the development of flow patterns in two basins of the same thickness (modified from Toth, 1962).

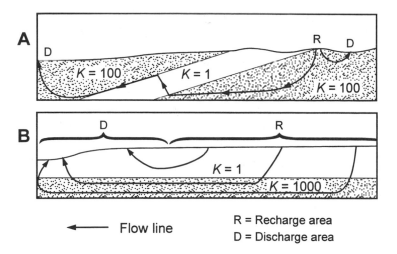

Figure 9.18.   The influence of geological complexities on flow patterns (modified from Freeze and Witherspoon, 1967).

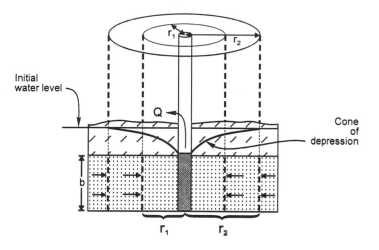

Figure 9.19.   Schematic diagram showing horizontal radial flow to a pumping well completed in a confined aquifer.

well, the larger of radius r_2 and the smaller of radius r_1. Assuming the response of the flow system to the pumping stress is steady-state and that the well discharges at a constant rate Q, then the quantity of water passing across the circumferential area represented by the larger cylinder must be equal to the quantity of water passing across the circumferential area represented by the smaller cylinder which must be equal to the pumping rate of the well. This mass balance can be expressed by Darcy's Law such that:

$$Q = (K) \ (2\pi r_1 b) \ (\frac{dh_1}{dr_1}) = (K) \ (2\pi r_2 b)(\frac{dh_2}{dr_2}) \qquad (9.10)$$

where dh_1/dr_1 and dh_2/dr_2 are the hydraulic gradients at r_1 and r_2, and $2\pi r_1 b$ and $2\pi r_2 b$ are the circumferential areas at r_1 and r_2. Cancelling terms and recognizing that if $r_1 < r_2$, then $dh_1/dr_1 > dh_2/dr_2$. Thus, as the distance to the pumping well becomes smaller and smaller, the hydraulic gradient becomes steeper and steeper, forming a cone of depression about the well, as shown on Figure 9.19.

Under transient conditions while the cone is still expanding under the pumping stress, the shape of the cone of depression is a function of the pumping rate, the hydraulic conductivity, and the coefficient of storage. In 1935, C.V. Theis, a geologist working in eastern New Mexico for the United States Geological Survey, published a paper presenting an equation describing transient changes in the shape (areal extent and depth) of the cone of depression based on certain simplifying assumptions regarding the characteristics of the pumped aquifer and the nature of the well and pumping stress. Theis' work relied on the analogy of groundwater flow in a porous medium to heat flow in a conductive solid. Thus, hydraulic head is analogous to temperature, hydraulic gradient to thermal gradient, hydraulic conductivity to thermal conductivity, and specific storage ($S_s = S/b$) to specific heat. The Theis equation assumes that the aquifer is isotropic, homogeneous, infinite, flat-lying, overlaid and underlaid by impermeable layers that provide an insignificant amount of water to the pumped aquifer, and water is removed instantaneously from storage. The well is assumed to fully penetrate the aquifer, to discharge at a constant rate, and to have a negligible amount of water stored in the borehole. Under these conditions the Theis equation can be used to compute the drawdown at any radial distance from the pumped well at any time, such that

$$s = \frac{Q}{4\pi T} W(u) \qquad (9.11)$$

and

$$u = \frac{r^2 S}{4Tt} \qquad (9.12)$$

where s is drawdown, Q is the pumping rate, T is transmissivity, t is time, S is the coefficient of storage, r is the radial distance to a point of interest, and W(u) is the well function of u. Any set of consistent units can be used in these equations. The value of the well function can be computed using a series expansion such that:

$$W(u) = [-0.5772 - \ln u + u - \frac{u^2}{2 \cdot 2!} + \frac{u^3}{3 \cdot 3!} - \frac{u^4}{4 \cdot 4!} + \cdots] \qquad (9.13)$$

Values of W(u) for various values of u are listed in Table 9.1 or can be computed using a spreadsheet program and expanding the well function to 9–10 terms.

Table 9.1.   Well function W(u).

u	W(u)	u	W(u)	u	W(u)	u	W(u)
1×10^{-10}	22.45	7×10^{-8}	15.90	4×10^{-5}	9.55	1×10^{-2}	4.04
2	21.76	8	15.76	5	9.33	2	3.35
3	21.35	9	15.65	6	9.14	3	2.96
4	21.06	1×10^{-7}	15.54	7	8.99	4	2.68
5	20.84	2	14.85	8	8.86	5	2.47
6	20.66	3	14.44	9	8.74	6	2.30
7	20.50	4	14.15	1×10^{-4}	8.63	7	2.15
8	20.37	5	13.93	2	7.94	8	2.03
9	20.25	6	13.75	3	7.53	9	1.92
1×10^{-9}	20.15	7	13.60	4	7.25	1×10^{-1}	1.823
2	19.45	8	13.46	5	7.02	2	1.223
3	19.05	9	13.34	6	6.84	3	0.906
4	18.76	1×10^{-6}	13.24	7	6.69	4	0.702
5	18.54	2	12.55	8	6.55	5	0.560
6	18.35	3	12.14	9	6.44	6	0.454
7	18.20	4	11.85	1×10^{-3}	6.33	7	0.374
8	18.07	5	11.63	2	5.64	8	0.311
9	17.95	6	11.45	3	5.23	9	0.260
1×10^{-8}	17.84	7	11.29	4	4.95	1×10^{0}	0.219
2	17.15	8	11.16	5	4.73	2	0.049
3	16.74	9	11.04	6	4.54	3	0.013
4	16.46	1×10^{-5}	10.94	7	4.39	4	0.004
5	16.23	2	10.24	8	4.26	5	0.001
6	16.05	3	9.84	9	4.14		

Example 9.3: The Theis equation can be used to answer many questions that arise concerning the relation of the cone of depression to pumping stress. For example, a power plant in Louisiana relies on two wells to supply water for fire protection. The wells are approximately 1800 feet deep and together, are capable of producing a sustained yield of 900 gpm ($\approx 173,000$ ft^3/day). As part of the licensing procedure for the plant, it is necessary to compute the effect of these wells on the nearest well completed in the same aquifer. A well survey was performed and it was determined that the nearest well is 7500 feet away. To determine the effect of pumping the fire protection wells on the potentiometric level at the nearest well also completed in the aquifer, it was necessary to perform a controlled field experiment called an aquifer test, to determine the transmissivity (T) and the coefficient of storage (S) of the pumped aquifer. The results of the aquifer test indicate that T equals 4680 ft^2/day and S equals 0.0007. For the purposes of making an overly conservative estimate of the effect of the pumping stress at the "nearest well," it is assumed that the fire protection wells discharge at 900 gpm for the entire 40-year design life of the power plant.

The first step in this computation is to determine u. Thus:

$$u = \frac{r^2 S}{4Tt} = \frac{(7500 \text{ ft})^2 (0.0007)}{(4)(4680 \text{ ft}^2/\text{day})(14,600 \text{ days})} = 0.000144$$

The value of the well function W(u) for this value of u can be interpolated from Table 9.1 as approximately 8.32. With this W(u) value, the drawdown at a distance of 7500 feet at the "nearest well" due to pumping the fire protection wells at a combined rate of 900 gpm for a 40-year period is computed as:

$$s = \frac{Q}{4\pi T} W(u) = \frac{(173,000 \text{ ft}^3/\text{day})(8.32)}{(4)(3.14)(4680 \text{ ft}^2/\text{day})} = 24.4 \text{ ft}$$

As the potentiometric level in the aquifer rises to within several hundred feet of the land surface and because the pump in the "nearest well" is located several hundred feet below the potentiometric level, the overly conservative amount of drawdown computed for the fire protection wells will not affect the operation of the "nearest well."

This is an example of a time-drawdown calculation made at a specific distance. Additional calculations can be made at different times to show the rate of spreading of the cone of depression. Alternatively, distance-drawdown calculations can be made at different locations at a specific time to show the shape of the cone of depression.

Answer: The drawdown due to pumping the fire protection wells will be 24.4 ft at the nearest wells.

In many problems it is necessary to account for the drawdowns produced by several wells. This is a common situation in the design of municipal wellfields, irrigation wells, contaminant pump-and-treat systems, and construction of dewatering/depressurizing systems for excavations. Because drawdowns are additive, the mathematical principle of superposition can be used to address these more complex problems. Thus, the composite drawdown from any number of wells at a particular location at a specific time is equal to the sum of the drawdowns from each individual well. Mathematically, this can be expressed simply as:

$$s_{total} = s_1 + s_2 + s_3 + \ldots. \tag{9.14}$$

where the subscripts refer to the component of the total drawdown due to an individual well. If the Theis assumptions are met, the following equation would apply to the calculation of the composite drawdown at a location between two pumping wells discharging at different rates:

$$s = \frac{Q_1}{4\pi T} W(u_1) + \frac{Q_2}{4\pi T} W(u_2) \qquad (9.15)$$

where

$$u_1 = \frac{r_1^2 S}{4Tt} \quad \text{and} \quad u_2 = \frac{r_2^2 S}{4Tt}$$

The assumptions required of the Theis equation are not met in all situations. In many geologic settings, aquifers are not infinite but are bounded on one or more sides by streams, which can provide an additional source of water in unconfined aquifers, or by bedrock valley walls or faults, which can preclude the flow of water. In an unconfined aquifer, the Theis equation will not make accurate predictions of the amount of drawdown near wells because the assumption of constant aquifer thickness is violated by the formation of the cone of depression in the water table around the well. (This is not a problem in confined aquifers because the cone of depression is formed in the potentiometric surface above the base of the overlying confining layer.) In the case of an unconfined aquifer, solution of the following equation using the quadratic formula enables drawdowns computed with the Theis equation to be corrected for the decrease in saturated thickness near the well and applied to an unconfined aquifer:

$$s^2 - 2bs = 2bs' \qquad (9.16)$$

where s is uncorrected drawdown computed using the Theis equation for a confined aquifer, b is the initial saturated thickness of the aquifer prior to pumping, and s' is the corrected drawdown in an equivalent unconfined aquifer. This correction factor is important near pumping wells but becomes increasingly less important with distance from the well. When the correction factor is less than 0.01 ft, it is below the precision of most methods of measuring hydraulic head and, therefore, can be ignored.

9.5. CAPTURE ZONES OF WELLS

Although wells most commonly are used to provide water for domestic, municipal, and industrial purposes, wells also are commonly used to extract contaminated groundwater from shallow aquifers. Once the contaminated groundwater is extracted, it can be treated in a variety of ways to remove the contaminants. This technology is called pump-and-treat. The duration that a pump-and-treat system needs to operate depends on the amount of contamination, the character of the groundwater flow system, the physical and chemical properties of the contaminants, the aquifer materials, and the desired level of decontamination. To design pump-and-treat systems in an efficient manner it is necessary to know the areal extent of the contamination so that the capture zones of the wells used to extract the contaminated groundwater do not "undershoot" the area of contamination and allow contamination to escape or "overshoot" the area of contamination and extract clean water that does not need to be treated.

The steady-state capture zone of a well completed in an isotropic, homogeneous aquifer with a uniform pre-pumping hydraulic gradient is shown in Figure 9.20. The capture zone extends upgradient to a regional groundwater divide. It extends downgradient only as far as a local groundwater divide created by the cone of depression of the well. In a Cartesian coordinate system, the bounding flowline within which all groundwater flows to the well can be estimated with the following equation:

$$x = \frac{-y}{\tan\left[\frac{2\pi Kbiy}{Q}\right]} \tag{9.17}$$

where x and y are coordinate values defined in Figure 9.20, Q is the pumping rate of the well, K is the hydraulic conductivity of the aquifer, b is the saturated thickness of the aquifer, i is the regional (pre-pumping) hydraulic gradient, and tan [ ] is in radians.

The distance to the downgradient groundwater divide created by the well is given by:

$$x_0 = \frac{-Q}{2\pi Kbi} \tag{9.18}$$

The maximum half-width of the capture zone as x approaches infinity is given by:

$$y_{max} = \pm\frac{Q}{2Kbi} \tag{9.19}$$

From these equations it can be seen that the distance to the downgradient groundwater divide is proportional to the pumping rate but inversely proportional to the hydraulic gradient and K. Thus, the greater the pumping rate or the flatter the regional hydraulic gradient, the greater the distance to the downgradient groundwater divide. These equations also show that the width of the capture zone is proportional to the pumping rate but inversely proportional to the hydraulic gradient and K. Thus, the greater the pumping rate or the smaller the regional hydraulic gradient, the wider the capture zone.

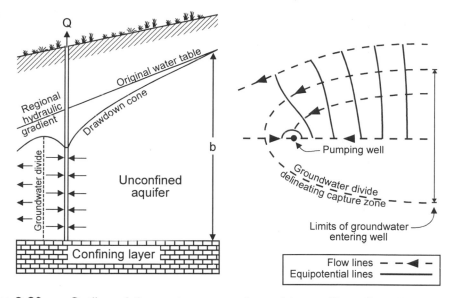

Figure 9.20.   Outline of the capture zone of a well in a uniform flow field.

Example 9.4: The following is an example of the calculation used to compute the x and y coordinates of bounding flowlines delineating the capture zone of an interceptor well constructed at a gasoline station to recover hydrocarbons that have leaked from an underground tank. Based on design considerations it is assumed the well pumps at a rate of 200,000 ft³/day. The geologic logs from a series of soil borings at the site indicate the aquifer is 30-ft thick. The hydraulic conductivity of the aquifer is estimated to be 1000 ft/day. Water levels measured in two wells completed in soil borings 700 ft apart indicate the (background) hydraulic head upgradient of the plume is 331 ft, whereas the hydraulic head downgradient of the plume is 327 ft. Substituting these values into Equation 9.17 produces

$$x = \frac{-y}{\tan\left[(2)(3.1415)(1000 \text{ ft/day})(30 \text{ ft})\dfrac{(331 \text{ ft})-(327 \text{ ft})}{(700 \text{ ft})}y\ 200,000 \text{ ft}^3/\text{day}\right]}$$

Substituting the values of y listed below into this equation, yields the values of x also listed below. When plotted on arithmetic paper (Figure 9.21) the x and ±y coordinates outline the bounding flowlines separating groundwater flow within the steady-state capture zone of the well from groundwater flow in the regional flow regime.

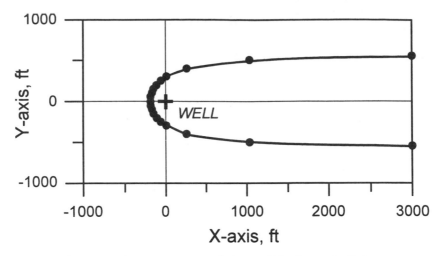

Figure 9.21.   Computed capture zone of the well in Example 9.4.

y	tan(argument)	x
± 1	0.005385	-185.7
± 50	0.269143	-181.3
± 100	0.538286	-167.5
± 150	0.807814	-143.4
± 200	1.077086	-107.6
± 250	1.346357	-57.07
± 300	1.615629	13.46
± 400	2.154171	264.0
± 500	2.692714	1038
± 550	2.961986	3029
± 583	3.139705	308825

The distance to the downgradient stagnation point is computed using Equation 9.18 and is:

$$x_o = \frac{-200,000 \text{ ft}^3/\text{day}}{(2)(3.415)(1000 \text{ ft/day})(30 \text{ ft})(0.0057)} = -186 \text{ ft}$$

and the *total* width of the capture at the upgradient divide is:

$$y_{max-total} = (2) \frac{200,000 \text{ ft}^3/\text{day}}{(2)(100 \text{ ft/day})(30 \text{ ft})(0.0057)} = 116 \text{ ft}$$

Answer: The capture zone is illustrated in Figure 9.21.

Many municipalities and private purveyors of water are being asked by state government agencies to delineate the capture zones of their wells in an attempt to encourage reasonable land-use practices within the capture zone as a means of pre-serving the quality of our groundwater supplies.

REFERENCES

Breen, K.J. 1988. Geochemistry of the stratified-drift aquifer in Killbuck Creek Valley west of Wooster, Ohio, in Regional Aquifer Systems of the United States—Northeast Glacial Aquifers, A.D. Randall and A.I. Johnson, Eds.: American Water Resources Association Monograph Series No. 11, p. 105–131.

Bureau of Reclamation. 1977. Ground Water Manual: U.S. Department of the Interior, 480 p.

de Roche, J.Z. 1993. An examination of the characteristics of seasonal recharge in Ohio based on well hydrographs: unpublished Senior Thesis, Department of Geological Sciences, The Ohio State University, 37 p.

Finton, C.D. 1994. Simulation of advective flow and attenuation of two agricultural chemicals in an alluvial-valley aquifer, Piketon, Ohio: unpublished M.S. Thesis, Department of Geological Sciences, The Ohio State University, 318 p.

Freeze, R.A., and P.A. Witherspoon. 1967. Theoretical analysis of regional ground-water flow: 1. effect of water-table configuration and subsurface permeability varia-tion: Water Resources Research, 3(2):623–634.

GeoTrans Newsletter. 1987. Woburn Toxic Trial: GeoTrans, Inc. Herndon, VA, June, p. 1–3.

Heath, R.C. 1983. Basic ground-water hydrology: U.S. Geological Survey Water-Sup-ply Paper 2220, 84 p.

Jenkins, C.T. 1970. Computation of rate and volume of streamflow depletion: U.S. Geological Survey Techniques of Water Resources Investigations, Book 4, Chapter D1, 17 p.

Johnson, A.I. 1967. Specific yield-compilation of specific yields for various materials: U.S. Geological Survey Water-Supply Paper 1662-D, 74 p.

Peters, J.G. 1987. Description and comparison of selected models for hydrologic analysis of ground-water flow, St. Joseph River Basin, Indiana: U.S. Geological Survey Water-Resources Investigations Report 86-4199, 125 p.

Price, M. 1985. Introducing Groundwater: George Allen and Unwin, Boston, Massa-chusetts, 195 p.

Toth, J.A. 1962. A theory of ground-water motion in small drainage basins in central Alberta, Canada. Journal of Geophysical Research, 67, no. 11, 4375–4387.

Problems

These problems are taken from Peters (1987) and relate to the diagram below, which shows a hydrogeologic cross section of an area in northwest Indiana. Site-specific field data indicate that the transmissivity of the confined aquifer is approximately 13,000 ft^2/day and the coefficient of storage is 0.000043. Well I6-1 is an irrigation well that is used periodically during the growing season.

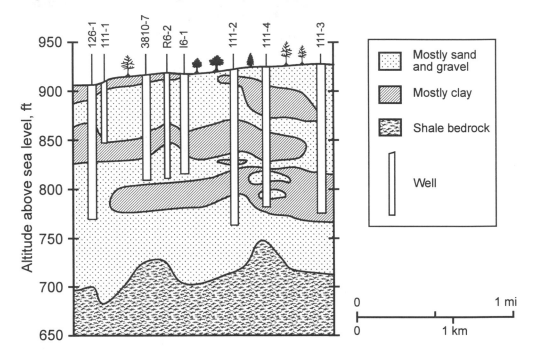

9.1   What would be the expected drawdown 1,000 ft away from well I6-1 if the well is pumped continuously at a rate of 110,000 ft^3/day for 30 days?

9.2   At what rate could well I6-1 be pumped for 30 days so that the resulting cone of depression would not exceed 5.0 ft of drawdown at a distance 1 mile from it?

9.3   How much time would be required for well I6-1 pumping continuously at 110,000 ft^3/day to cause 1.0 ft of drawdown 1 mile away?

9.4   After 1 day of pumping at 110,000 ft^3/day, at what distance from well I6-1 would drawdown be 1.0 ft?

9.5   Well 3810-7S is located 1,800 ft away from well I6-1. What would be the drawdown at a point halfway between the two wells after 30 days of continuous pumping assuming well 3810-7S pumps at 70,000 ft^3/day and well I6-1 pumps at 110,000 ft^3/day?

CHAPTER 10

Water Quality

Terry J. Logan

10.1. INTRODUCTION

An important component of any modern text on hydrology is water quality. We are not only concerned with problems of water supply, but on the quality of that water, which will determine the extent to which our water must be treated before use, or before discharge into the environment. Whereas, in the early parts of the 20th century in North America we were concerned with allocation of water resources among states and nations, today water quality issues are predominant. From the 1970s to the present, the U.S. and Canada entered into a number of agreements to reduce the discharge of contaminants to the Great Lakes, a shared water resource, and for much of the last decade, the U.S. has been constructing one of the world's largest water treatment plants to reduce pollutants in the Colorado River before it enters Mexico. Likewise, European nations in the Rhine River watershed are cooperating to improve the water quality of this important resource which serves both potable and industrial water needs. Humans have always recognized that water carries dissolved and suspended constituents, that it is not pure water. Early humans probably took the cue of wild animals in avoiding contaminated waters and learned to identify some dissolved constituents by taste. The early Egyptians certainly recognized the benefits of suspended matter from Nile flooding, and in a remarkable paper, published by the Royal Society of England in 1699, John Woodward found that spearmint grown in Thames River water grew more than when grown in rain water. Woodward correctly identified the source of the increased growth as impurities in the river water.

In this chapter we will consider water quality from the standpoint of both natural and anthropogenic (human caused) sources, and we will distinguish water quality issues associated with surface and ground water.

10.2. DEFINITION OF TERMS

Any discussion of water quality will involve terms such as *pollutant, contaminant, dissolved, sediment, sediment-bound, particulate, point source,* and *non-point source.* These are defined below:

311

Pollutant—a substance whose presence has the potential to cause harm through environmental degradation, or to human and animal health.

Contaminant—a substance whose presence is not normally expected. In the case of water, any substance other than the chemical compound H_2O might be considered a contaminant. A more reasonable definition would exclude the presence of constituents in equilibrium with the natural system; e.g., dissolved O_2 and CO_2 gases in equilibrium with the atmosphere, dissolved bicarbonate (HCO_3^-) in equilibrium with carbonate rocks.

Dissolved—in chemistry this normally refers to ions that have dissociated and are coordinated with water molecules. It can also include ion pairs like $CdCl^+$, organic matter as humic and fulvic acids with molecular weights in the hundreds and thousands, metal-organic matter complexes, water-soluble pesticides like atrazine, and colloidal matter that includes organic matter, clays, and polymeric iron (Fe) and aluminum (Al) oxides. Dissolved matter is *operationally defined* by the American Public Health Association as that material that passes through a 0.45 μm pore diameter filter. This size pore was originally developed for trapping bacterial cells in water and wastewater treatment testing, but has been adopted for water quality testing in general. Considerable colloidal matter can pass through a 0.45 μm diameter pore, particularly if the colloids are dispersed, and it is, therefore, important to recognize that the *reactivity* of a dissolved substance may vary considerably depending on its chemical form. For example, an important water pollutant is the nutrient element phosphorus (P) which in the dissolved state can occur as the ion $H_2PO_4^-$, as detergent-based polyphosphate, as mono- and di-ester organic phosphates, adsorbed to clay particles, and precipitated as colloidal Fe and Al phosphates. The term *soluble* is often used instead of dissolved, but should be avoided unless it is used in the strict chemical sense in which the solubility of a solid substance is given as the amount in grams that will solvate into 100 cm^3 of pure water.

Sediment, sediment-bound, particulate—these terms are used to describe waterborne constituents that are *not* dissolved, i.e., they are retained on a filter. Sediment is typically used to denote the larger particles entrained by moving water that will settle out when water loses its velocity and kinetic energy, and sediment-bound substances are elements or compounds that are physically and chemically associated with sediment. Particulate is a more general term that can include settleable sediment as well as finer colloidal material that can only be removed from the water by filtration, centrifugation or coagulation (flocculation).

Point Source—contaminants that are discharged to surface or ground water from a defined source. This would include industrial discharges, municipal effluents, concentrated feedlot runoff, etc.

Non-Point Source—Also described as "diffuse." These are watershed sources of contaminants that are contributed by runoff or percolation from land as a result of different land use activities. While the flux or load from a small area of the watershed may be low, the overall loading to the watershed may be large if the total area contributing contaminants is large.

10.3. WATER SAMPLING

Water can be sampled for analysis by a variety of manual or automatic techniques. Instantaneous "grab" samples can be taken when only periodic analyses are needed and where water quality does not change rapidly. For more intensive sampling of rapidly changing water systems, automated sampling is usually required. Automatic sampling devices can vary from passive, water driven devices like the Coshocton wheel or multislot divisor (Figure 10.1), to pump samplers (Figure 10.2). Because of the dependence of water quality on water flow, samples are synchronized with flow measurements (see chapter on flow measurement) to yield a concentration-flow relationship known as a *chemograph*. The chemograph can be used to calculate fluxes of waterborne chemicals (see Section 10.5).

Lakes and Ponds. These are relatively quiescent waters that can be sampled manually using depth stratified or depth integrated samples on some kind of surface grid. Samples can also be taken at inlets and outlets to measure effects of lake processes on water quality.

Rivers and Streams. Larger streams and rivers that have sustained base flows are easier to sample than ephemeral streams in which flow is intermittent and unpredictable. Base flow sampling is important in determining in situ water quality; for example, the dissolved O_2 content of streams in the summer can greatly affect fish survival, and can be monitored with regular, periodic grab sampling. If the objective of water quality monitoring is to determine the discharge of contaminants from the river to a lake or other water body, then storm hydrograph monitoring is essential. Because contaminant concentrations often increase with flow in the river (see Section 10.5), storm events may only represent a small percentage of time but a disproportionate percentage of the contaminant load. Storm hydrographs can be sampled manually from large rivers where the total hydrograph may occur over a period of several days; for smaller streams, flow activated pump samplers are used and the time of sampling is recorded on the hydrograph so as to calculate the contaminant load from the chemograph.

Drainage. Drainage by buried tile or open channel in areas of poorly drained soils or where irrigation is used can be sampled together with flow measurement by both manual and automatic means. Tile flow can be sampled by intercepting the tile line with a vertical sump with enough head fall for the tile flow to enter the sampler. The sampler can be a sump pump, tipping bucket, Coshocton wheel, or commercial pump sampler. In open channels, a pump sampler can be used with a Parshall flume or V-notch weir.

Ground Water. Shallow ground water above the water table is usually sampled with some kind of lysimeter. The simplest type is a plastic tube of 4 to 10 cm diameter capped at the sampling end with a porous ceramic cup. Suction is applied at the upper end via a rubber stopper and water is drawn through the cup into the tube in response to the applied vacuum at soil matric potentials < 1 bar. Pan and monolith lysimeters involve some soil disturbance to install and usually only can sample saturated flow; however, they sample a larger cross-sectional area than the suction lysimeter and vertical chemical fluxes can be estimated. Deeper ground water is sampled by installing a tube well to the depth of interest, usually the depth of maximum water table recession. The open end of the well is filled with a few centimeters of sand to prevent sediment movement into the well. Water moves into the well in response to the hydraulic head of the water table. Samples are removed with air-driven or peristaltic pumps. An advance type of ground water well sampler is the Solist system (Figure 10.3). Sections of the well are made of perforated stainless steel to allow water entry from a given depth; sections can be separated

Figure 10.1.   Proportional samplers used in field water sampling. Top: Coshocton
wheel; bottom and next page: multislot divisors.

from each other with inflatable balloons. Samples are withdrawn via a series of
small-diameter tubes inserted to each depth of sampling, and a manifold at the surface permits easy switching of the sample pump from one depth to another.

Figure 10.1.   Continued.

10.4. WATER ANALYSIS

Water quality can be evaluated by subjective and qualitative taste and odor tests. Human taste and smell is incredibly sensitive and accurate, and humans can be calibrated to perform as well as some sophisticated analytical instruments in detecting the presence and concentration of some constituents. Only volatile substances can be smelled, but important water contaminants like gasoline, hydrogen sulfide (H_2S), and ammonia (NH_3) can be detected by the nose. Wine, tea and coffee tasters are highly skilled in detecting the subtleties in these beverages, but I once saw a man on a late-night talk show identify bottled samples of water from the five Great Lakes in about five minutes by tasting and smelling them. Waterborne constituents are normally identified and quantified by chemical analysis following sampling preparation. Samples can also be analyzed for pathogenic organisms by culturing them on growth media or by microscopic observation. Samples are normally prepared by filtering or centrifuging to separate the dissolved and filterable (sediment-bound, particulate) fractions. Following separation, the dissolved fraction can be analyzed by a variety of gravimetric, potentiometric or spectroscopic techniques that have evolved in the last two decades to the point where concentrations in the parts per billion (10^{-9}) or even parts per trillion (10^{-12}) can be measured. Different chemical species can be separated using gas, ion and liquid chromatography, and chemical speciation models can be used to estimate the concentrations of complex chemical species that are not easily measured analytically. The filtered material can be digested in strong acid to give the total chemical content, or selective extractions can be used to yield information about the relative solubility of particulate forms of an element. Chemical composition of the filtered material can also be inferred from X-ray diffraction, infrared, UV-visible, and electron spectroscopy.

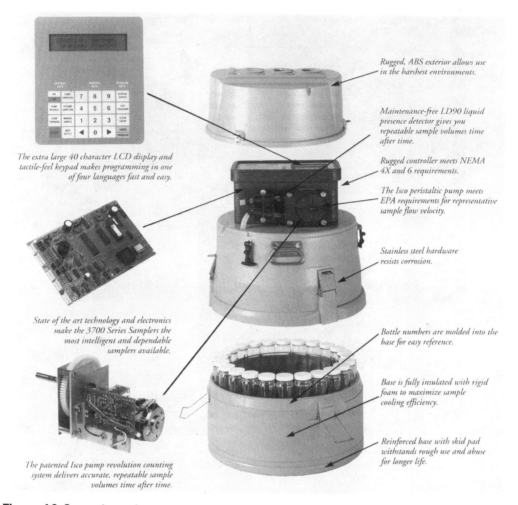

The extra large 40 character LCD display and
tactile-feel keypad makes programming in one
of four languages fast and easy.

Rugged, ABS exterior allows use
in the harshest environments.

Maintenance-free LD90 liquid
presence detector gives you
repeatable sample volumes time
after time.

Rugged controller meets NEMA
4X and 6 requirements.

The Isco peristaltic pump meets
EPA requirements for representative
sample flow velocity.

Stainless steel hardware
resists corrosion.

State of the art technology and electronics
make the 3700 Series Samplers the
most intelligent and dependable
samplers available.

Bottle numbers are molded into the
base for easy reference.

Base is fully insulated with rigid
foam to maximize sample
cooling efficiency.

Reinforced base with skid pad
withstands rough use and abuse
for longer life.

The patented Isco pump revolution counting
system delivers accurate, repeatable sample
volumes time after time.

Figure 10.2. A modern automated pump sampler. Reproduced with permission of
Isco Environmental Division, Lincoln, Nebraska.

For specific analytical methods for water quality analysis, refer to "Standard
Methods For The Examination of Wastes and Wastewater," published by the Ameri-
can Public Health Association (1971), "Methods For Chemical Analysis of Water
and Wastes" by U.S. Environmental Protection Agency (1983), and "Methods For
Determination of Inorganic Substances in Water and Fluvial Sediments" by U.S.
Geological Survey (1989).

Sampling Techniques. The information gained from a water analysis is only as
good as the sample that was analyzed. Considerations in sampling include sample
size, timing of sampling, containers used, sample preservation and handling. Samples should be protected from contamination arising from the sampler itself, and
from improperly washed and sealed containers. Sample size should be at least
double that needed to perform all anticipated analyses, but sample size may be
limited if an automatic sampler is used. Sample timing is selected to coincide with
anticipated events, e.g., storm hydrographs, rising or falling water table, tile discharge, application of agrichemicals. Samples for analysis of inorganic constituents

Figure 10.3.   The Solist multistage ground water sampler. Each tube is inserted through the casing to a different depth.

are taken in plastic bottles to avoid contamination by metals that can leach from glass, while samples for organics analysis are stored in glass to avoid sorption of dissolved organics by plastic bottles. Samples containing significant suspended sediment are usually preserved by refrigeration at 4°C; filtered samples for metals and nutrient analysis can be preserved in dilute acid. Samples for analysis of pathogens must be taken with presterilized sampling equipment and immediately placed in presterilized containers and sealed tightly. Samples should be transported from the sample site to the laboratory as soon as possible, but should not be kept in the field for more than a week.

10.5. UNITS IN WATER QUALITY DETERMINATION

Water is a liquid and is sampled volumetrically. The constituents in water are usually expressed in terms of their mass concentrations on a volumetric basis, and watershed loads are given in areal units. Although hydrologic parameters are often given in the U.S. in English units (gallons, ft^3/sec, etc.), metric units are used in analytical chemistry. The following mass/volume concentration units are preferred in water quality determinations:

grams per liter	g/L	10^{-3}	parts per thousand
milligrams per liter	mg/L	10^{-6}	parts per million (ppm)
micrograms per liter	μg/L	10^{-9}	parts per billion (ppb)
picograms per liter	pg/L	10^{-12}	parts per trillion (ppt)

Under the Systeme Internationale (SI) for chemical and physical units, only liter (L) can be used in the denominator for volume, while any multiple of gram (g) can be used in the numerator for mass. Ratio units like parts per million are unacceptable because they are usually incorrect and ambiguous. For ratio units to be used, the volume of water would have to be expressed as a mass; with a density of 1 g/cm^3 and with 1 liter = 10^3 cm^3, a concentration of mg/L would give a mass ratio of mg/10^6 mg, or parts per million. However, the density of water is not exactly equal to 1 at all temperatures, and there is still the additional problem that hydrologic units are based on water volume, not mass.

When the concentration of an element is measured, the chemical form is often denoted even though the mass concentration is expressed on an elemental basis. For example, the nutrient element nitrogen (N) can exist in water in a number of chemical forms, two important pollutant forms of which are nitrate (NO_3^-) and ammonia (NH_3). The concentrations of these chemical species would be designated as mg/L NO_3-N and mg/L NH_3-N, meaning that the concentration is of the element not the compound mass. This has led to confusion in the past for the drinking water standard (today referred to as the maximum concentration limit, MCL) for nitrate, which is 10 mg/L NO_3-N but is still found in some older texts as 45 mg/L NO_3^-. The two numbers are approximately equal since the molar ratio of NO_3 to N is 62:14 = 4.42.

Where only ionic constituents are being considered, as in the case of salinity or sodicity (sodium, Na, content), ionic concentrations are expressed in terms of ionic charge. Ionic charge is determined from the molar concentration (moles/L, M) by multiplying by the valence of the ion. For example, the ionic charge concentration of calcium (Ca^{2+}) would be twice its molar concentration. Charge concentration is expressed as equivalents/liter (equiv/L) or as mole charge/liter (mol_c /L), the latter more commonly used today.

Example 10.1: A water analysis is returned to you from a testing lab with a concentration of sulfate (SO_4^{2-}) of 25 ppm. Convert this result to millimole charge per liter ($mmol_c$ /L).

Solution:

25 ppm SO_4^{2-} = 25 mg SO_4^{2-}/L
Molecular weight of SO_4^{2-} = 32 (S) + 4x16 (4 O) = 96 g/mole = 96 mg/millimole
millimole SO_4^{2-}/L = (25 mg SO_4^{2-}/L)/(96 mg/millimole) = 0.26 millimole SO_4^{2-}/L
Valence of SO_4^{2-} = 2 mol_c /mole = 2 $millimol_c$ /millimole

$mmol_c$/L SO_4^{2-} = 2 millimol$_c$/millimole $\times$ 0.26 millimole SO_4^{2-}/L = 0.52 $mmol_c$/L SO_4^{2-}

Answer: There are 0.52 $mmol_c$/L SO_4^{2-}.

10.6. FLOW CHEMOGRAPHS

The concentrations of dissolved and suspended constituents in surface waters vary in response to processes occurring on the watershed and within the channel. These processes are primarily driven by storm events, but also include in-stream sedimentation and resuspension, point source discharges, and biological immobilization and mineralization. These changes manifest themselves at the monitoring point in the channel as time-series concentration changes associated with the flow-discharge hydrograph. There are two basic types of chemograph-hydrograph relationships, and both are illustrated for an intensively farmed watershed in northern Ohio (Figure 10.4). The first type of chemograph (Figure 10.4a) is characteristic of sediments, or sediment-bound constituents like P, which are entrained in the water column by the kinetic energy of the storm event as it passes downstream. Concentrations increase and fall as the flow velocity increases and ebbs. In most cases, it has been observed that the chemograph peak slightly precedes the hydrograph, and this has been interpreted to occur because the kinematic wave moves through the water column faster than the water moves down the channel. This phenomenon is similar to that observed with ocean and lake waves in which the wave is moving independently of the mass of water. The second type of chemograph is characteristic of dissolved constituents that are not bound to sediments. This would include nitrate (Figure 10.4b) and water soluble pesticides like the herbicides atrazine and alachlor (Figure 10.4c). In the case of nitrate in northern Ohio, the delay in the chemograph with respect to the hydrograph has been interpreted as being due to tile drainage, high in nitrate relative to surface runoff, that increases as a percentage of total flow as water levels fall in the tributaries and main channel of the river. The pesticides appear to have behavior intermediate between that of sediment and nitrate because they are sufficiently biodegradable that their concentrations in tile drainage are low.

Example 10.2: The flow-sediment rating curve for a river monitoring station can be expressed by the linear equation:

Sediment conc. (mg/L) = 10 + 0.50 flow (L/sec)

Total P concentration of sediment in the river is 1200 mg P/kg, and watershed area is 1500 hectares.

Calculate the watershed total P load in kg P/ha for a 24-hour storm with a flow of 600 L/sec.

Solution:

Sediment concentration = (10 + 0.50 $\times$ 600 L/sec) = 310 mg/L

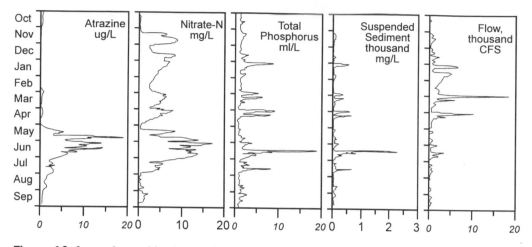

Figure 10.4.   Annual hydrograph, sediment graph and chemographs for the Sandus-
ky River monitoring station during the 1985 water year (Baker, 1993).

Total flow = 600 L/sec × 24 hours × 3600 sec/hour = 5.184×10^7 L
Sediment load = 5.184×10^7 L × 310 mg/L × 10^{-6} kg/mg = 1.61×10^4 kg
P load = 1.61×10^4 kg sediment × 1200 mg P/kg sediment × 10^{-6} kg/mg
    = 19.32 kg P
Watershed P load = 19.32 kg P/1500 hectares = 0.013 kg P/ha

Answer: The watershed P load is 0.013 kg P/ha.

10.7. SURFACE WATER QUALITY PROCESSES

As discussed in Chapter 5 on surface hydrology, surface water is the result of the complex hydrologic cycle in which precipitation intercepted by land surfaces is shed (hence the name watershed) by the land and proceeds downslope by gravity until it enters some confined water body (lake or reservoir) and ultimately the ocean. Much of the precipitation on the upper reaches of the watershed interacts only with the soil surface, particularly where there are impermeable soils, while lower reaches will have increasing contributions from ground water. Human intervention has modified this natural process through the installation of drainage and irrigation systems, ground water extraction, upground reservoirs, and liquid waste disposal, with important consequences for water quality. The use of irrigation may result in runoff at a time when runoff would not normally occur, and tile drainage alters the hydrologic cycle by rapidly returning to the river water that would have percolated down to the ground water.

When water contacts the land, it interacts physically, chemically, and biologically with the rocks, soils, and biota (in the form of growing vegetation, decaying organic matter, microorganisms, soil arthropods, and other animals) so as to release particulate matter and dissolved constituents into the water. To this must be added the deliberate or inadvertent discharge of wastes into surface and ground waters. The major processes by which material is *released to water* include:

Water Erosion. The detachment and transport of rock, soil, and biota by the kinetic energy of flowing water.

Decomposition. The degradation of complex organic molecules, such as in decaying plants and animals and wastes, by microorganisms, into CO_2, H_2O, ions like Ca^{2+}, $H_2PO_4^-$ and NO_3^-, and smaller organic molecules like fatty acids and amines.

Dissolution. This is normally a slow process by which solids in rocks and soils dissolve in water to produce ions. For example, limestone-derived rocks dissolve to produce Ca^{2+} and CO_3^{2-} ions. The ionic composition of natural waters is determined to a large extent by this process, but this like many other natural processes can be greatly affected by human activity. An example would be the widespread use of salt (NaCl) in northern U.S. states for road deicing and snow control. This has resulted in elevated concentrations of Na^+ and Cl^- in surface waters during spring runoff.

Desorption. The transfer of sorbed constituents from exchange sites on soil minerals and organic matter. This is a rapid process and is the primary mechanism by which soil constituents like exchangeable cations (Na^+, K^+, Ca^{2+}, Mg^{2+}) are buffered in soil water.

Oxidation-Reduction. Both oxidation and reduction reactions in rocks and soils can result in water solubilization of constituents that have multiple oxidation states. An important example is the oxidation of the mineral pyrite (FeS_2) found in high-S coals in the Appalachian region of the U.S. to Fe^{3+} and sulfuric acid (H_2SO_4). This process releases large amounts of soluble iron and sulfur into surface waters draining areas of abandoned surface mines. Similar situations occur in areas of gold, nickel and copper mining where the ore deposits are sulfides that oxidize on exposure to the air to form H_2SO_4 and high concentrations of dissolved metals that can be toxic to downstream biota.

The major processes by which material is *removed from* surface water include:

Sedimentation. Sediments settle out of water in channels, flood plains and behind dams as water velocities slow. Storage of sediments and sediment-bound constituents may be permanent or transitory. Sediment concentrations in the Genessee River in New York at Rochester as it enters Lake Ontario were lower in storm events for some time following the large event that occurred with Hurricane Agnes in 1972. Even if sediments are resuspended and transported out of the watershed, temporary storage can result in significant water quality changes. Bottom sediments in the Maumee River of northwestern Ohio had lower total P contents than suspended sediments, presumably because of reduction of iron oxides and iron phosphates and their mobilization into the water column.

Immobilization. The uptake of dissolved constituents from water by biota and conversion into biomass compounds. Phytoplankton (algae) and rooted macrophytes (such as cattails) can remove significant quantities of nutrients like N and P from water in river channels. Riparian systems (vegetation growing at the verge between the land and water) and wetlands connected to surface water channels are being engineered in some areas to enhance the immobilization of waterborne pollutants into biomass.

Precipitation. Chemical interaction of dissolved ions to form a solid compound. Precipitation will occur when the solubility product of a given compound is exceeded. This rarely occurs in open waters because ion concentrations are usually low compared to the solubility products of most compounds. It may occur in concentrated drainage, such as from an industrial discharge or from mine tailings. For example, a common sight in the surface coal mining areas of the Appalachian region of the U.S. is the formation of a reddish-yellow precipitate known locally as "yellow boy." This material is a complex mixture of Fe oxides and sulfates which precipitate when the strongly acid mine drainage is neutralized when it comes in contact with more neutral stream waters.

Sorption. The rapid removal of dissolved ions, complexes and organic compounds from solution by binding to the reactive surfaces of suspended and settled sediment. This is probably the major mechanism that regulates the dissolved concentration of those chemical species which react strongly with sediment. These include P, all of the metals, and organics with low water solubility or with reactive functional groups.

10.8. PARTITIONING

The tendency of a chemical species to sorb to sediment is termed *partitioning.* This general term does not imply the mechanism by which sorption occurs, only the extent to which it occurs. Partitioning is usually assumed to be an equilibrium process because, at equilibrium, it is possible the describe the extent of partitioning by an equilibrium constant, and this constant can be used with knowledge of the sediment concentration to estimate the distribution of the species between the dissolved and sediment-bound phases (see Example 10.3). In open water systems, partitioning is rarely at thermodynamic equilibrium, but may be close enough to provide useful information on the distribution of dissolved and sediment-bound species.

The partition function is normally determined for a chemical species and a specific sediment by shaking water samples, to which increasing amounts of the chemical species in question are added, with sediment at a fixed sediment concentration. At near equilibrium (usually after 24 hours), the concentrations of the chemical species in solution are determined and the amount of the species sorbed by the sediment is calculated by difference from the amount added. A plot of sorbed chemical species versus equilibrium aqueous concentration at fixed temperature (usually room temperature) is called a *sorption isotherm.* The plot can then be described mathematically by one of several equations, the two most common of which are the Langmuir and Freundlich. The Langmuir equation is the most useful because it allows for the calculation of sediment sorption capacity, but its assumptions are rarely appropriate for sediment systems. The Freundlich equation is most commonly used and is:

$$Q/X = KC^n \qquad (10.1)$$

where C is the equilibrium aqueous concentration of the sorbed chemical species, Q is the mass of sorbed chemical, X is the mass of sediment, and K and n are constants. K is a measure of the overall partition tendency (chemical species that bind strongly to sediments have large K values), and n is a measure of nonlinearity of the isotherm (a value of n of 1.0 means that the isotherm is perfectly linear). Because this is an exponential function, it is most conveniently used in its linear, log-transformed form:

$$\log (Q/X) = \log K + n \, \text{Log} \, C \qquad (10.2)$$

For chemical species like P which only bind to a limited number of reactive sites on the sediment particle, n is less than 1 and the isotherm is curved downward as C increases (Figure 10.5). For pesticides and other organic compounds, sorption is onto the organic matter in the sediment and occurs by entering into the organic matter mass rather than being held only on the particle surface. Because of this, the available sites for sorption are essentially limitless and the n constant in the Freundlich isotherm will have a value close to 1.0. The Freundlich equation then becomes:

$$C/X = KC \qquad (10.3)$$

and K is known as the *partition coefficient*, K_D.

Sorption isotherms are used to estimate the distribution of chemicals between soil and water. This knowledge is very useful in predicting if a chemical will be easily leached or will move in surface water primarily in the aqueous phase (chemicals with low partition coefficient), or will be tightly bound to soil and sediment and move only in response to soil erosion or sediment suspension (chemicals with high partition coefficient)(Figure 10.6). Approximate partition coefficients for common water quality chemicals are summarized in Table 10.1.

Example 10.3: A runoff water sample has a sediment concentration of 500 mg/L. Determine the fraction of pesticide A that is in the sediment and water phases if pesticide A has a partition coefficient of 200 (mg/kg)/(mg/L).

Solution:

Assume a volume of 1.0 L.

Mass of sediment (kg) $= 500 \text{ mg/L} \times 1.0 \text{ L} \times 10^{-6} \text{ kg/mg} = 5 \times 10^{-4} \text{ kg}$

Partition of pesticide A $= \dfrac{200 \text{ mg/kg} \times 5 \times 10^{-4} \text{ kg}}{1.0 \text{ mg/L} \times 1.0 \text{ L}}$

$= 0.1 \text{ mg/1.0 mg} = 0.1$

Answer: 10% of the pesticide is in the sediment phase and 90% is in the aqueous phase. The K_D units of (mg/kg)/(mg/L) can be shortened to L/kg and K_D is normally given in these units (see Problem 10.4).

10.9. GROUND WATER QUALITY PROCESSES

The processes that control ground water quality are the same as those previously discussed for surface water. The important difference, however, is in the relative ratio of soil to water in the two systems. In surface water processes, chemicals are more likely to be in contact with water than with soil, and processes that transfer

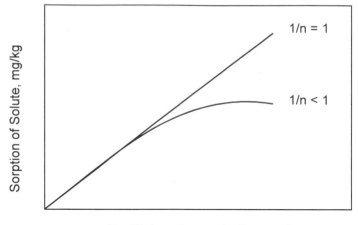

Figure 10.5.   Freundlich adsorption isotherm with (1/n <1) and without (1/n = 1) surface limitation.

chemicals from soil or sediment to water are favored. As the example above showed, even though the partition coefficient for pesticide A was positive (the *concentration* of the pesticide higher in the sediment than in water), most of the pesticide was in the aqueous phase because there was a limited mass of sediment (500 mg) relative to the volume of water (1 L, which has an approximate mass of 1000 g or 10^6 mg). In the case of ground water, chemicals enter ground water through transfer processes between subsurface strata containing the ground water, and via transport of percolating water from the surface. In these cases, the mass of soil is large relative to the volume of water and transfer processes will favor the soil phase. Therefore, chemicals that are strongly bound to soil or subsurface rock strata are found in ground water in very low concentrations unless the subsurface rock strata itself maintains aqueous concentrations by mineral dissolution.

Two other important differences between surface and ground water processes affecting water quality are redox status and biological activity. Surface waters are highly oxygenated and reductions in dissolved oxygen are usually temporary unless there is a strong continuous oxygen demand, such as in a marsh or bog. In ground water systems, however, oxygen must diffuse from the surface and any oxygen demand as a result of biological activity will rapidly deplete dissolved O_2. When this occurs, some bacteria are able to utilize alternative electron acceptors. These chemical species are reduced in the process, greatly changing their chemical character. The alternative electron acceptors are utilized in a sequential manner beginning with nitrate (NO_3^-) and eventually water itself (Table 10.2). Nitrate is readily reduced under ground water conditions if there is any biological activity, but may become quite stable in deeper ground water if there is insufficient available carbon to support growth of heterotrophic bacteria.

Extensive ground water studies in the last decade have dispelled a commonly held belief that ground water is low in microbial activity because of low numbers of organisms. In fact, numbers of organisms are only about one to two logs lower than in surface soil (e.g., 10^6/g vs. 10^8/g).

Agricultural Contaminants

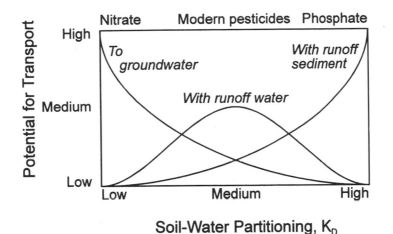

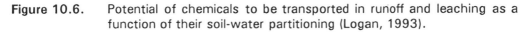

Figure 10.6.   Potential of chemicals to be transported in runoff and leaching as a function of their soil-water partitioning (Logan, 1993).

Table 10.1. Soil-water partition coefficients for various chemicals.

Chemical	Formula	Partition Coefficient (L/kg)[1]	Assumptions[2]
Nitrate	NO_3^-	~0	No precipitation or sorption
Phosphate	$H_2PO_4^-$	25–100	Sorption only
Trace elements	M^{n+}, MO_x^{n-}	1–500	Low values for oxyanions like CrO_4^{2-}; high values for precipitation of metals like Pb^{2+}
Organics	Various	10–10,000,000	Organic carbon content of soil of 2–3%

[1] K_D = Soil concentration (mg/kg)/Solution concentration (mg/L) = L/kg
[2] Soil: solution ratio in the range of 5–25

Ferric iron (Fe^{III}) and oxidized manganese (Mn^{IV}) oxides are the next to be reduced. Because they are quite extensive in soil (occurring in percentage quantities by mass as compared to parts per million for NO_3^-), reducing conditions must be severe (large amounts of decomposable organic matter) before the redox status shifts to sulfate (SO_4^{2-}). Sulfate and water reduction are usually indicative of prolonged reducing conditions in which degradable carbon is continually added. These conditions occur in stratified lakes and bogs, but are unusual in ground water strata.

Movement of chemicals to ground water by percolation was long viewed as a miscible displacement process in which pore water was displaced by water infiltrating from the surface. The conductive-dispersion equation combines Darcy's Law for

Table 10.2. Oxidation-reduction reactions in soil-water systems (Sposito, 1989).

Reaction	Oxidized Species	Reduced Species	Redox Potential (pe)[1]
$O_{2(g)} + H^+ + 4\ e^- = 2\ H_2O$	O_2	O^{2-}	5.0–11.0
$2\ NO_3^- + 12\ H^+ + 10\ e^- = N_{2(g)} + 6\ H_2O$	NO_3^-	N_2	3.4–8.5
$MnO_{2(s)} + 4\ H^+ + 2\ e^- = Mn^{2+} + H_2O$	MnO_2	Mn^{2+}	3.4–6.8
$FeOOH_{(s)} + 2\ H^+ + 2\ e^- = Fe^{2+} + H_2O$	$FeOOH$	FE^{2+}	1.7–5.0
$SO_4^{2-} + 10\ H^+ + 8\ e^- = H_2S_{(g)} + 4\ H_2O$	SO_4^{2-}	H_2S	-2.5–0.0

1 pe = -log (e⁻); (e⁻) = electron activity, calculated from the redox potential using the Nernst equation.

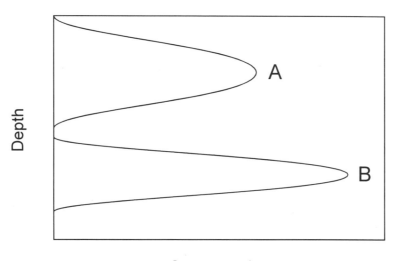

Figure 10.7.   Migration of chemicals by leaching. Chemical A is retained better by the soil and is less mobile than chemical B.

saturated flow in porous media (see Chapter 3) with a retardation coefficient that reflects partitioning behavior between the aqueous chemical species and the soil matrix. This equation predicts that the mass of chemical will move vertically and laterally with the water front in response to hydraulic gradients, but will be smeared (Figure 10.7) by the retardation process. The larger the retardation coefficient, the broader is the distribution of chemical as it migrates to ground water. A recent development in the theory of contaminant transport to ground water is the recognition that some of the percolating water and associated chemicals can move preferentially in large connected channels. These channels may be caused by soil cracking in the case of soils with swelling clays, dissolution channels in limestone (karst), decaying roots, and earthworms or other animal burrows. Flow is more rapid in these channels than would be predicted by miscible displacement, and may account for the presence in ground water of chemicals that should have been attenuated by the soil matrix. On the other hand, the presence of preferential flow channels might cause chemicals in smaller discontinuous pores to be bypassed during percolation and to be retained in the soil.

10.10. SEDIMENT

Sediment is probably the most significant of all surface water pollutants in terms of its concentration in water, its impacts on water use, and its effects on the transport of other pollutants. Sediment is generated in a watershed as a result of water erosion (see Chapter 6). Most eroded soil is rapidly deposited on the land surface, with only a small percentage entering the surface drainage system. There is further deposition of sediment within stream channels and flood plains, and there are a number of general relationships between sediment delivery and drainage area. For example, Renfro (1975) developed the empirical equation:

$$\log DR = 1.87680 - 0.14191 \log (10 \ A) \qquad (10.4)$$

where DR is the sediment delivery ratio (in percent of annual erosion), and A is the drainage area in square miles ($A < 100$ mi^2).

The sediment deposition process results in a selective removal of clay-sized particles and organic matter. Because these are the most chemically reactive materials in soil, sediment often has higher concentrations of chemicals like trace elements and phosphorus (P) than the whole soil. The ratio of the concentration of a chemical in sediment to that in surface soil is termed the *enrichment ratio.* The selective removal of finer and lighter clay and organic matter during runoff is inversely related to the energy of the runoff. As kinetic energy in the form of runoff velocity and turbulence increases, progressively larger and heavier particles are entrained in the runoff and, in the limit, all particles in the surface soil are transported. This means that the enrichment ratio decreases as sediment load increases, approaching 1.0 (Figure 10.8). The clay enrichment ratio will be greatest for coarser textured soils and smaller for soils that are high in clay content.

Once entrained in the stream system, suspended sediments can greatly affect water quality by acting as either a *source* or *sink* for chemicals. When sediment, enriched in nutrients like P from fertilization or waste application, enters the stream or lake system, P desorbs or dissolves from the sediment in response to the lower dissolved P concentrations in the stream relative to those in soil pores. On the other hand, sediment has been shown to scavenge dissolved chemicals discharged to the stream as point sources. Sediment-bound chemicals may be permanently stored in sediments as flood plain deposits or as buried sediment in lakes. Much of the sediment in stream systems, however, is temporarily stored as bottom sediments and is resuspended during larger storm events. A sediment particle may take several years to move from the edge of the field to the outlet of the river.

10.11. NITRATE

Nitrate (NO_3^-) is a major water pollutant. It is toxic to humans and animals because, in their digestive systems, NO_3^- is reduced to nitrite (NO_2^-) which binds with

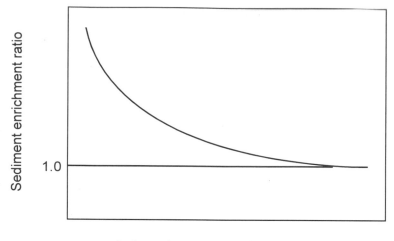

Figure 10.8.   Relationship between soil erosion rate and sediment enrichment ratio.

hemoglobin in the blood to restrict the transfer of oxygen from the lungs. This effect is termed *methemoglobinemia* or Blue Baby syndrome (because of the bluish color of deoxygenated blood). Nitrate has many sources including industrial and municipal waste discharges, but a major source of concern in this text is agricultural. It is a non-point source, and is associated with excess fertilizer use and improper management of livestock wastes. Nitrate is highly water soluble and forms no insoluble minerals. It is an anion and thus is repelled by the net negative charge in most soils, and it does not form strong chemical bonds with mineral surfaces, as does P. It is, therefore, highly mobile and is more likely to be a ground than surface water contaminant. Nitrogen is a non *conservative* element, meaning that it can readily change form in the environment and move between the soil-water-air compartments of the earth. Nitrogen exists in many oxidation states, from NH_3 (-3) to NO_3^- (+5). It can also exist in gaseous (N_2, N_2O, NO, NH_3), aqueous (NO_3^-, NO_2^-, NH_4^+) and solid (organic matter bound) forms. Most N transformations are biologically driven and are summarized in the generalized N cycle (Figure 10.9). Nitrate primarily enters the system by nitrification of NH_3 sources, like fertilizers and organic wastes, and is lost from the system by biological uptake (crop and animal removal) and denitrification.

In natural ecosystems, N is usually in a very tight cycle in which the major sources are rainfall and biological N_2 fixation. Nitrate is rapidly assimilated by the standing biomass, there is little change in N stored in soil organic matter, and there is little loss of N from the system by denitrification and NO_3^- leaching. Nitrate concentrations in surface waters are rarely > 1 mg NO_3^--N/L, and subsurface drainage waters are usually < 1-2 mg NO_3^--N/L. In fertilized systems, particularly where annual crops are grown, the N cycle is not tight. Nitrogen fertilizer or organic N source is often applied at periods before the crop is growing. Crop yields may not be as high as anticipated because of unrealistic yield goals or adverse weather. This can result in runoff and leaching of fall through spring applications of N fertilizer, and leaching of residual NO_3^- in the soil profile following crop

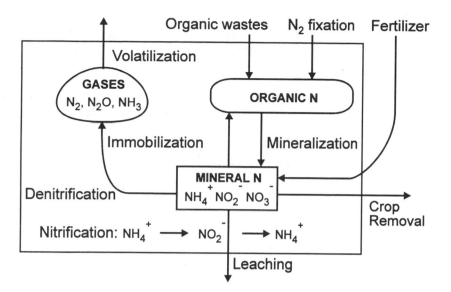

Figure 10.9.   The nitrogen cycle in soil.

harvest. In temperate climates such as in the central U.S., NO_3^- concentrations in streams peak in the spring and are lowest in midsummer. The spring highs are a result of carryover from residual NO_3^- in the soil profile following crop harvest and not from spring applied fertilizer.

Much of the NO_3^- lost from agricultural watersheds in surface drainage is in the form of return flow from tile or surface drains. Nitrate is so soluble that little is lost in direct surface runoff but rather as percolate that is intercepted by tile or surface drains. Where soils are permeable enough not to require artificial drainage, NO_3^- is primarily lost from agricultural watersheds as percolation to ground water. Concentrations of NO_3^- from intensely fertilized agricultural land, particularly for high N consuming crops like maize and potatoes, may exceed 50 mg NO_3^--N/L, and may average higher than the U.S. Environmental Protection Agency Maximum Concentration Limit (MCL) of 10 mg NO_3^--N/L at the edge of the field. Watersheds as large as 17,000 ha in northwestern Ohio, draining intensively farmed land, have had NO_3^- concentrations in excess of the MCL for several weeks of the year. Irrigated maize production in the sandy soils of western Nebraska has resulted in shallow ground water NO_3^- concentrations in excess of 100 mg NO_3^--N/L.

Example 10.4: A maize crop removes 100 kg N/ha in harvested grain. There is a carryover of 25 kg N/ha from the previous year's crop. The recommended N fertilizer rate of 150 kg N/ha is applied. Denitrification losses of all sources is estimated to be 20%. The soil water content at crop harvest is 25% by volume. Calculate the NO_3^- concentration in soil solution at harvest as mg NO_3^--N/L. Assume that all residual N is in the form of NO_3^- and that residual N is distributed uniformly in the top 1 m of soil.

Solution:

Residual NO_3^--N at harvest $= (150 + 25) \times 0.8 - 100 = 40$ kg NO_3^--N/ha
$\times 10^6$ mg/kg
$= 4 \times 10^7$ mg NO_3^--N/ha

Volume of water $= 10^4$ m^2/ha $\times 1$ m $\times 10^6$ cm^3/m$^3 \times$ L/10^3 cm$^3 = 10^7$ L/ha

Nitrate concentration in soil solution $= (4 \times 10^7$ mg NO_3^--N/ha)$/10^7$ L/ha $=$
4.0 mg NO_3^--N/L

Answer: The NO_3^- concentration in soil solution is 4.0 mg NO_3^--N/L.

10.12. PHOSPHORUS

Phosphorus is a naturally occurring element in soil with a typical total P concentration of 300 to 1200 mg/kg. Phosphorus is one of the three elements (the others are N and potassium, K) required by plants in the largest amounts. Phosphorus in soil is composed of solid and solution phases, with the majority (> 99%) in solid phase (Figure 10.10). The solid phase is composed of organic P, Fe, Al and Ca phosphates, and P sorbed to the surfaces of Fe and Al oxides. The solution phase is mostly orthophosphate ($H_2PO_4^-$) and small amounts of dissolved organic P or P bound to colloidal organic matter and Fe oxide. A fraction of the solid phase, usually < 25%, is in dynamic equilibrium with the solution phase through sorption-desorption, precipitation-dissolution, or immobilization-mineralization reactions. This fraction of the solid phase P is termed *labile, phytoavailable* or *bioavailable,* meaning that it is readily transferred to the solution phase and is available to plants that include crops as well as algae. Bioavailable P can be estimated by a number of methods, including direct bioassay, but a preferred method is to use a dilute chemical extractant or an exchange resin to mimick P removal by the plant. It is not surprising that methods to estimate crop available P give results that are positively correlated with those that estimate algal available P. The total bioavailable P in a water sample is the sum of the dissolved inorganic P and the bioavailable sediment P.

Phosphorus is strongly bound to soils and sediments by the reactions described above. For this reason, P movement by leaching is not common. The exceptions are instances where septic tanks are placed over very sandy soils or fractured bedrock, or where organic wastes are applied at excessive rates on similar soil conditions. Some dissolved P may be in the form of organic P from organic wastes, and these compounds are more mobile in soil than orthophosphate because they are less reactive with soil minerals. Contamination of ground water by P is not a major environmental problem unless the ground water contributes significantly to surface water flow from a watershed; the major environmental problem associated with P in water is eutrophication which can only occur in surface waters rather than ground waters.

Eutrophication, the accelerated growth of phytoplankton (suspended and rooted algae) in ponds, rivers, and lakes from increased nutrient input, is usually limited by the addition of P to the system. The major sources of P to water bodies are point source discharges from wastewater treatment plants and industry, and non-point

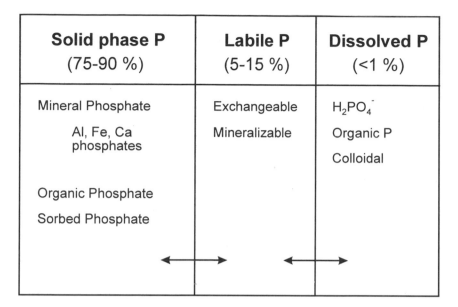

Solid phase P (75-90 %)	Labile P (5-15 %)	Dissolved P (<1 %)
Mineral Phosphate Al, Fe, Ca phosphates Organic Phosphate Sorbed Phosphate	Exchangeable Mineralizable	$H_2PO_4^-$ Organic P Colloidal

Figure 10.10.   Forms of phosphorus in soil-water system.

source discharges from land runoff. Important sources of P in land runoff are intensive agricultural production and urban runoff. Losses from forested or well managed pasture are generally lower.

Non-point source runoff of P from land occurs as a complex interaction of soil and water in the rainfall-erosion process (Figure 10.11). As rainfall strikes the land surface, runoff is generated and soil particles are detached and suspended in the flowing water. Total particulate P is transported with the eroded sediment, and total sediment P concentration decreases as the sediment concentration increases (See 10.10 for a discussion of sediment enrichment ratio). In addition, there is an even greater decrease in sediment labile P as sediment concentration increases. As runoff water moves over the land surface, inorganic and organic P are dissolved from the soil, and plant residues, manure, fertilizer or other P containing materials on the soil surface. The interaction between dissolved P and sediment is rapid and near equilibrium is achieved by the time the runoff leaves the edge of the field. The proportion of total P load in runoff between the sediment and water phases will be determined by the total and labile P content of the eroded sediment, and the sediment concentration. The partition coefficient for P is very large and, if sediment concentrations are much higher than 100 mg/L, most of the P load is in the form of particulate P. Under forest, pasture, or no-till conditions, where sediment concentrations are low, most of the P load may be dissolved.

The implications of these phenomena are that, while erosion control is effective in reducing total P losses in runoff because most of the P loss is associated with sediment at high to moderate erosion rates, control of dissolved P losses in runoff will require control of labile P content of the surface soil. This can only be achieved by limiting labile P levels to those needed to satisfy crop requirements, or by subsurface injection of P fertilizer. Even where P is subsurface injected into no-till crops or pasture, labile P concentrations at the soil surface will increase with P fertilizer application by the decay of crop residues at the surface.

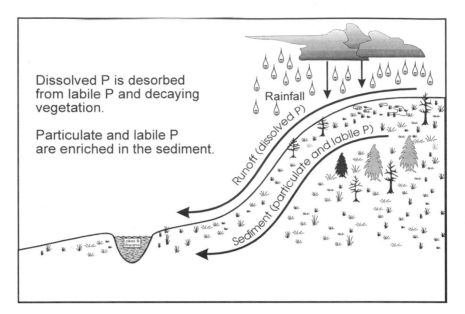

Dissolved P is desorbed
from labile P and decaying
vegetation.

Particulate and labile P
are enriched in the sediment.

Rainfall

Runoff (dissolved P)

Sediment (particulate and labile P)

Lakes & Streams

Figure 10.11.   Generation of dissolved, labile, and particulate phosphorus in run-off/erosion process.

10.13. XENOBIOTICS

Xenobiotics are synthesized organic compounds. They are characterized as having relatively low water solubilities and strong binding to soil organic matter. They also have greatly varying persistence in soil with half-lives (see Example 10.5 below) of days to years. Partition coefficients vary from low to very high, and this factor, combined with persistence, determines the degree to which xenobiotics are transported in surface or ground water. In the case of pesticides which are widely applied to land and thus particularly susceptible to losses by rainfall, the properties of water solubility, persistence, and partitioning are often selected by the manufacturer to enhance efficiency. For example, nematicides like aldicarb are among the most widely reported ground water pesticide contaminants because of their relatively high water solubility, long persistence, and low soil binding. These are the properties required of compounds which must be mobile enough to move to the subsoil where nematodes are found and persistent enough to maintain efficacy.

The partition coefficient for xenobiotics between soil and water can be related to intrinsic properties of the compound itself by correcting partitioning for the organic matter content of the soil, thereby removing the effect of the soil. The corrected K_D, termed the K_{OC}, can also be estimated by linear regression relationships with properties of the compound that include water solubility and molecular weight, among others.

The partition coefficients of the older, chlorinated insecticides like DDT, and other xenobiotics like PCBs, are large enough that they are transported mainly with eroded sediment. Because of their long half-lives, they continue to be detected in sediments years after their use was banned. Modern pesticides, particularly widely used herbicides like atrazine, alachlor, metribuzin, metolochlor and cyanazine, are more water soluble, and are far less persistent. Although detected in ground waters, they are more likely to be seasonal surface water contaminants because they degrade rapidly relative to water percolation from the surface to the water table. In surface water, they are mainly in dissolved form and are found in surface waters for a few months following their application.

Example 10.5: Persistence of xenobiotics in the environment is often expressed as a first-order decay according to the equation:

$$P = P_0 e^{-kt} \qquad\qquad (10.5)$$

where P is the amount of pesticide at time t, P_0 is the initial amount of pesticide, and k is the first-order rate constant. When P is expressed as a fraction or percentage of P_0, the half-life ($t_{1/2}$), the time required for 50% of the pesticide to degrade, is given by 0.693/k.

Calculate the number of half-lives and time required for more than 99% of the chemical to degrade if $t_{1/2}$ is 30 days.

Solution:

Half-Lives	Percent of Pesticide Degraded
1	50
2	75
3	87.5
4	93.75
5	96.875
6	98.4375
7	99.21875

Answer: A minimum of 7 half-lives are required. Time required = 7×30 days = 210 days.

10.14. TRACE ELEMENTS

Trace elements are generally defined as those that occur in rocks and soils in concentrations < 100 mg/kg. They are often mistakenly called "heavy metals," but this term is incorrect because some of the trace elements have low densities (e.g., Be or B), and many are not metallic (e.g., B). The trace elements can exist primarily as cations (Cd, Cu, Cr^{III}, Co, Hg, Ni, Pb, Zn), as oxyanions (AsO_4^{3-}, CrO_4^{2-}, MoO_4^{2-}), or neutral species ($B(OH)_3$). The cations form strong coordinated complexes with soil organic matter and Fe and Al oxides and are also held on the predominantly negatively charged clay minerals in temperate region soils. The oxyanions are more mobile in soil than the cations but can form coordinated bonds with Fe and Al oxides by displacing -OH groups, or electrostatic interactions with positively charged sites on the Fe and Al oxides. In addition to sorption reactions, the trace metals form coprecipitates with soil minerals and they generally have very low water solubilities. The solubility of cationic trace elements increases as pH decreases, particularly for pH < 5. The opposite is true for the oxyanions.

Trace element contamination of surface waters is limited to point source discharges, seepage from mine tailings, or runoff of eroded sediment from waste disposal sites. Even where acidic mine drainage with high concentrations of trace elements is discharged into surface waters, the higher pH of the receiving waters results in precipitation of solids in which the trace elements are coprecipitated. Ground water contamination by trace elements is rare. Where ground waters are reduced, the conversion of sulfate to sulfide results in precipitation of very insoluble metal sulfides, and percolation of trace elements from the surface is limited by their very high partitioning to soil. A noted exception is the presence of high concentrations of CrO_4^{2-} in ground water at the Hanford, Washington DOE site where chromate was used to process nuclear fuel rods and then disposed of in surface impoundments. Chromate, as an oxyanion, is only weakly retained by soil. Boron, which occurs at normal pHs as a neutral species $(B(OH)_3)$, is very mobile in soils.

Recent developments in our ability to understand aqueous chemistry include analytical and modeling approaches that permit us to determine the actual form of an element in the waters being studied. This knowledge has led to greater understanding of how trace element chemistry affects mobility in aqueous systems. Dissolved chemicals, rather than existing as independent ions (e.g., NaCl dissociation to Na^+ and Cl^- ions), form solution complexes with each other. These complexes may be cationic, anionic, or neutral, and may dramatically alter the mobility of the element. A good example is Pb^{2+} which is very insoluble in soil but complexes with Cl^- to form the anion $PbCl_4^{2-}$, a highly mobile species. The use of road salt (NaCl) could result in the mobilization of Pb from roadside soils contaminated by burning of leaded gasoline.

REFERENCES

American Public Health Association. 1971. Standard Methods for the Examination of Water and Wastewater. 13th Ed. Washington, D.C.

Baker, D.B. 1993. The Lake Erie agroecosystem program: water quality assessments. Agric., Ecosystems and Environ. 46:197-215.

Logan, T.J. 1993. Agricultural best management practices for water pollution control: current issues. Agric., Ecosystems and Environ. 46:223-231.

Renfro, G.W. 1975. Use of erosion equations and sediment-delivery ratios for predicting sediment yield. In Present and Prospective Technology For Predicting Sediment Yields and Sources. U.S.D.A. Agricultural Research Service. ARS-S-40.

Sposito, G. 1989. The Chemistry of Soils. Oxford Press, New York.

U.S. Geological Survey. 1989. Techniques of Water-Resources Investigations of the United States Geological Survey. Federal Center, Denver, CO.

U.S. Environmental Protection Agency. 1983. Methods For Chemical Analysis of Water and Wastes. Office of Research and Development. Cincinnati, OH.

Problems

10.1 A laboratory analysis for a water sample lists the results as follows:

NO_3-N   5 ppm
PO_4-P   3 ppb
Al   25 ppt

Calculate the concentrations of each chemical *species* in units of moles/L and $mole_c$ /L.

10.2 The flow weighted mean concentration of nitrate in runoff from a watershed is 5.0 mg NO_3-N/L. If the runoff flow from the watershed is 15 cm, calculate the nitrate load in kg NO_3-N/ha. (Hint: the runoff flow number is actually a volume.)

10.3 Use the data below to calculate the total flow, annual load, and flow weighted mean concentration of total P from a watershed. Plot runoff, concentration, and monthly load versus time on the same graph and determine the periods of maximum P loss. Explain this seasonality. Regress flow versus P concentration and develop a predictive linear equation. Is this a good relationship? Explain the physical basis for this relationship.

Month	Runoff Flow (cm)	Concentration (mg PO_4-P/L)
January	0.001	0.005
February	0.005	0.007
March	3.00	0.250
April	5.00	0.500
May	6.00	0.550
June	3.50	0.300
July	2.00	0.150
August	0.10	0.100
September	0.25	0.125
October	1.00	0.080
November	4.00	0.350
December	0.005	0.010

10.4 A pesticide has a log K_D of 2.5 L/kg sediment. Calculate the sediment concentration (mg/L) in a runoff sample that would be required in order for 25% of the pesticide in the sample to be associated with the sediment. (Hint: assume a sample volume of 1 L.)

10.5 Total phosphorus enrichment ratio (PER) is related to sediment load by the equation:

ln PER = 2.00 - 0.20 ln (sediment load) (kg/ha)

If an initial sediment load of 10 mt/ha is reduced by 90% by conservation practices, what will be the corresponding percent reduction in total P?

10.6 Determine the minimum overapplication to a corn crop of nitrogen fertilizer in kg N/ha that can be tolerated so as not to exceed the drinking water standard for NO_3-N of 10 mg/L. Assume that all of the fertilizer N is converted to nitrate, that 25% of the applied N is lost by denitrification, and that all of the excess nitrate is dissolved in 20 cm of rainfall.

10.7 A pesticide has a field first-order half-life of 20 days. It is surface-applied to a field in a watershed on April 1. A runoff event occurs on June 1 in which 5% of the remaining pesticide is removed in runoff water. If it takes 7 days for the runoff to be discharged into a receiving body (e.g., a lake), what percentage of the applied pesticide is delivered to the receiving body?

CHAPTER 11

Remote Sensing and Geographic Information Systems in Hydrology

John Grimson Lyon

11.1. WHAT ARE REMOTE SENSING AND GEOGRAPHIC INFORMATION SYSTEMS (GIS)?

There exists a variety of technologies which offer great opportunities for evaluations of general hydrology and related issues. The last twenty or so years have seen their application to experimental and operational activities. The utility of these technologies make them very appropriate to supply information and to facilitate important work related to hydrological problems (Lyon and McCarthy, 1995).

Modern technologies have historical antecedents. The technologies of remote sensing and Geographic Information Systems (GIS) have a number of antecedents because they bring together a number of early technologies, and they are useful in many different applications. Remote sensing and Geographic Information Systems (GIS) were developed from earlier and still useful technologies such as surveying, photogrammetry, photointerpretation and the like.

Remote sensing is the practice of measuring an object from a distance or remote location using an instrument. In essence, remote sensing is instrumental sensing that is done without touching the object. This may be performed with an aircraft and a camera or sensor, or with a spacecraft and sensor instrument system.

Geographic Information Systems (GIS) are technologies with antecedents in surveying, mapping, cartography, and information management technologies. GIS are technologies that allow the storage and processing of data in a spatial or map-like reference system. The term is but ten years old, yet elements of GIS technologies have been applied for many years. The advent of lower cost computing capabilities has made the discipline grow in a rapid fashion.

11.1.1. Mapping Science and Engineering Technologies

A very good way to conceptualize the common elements and heritage of these technologies is to view them with a broad look. It's valuable to organize these technologies under the heading of the Mapping Sciences. This concept allows one to address the variety of technologies without excluding necessary examples. It also recognizes that the Mapping Sciences is like a "tool box" of methods, that can be applied when the need arises. And like the old adage, one applies the "right tool for the job."

11.1.2. Geographic Information Systems

Geographic Information Systems (GIS) are databases that usually have a spatial component in the storage of the data. Hence, they have the potential to both store and create map-like products. They offer the potential for performing multiple analyses or evaluations of scenarios of model simulations (Kessler, 1992).

Data are stored in multiple files. Each file contains data in a coordinate system that identifies a position for each data point or entry. Characteristics of the data point are stored as "attributes." A database of individual files is developed and may contain files with characteristics or attributes such as stream locations, topography, chemical sampling, and management practices.

The strength of the GIS approach lies in the quality of the database, and how it can be used to address the application of interest. Each variable or "layer" in the GIS will hopefully be useful in the application by supplying information on physical or chemical processes, or characteristics of the features.

An important capability of GIS is the simulation of physical, chemical, and biological processes using models. GIS can potentially be used with deterministic or complex models based on algorithms simulating processes, or they can be applied with statistical models. The requirement is that the model to be applied has the capability to take spatial and/or multiple file or "layer" data as input to computations.

GIS databases and products are also amenable to evaluations as to quality. It is an important element of these efforts to document the quality of the data, and determine the accuracy of the product. An number of methods can be used to conduct an assessment of accuracy (Congalton and Green, 1992; Lyon and McCarthy, 1995), and these approaches can be implemented in the experimental design (Bolstad and Smith, 1992).

11.1.3. Remote Sensing

Remote sensor technologies can be used to acquire a variety of data for applications. A number of these technologies can supply data to help to solve problems, and can often be accomplished at a lower relative cost than many other technologies. These advantages have attracted great interest in the scientific and engineering community.

To apply remote sensing technologies, it is necessary to identify the application problem and the characteristics that may be measured by the remote sensor. These characteristics or variables may be measured directly by the sensor or indirectly through measurement of a "surrogate" variable, whose behavior is correlated to the variable of interest. Then, one can select the appropriate technology or tool from the suite of technologies. This selection is made with the knowledge of the characteristics or variables to be measured, the available budget and resources, and the "time-frame" of the project.

11.1.4. Photointerpretation

Photointerpretation is the science of deriving or interpreting information from the characteristics of features recorded on photographs. Generally, the term refers to interpreting photographs taken from aircraft, though the same skills can be used to obtain information from any type of photograph or image.

Photointerpretation utilizes several characteristics or elements of photographs to supply information. These elements include characteristics such as: tone or color, shape, size, texture, pattern, shadow, and associations.

The tone refers to the absolute or relative grey tone of features within the photo. For example, "clean" or "optically clear" water may be dark-toned, while soil is light-toned on black and white aerial photographs. Tone can also be applied to interpretations of color photographs and to the description of the relative or absolute reflectance or tone. One can describe a scale of, say, dark green approaching black and continuing onto bright green tones. This may also be done for the blue and red scales of the additive color primaries.

Shape refers to exterior configuration of materials or features. Many features or objects have distinct regular or irregular shapes. Shape can be a unique clue as to the identity of the feature. For example, most people recognize the shape of certain buildings, such as the five-sided shape of the Pentagon, the horseshoe shape of the Ohio State University Stadium, the shape of the Arch in St. Louis, the pyramids of Egypt, the pyramid-shaped Hotel and Casino Luxor, in Las Vegas, and so on.

Size refers to the absolute or relative dimensions of the object or feature. Size can be very important as features are often indistinct on photographs and knowledge of size can be a big clue as to the identity of the feature.

The texture refers to the regular or irregular organization of features within the photo. Usually, the features are small relative to the resolution of the photo. For example, one may view a distinct texture resulting from small rocks breaking the surface of a lake yet the rocks may be too small to view directly. Their presence results in a different texture from that of lake waters without the small rocks.

Pattern, conversely, is the regular or irregular distribution of relatively large features on the earth's surface. An example is the regular pattern of field boundaries in an agricultural area. When this pattern is disturbed, say by a stream course, one can infer certain characteristics of the stream channel such that it may be too big for farm equipment to traverse through it.

The shadow of a feature can be an important clue as to its size and shape, and may help identify the feature as to type. Shadows cast by a feature can provide important shape and size information. A common example is that of interpreting the name of a business or building from the shadows of individual letters that form a sign. The individual letters "blend" into the building, but the shadows are distinct to the interpreter.

Conversely, shadows can obscure detail. It is difficult to view a feature hidden by a shadow, due to the lower relative illumination in the shadow area as compared to the overall illumination of the sunlit areas.

The "association" of a feature is the other characteristic or clue that is found together, in association, with the feature of interest. For example, the course of a stream is usually evident by shape, but often the tone of the water is hidden from view by trees. Stream courses are usually found in association with other clues such as the following: the meandering stream pattern; the branching drainage pattern; automobile and railroad bridges; ponds or lakes; stream-side vegetation; and lower relative elevation and a downhill course as compared to the surrounding landscape (Argialas, Lyon, and Mintzer, 1988).

Figures 11.1 and 11.2 display many of the characteristics that can be used to derive information on an area using photointerpretation techniques. In Figure 11.1 one can see details of the nearshore topography and shape of coastal feature due to the penetration of visible light into the shallow water column. Note the very light tone of bottom sediments, roads and beach areas. In comparison, the vegetation of the swamp areas and beach ridges is dark. On the left side, the grey areas within the water are emergent and submergent wetland plants that retain the low reflectance or grey tone of vegetation. The low grey tone is due to the lack of reflectance by plants of blue light, low reflectance of green light as compared to soil, and lack of red reflectance due to the presence of chlorophyll in plants.

Figure 11.2 is a black and white infrared image of same location. Here, the areal extent of surface water can be seen by the very dark tone of water. Note the north flowing stream is much more obvious, as are the general wetlands areas with standing water.

Figure 11.3 is an additional black and white image of the area. The use of a number of images and other different film types can be a great help in photointerpretation (Lyon, 1987; Williams and Lyon, 1991; Lyon and Greene, 1992; Lyon, 1993; Lyon and McCarthy, 1995). This is particularly true of hydrological characteristics on areas which change follow rainstorms and other hydrological-related events. It is always useful to obtain many aerial photos of a given study area in support of analyses. This is because each photo supplies unique information, and repetitive coverage adds the value of multiple samples and statistical capabilities. Photos may also be inexpensive in comparison to the costs of obtaining the similar quality and quantity of field data using traditional methods.

11.1.5. Photogrammetry

Photogrammetry is a very valuable technology which has provided many years of service. Almost all maps showing horizontal positions, point elevations, elevation contours, and/or topographic maps are made using photogrammetric technologies. Photogrammetry is the science of obtaining precise and accurate measurements from overlapping or stereoscopic photographs (USACE, 1992; Falkner, 1994).

Commonly, these photographs are taken from an aircraft platform, and individual exposures record the ground under the aircraft overflight. The photos are taken such

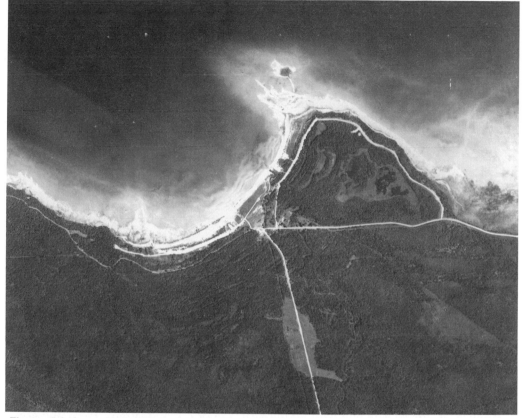

Figure 11.1.   This is a typical black and white aerial photo of coastal Lake Michigan (July 12, 1965).

that an individual photo includes ground coverage of the previous and next exposure in a series or flight line. These photographs are called stereoscopic pairs. The difference in perspective of a common objective photographed from different positions is called parallax. The relative parallax of the pair of exposures creates the stereoscopic effect, and allows one to determine the elevation of features in addition to their horizontal, relative positions.

Photogrammetry is important because it is the most cost-effective method to make topographic maps. Almost all topographic maps are made from aerial photographs using photogrammetric technologies. Most large scale engineering maps are also made from aerial photographs. Hence, most of the products that are used in hydrological analyses are made from photogrammetric analysis of aerial photographs (USACE, 1992; Falkner, 1994).

Vertical, overlapping photographs can be valuable for photogrammetric measurements and for photointerpretation. Figure 11.4 demonstrates a stereo triplet of a coastal area near the location of Figures 11.1 to 11.3 in the Straits of Mackinac, Lake Michigan.

Photogrammetric measurements from aerial photographs can document historical conditions and change over time (Lyon and Drobney, 1983; Lyon, 1987). This is particularly true of general hydrological events that cause change. Figures 11.5,

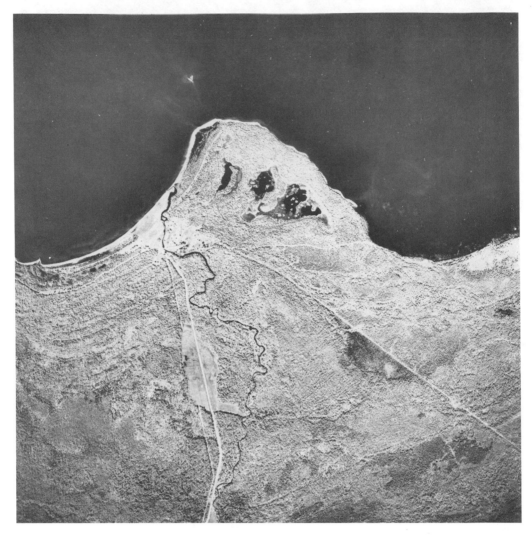

Figure 11.2.   The difference in this image (August 23, 1952) of the same location as Figure 11.1 is due to the fact that the film is black and white infrared. Note the lack of detail within the water due to the absorbance of infrared light at the surface and near surface of the water.

11.6, 11.7, and 11.8 show the remarkable change in emergent wetlands over a 27-year period. This information is very valuable for scientific and engineering studies.

Color and color infrared photographs are also very valuable for hydrological-related analyses. The CIR image can facilitate interpretation of water resource characteristics and general vegetation type and condition (Figure 11.9). The stereoscopic coverage allows viewing of relative topography, and the shape of tree canopies.

11.1.6. Surveying

Surveying is the science of obtaining precise and accurate measurements of the earth's surface, or its waters, using instruments. These measurements include hori-

Figure 11.3.   A black and white photo (June 12, 1958) from another date can supply additional details that may not be observable in other photographs.

zontal locations as well as vertical or elevational locations. Commonly, these measurements are tied to some absolute reference to characterize the location of earth features in an absolute sense, and to allow one to later relocate these features using the original measurements or map products.

Surveying is important in hydrology, as these measurements form the basis of many hydrological calculations. The surveyed positions of basins, sub-basins, channels, control structures and the like all form important inputs to calculations. These positions may also provide the basis for GIS databases, and for GIS model calculations.

Surveying is also important in the production of photogrammetric products (USACE, 1992). To make photogrammetric calculations in absolute units and to reference measurements absolutely, it is necessary to collect ground-surveyed positions. These positions or ground control points (GCP) allow the stereo model to be tied into absolute ground positions.

Figure 11.4.   Low altitude, stereoscopic coverage (August 1980) acquired with a 35mm camera and light aircraft system.

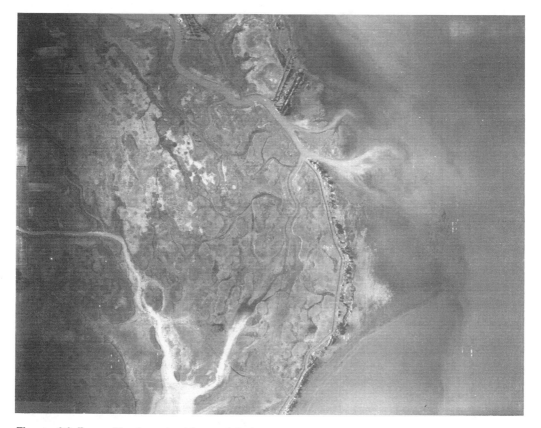

Figure 11.5.   Black and white aerial photograph (July 27, 1937) of the Pointe Mouillee area of Lake Erie in Michigan. Note that aerial photos are available from the late 1930s.

Surveying is also used to mark the positions and follow the progress of field construction. These measurements help to assure that the engineering design specifications are met.

The advent of more capable instruments and computers has greatly influenced the contributions of surveying. The use of total station instruments has allowed distances to be measured with laser Electronic Distance Measurement (EDM) devices, as well as angles to be turned. Data can be stored directly to an "on-board" computer system to facilitate record keeping, and to expedite calculations of positions. This tool and its application has allowed surveyors to be more efficient and to work with fewer crew members. It has also made the data readily available for computer analyses and production of final products using computer output devices.

Global Positioning Systems (GPS) have now made position information available to most users at minimal cost. This technology has brought general surveying information to the general public, and has really contributed to the lay person. Outdoor pursuits, general navigation, safety, and science and engineering have all benefited.

GPS is a system of satellites transmitting a microwave signal to earth that can be received by GPS instruments on the surface. The signal is a timing code such that the receiver can measure time at the speed of light, or convert the timing information

Figure 11.6.   Black and white aerial photograph (November 5, 1955) of the Pointe Mouillee area which demonstrates a loss of emergent wetlands.

to a distance or position. Receiving a number of signals from different satellites allows one to position the receiver with multiple time or distance measurements.

To better define the position, many civilian applications make use of more than one receiver. This allows the user to compare the timing signal received from one instrument to another. The comparison and use of correction algorithms removes sources of error that can result from the atmosphere or other conditions. This differential GPS approach can be implemented in a number of manners based on position accuracy and precision requirements.

11.2. PRODUCTS

Products are the media that we use to handle the information. Often the products are helpful in many aspects of the work. They may form original examples of products or may be derivative products optimized for a certain application.

Figure 11.7.   Black and white aerial photograph (May 27, 1964) of the Pointe Mouillee area demonstrating further loss of wetlands, and the construction
of a dike area to maintain the shrinking wetlands (Lyon and Greene,
1992).

11.2.1. Photographs

Photographic products have well known characteristics and can be used in a
variety of forms to collect and store original data. Photographs use silver-halide
chemistry to capture black and white or color images on a film transparency or on
paper. As such, photography using cameras is a remote sensor technology. It has
a number of advantages in that the products are usually available, and are often low
in relative cost to other sensor products.

Figure 11.8.   Smaller scale aerial coverage of the areas shown in previous figures. Note the regional view and the information supplied by stereoscopic viewing (Lyon, 1981).

11.2.2. Maps

Maps are often the starting point in an analysis and often are used in the presentation of analyses results in reports. U.S. Geological Survey (USGS) and other original source maps may be used to obtain a variety of information. They can be used to collect: point-position information; topographic contour information; cultural or planimetric details such as roads, waterways, dwellings; or public land ownership boundaries. The maps can provide the starting point to form a variety of land cover themes or layers in a GIS.

Engineers often have large scale maps available for site planning and design. The scale is commonly: one inch is equal to one-hundred feet (1"=100') or one inch is equal to two-hundred feet (1"=200'). The contour interval is often one-foot of elevation for engineering design, and a two-foot interval for planning purposes. These products are especially useful in hydrological analyses, and they are created through surveying or a combination of surveying and photogrammetric technologies.

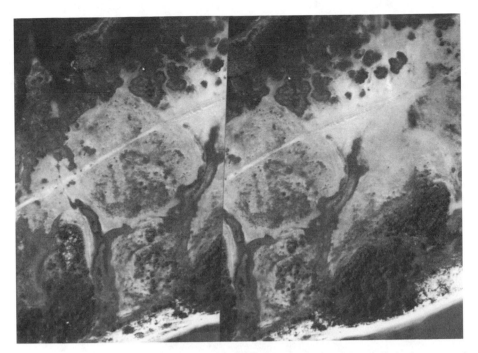

Figure 11.9.   Low altitude, small format (35mm) color infrared aerial photographs of coastal vegetation and land form. This photo was taken in the same mission as Figure 11.4.

11.2.3. Images

Images are defined as two-dimensional, pictorial representations of data. Images are often presented as non-photographic data, and they are usually the result of computer systems. The word "image" helps to identify and separate the photographic-image products that are so common, from the pictorial presentation of data or results from computer processing systems.

Computer data can be produced in either image form on a television monitor, on paper, or as output from a printer, or as a computer data result stored on a photographic image. Computer data format has the advantage of being an excellent medium to process hydrological and other calculations, and to also facilitate the interpretation of results through digital imagery display of data and results.

Landsat satellite data have been acquired since 1972. These data are in the form of computer image products, and much information can be developed from analyses of digital or image data.

Landsat Multispectral Scanner (MSS) images demonstrate some of the general capabilities of these data. Figures 11.10 through 11.13 are MSS images of the Portland, Oregon and south central Washington abutting the lower Columbia River (Lyon, 1983).

Figure 11.11 is a band-5 or red reflectance image of the area. The dark-toned areas are forest cover, and the white toned parts are urban and snow covered areas.

Figures 11.12 and 11.13 hold some interesting, general hydrological information. They are images in the infrared portion of the spectrum, and water resources appear

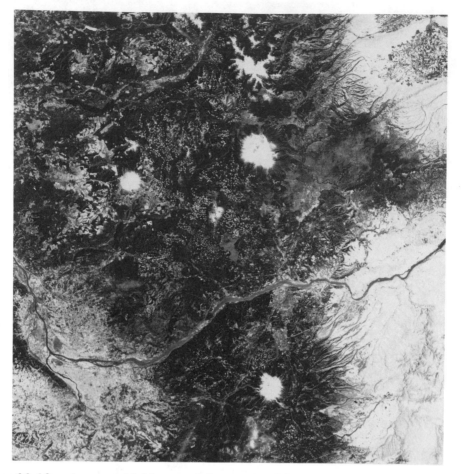

Figure 11.10. Landsat Multispectral Scanner (MSS) image of the Portland, Oregon area and surroundings.

dark black. Figure 11.12 is from May 5, 1975 and Figure 11.13 is from September 14, 1975. The fall image has smaller areas of black, such as a narrower black Columbia River, smaller lakes and reservoirs, and narrower black, general wetland areas found along the Columbia at Sauvie Island, OR and Ridgefield, WA. The extent of snow cover (white) on Mounts Hood, OR and Adams and St. Helens, WA can be identified.

11.3. PARTS OF THE SPECTRUM AND RADIATION CHARACTERISTICS

Remote sensing and products derived from remote sensing supply measures of light. Light or electromagnetic radiation emanates from the sun and is propagated through the atmosphere, striking and interacting with the earth's surface. Subsequently, the light returns through the atmosphere and is ultimately measured by the sensor above the earth. The study of many of these phenomena is remote sensing.

Figure 11.11.   Landsat MSS image of Portland, Oregon.

There are many valuable portions of the spectrum that can be sensed. In particular, water characteristics are identified using visible, near and thermal infrared, microwave or radar, and sonar sensors. The selection of the appropriate portion of the spectrum and sensor is based on the application and the cost of using the sensor.

There are a variety of types of light and these are usually measured based on the perceived information that can be obtained about a feature such as water. Hence, only a portion of the total field of electromagnetic energy is measured and this is called a spectral measurement.

11.3.1. Visible

The visible portion of the spectrum is found approximately from wavelengths between 0.3 μm and 0.7 μm. A micrometer, μm, is 1 x 10-6 m. This is the part of the spectrum that the human eye is sensitive. Unsurprisingly, it is the same part that black and white and color film sense to produce photographs that capture images in the ways human see them.

Figure 11.12.   Landsat MSS image of Portland, Oregon.

11.3.2. Infrared

The infrared portion of the spectrum is just beyond what a human may see. The "reflected infrared" is found approximately between 0.7 μm and 1.4 μm, and the "middle infrared" is found approximately between 1.5 and 2.8 μm. Figure 11.9 is an example of a color infrared image (CIR) which is sensitive to the green, red, and near infrared portions of the spectrum. Figures 11.11-11.12 are near infrared images from the Landsat Multispectral Scanner (MSS) sensor.

11.3.3. Thermal Infrared

The thermal infrared (TIR) is a measure of the heat emitted from features. TIR is found approximately between 3.0 and 14.0 μm, and it provides very good information on features resulting from their emission of heat. All features emit thermal infrared radiation, and as such they can be sensed both night and day and in any season, depending on cloud cover. Sensors that record the passive emission of light energy from a feature are called passive sensors.

11.3.4. Radar

Radar works in the microwave region of the spectrum, and has wavelengths that are relatively long compared to the visible and infrared. Radar systems generally operate between millimeter wavelengths, and wavelengths 24 centimeters and longer. Radar systems generate their own microwave wavelength electromagnetic radiation and can be used during night or day, and in most weather. Sensors that create their own radiation are called active sensors.

11.3.5. Radiation

One of the greatest capabilities of remote sensing is the fact that light can be measured or modelled in a quantitative manner. Photointerpretation is inherently qualitative, though the photographic process may be the subject of quantitative analyses. Remote sensor data are often digital as they come from the sensor, and can be processed in the same fashion as any digital product. With these characteristics it is possible to model the characteristics of light and sensor, and predict the behavior of measurement variables in experiments.

11.3.6. Relative Units

In many applications using measurements or calculations it is desirable to work in relative units. The relative units are usually simple ratios of some absolute measure of incoming radiation and some portion of the outgoing radiation.

Upon striking a surface, the incoming radiation is proportioned into three components. The sum of the components is equal to 1.00. The light components sum to 1.00 because energy is conserved as per the First Law of Thermodynamics. The upwelling component is known as reflectance, or the light that upwells or is returned from the surface. Radiation also travels through materials and this concept is called transmittance. Radiation that is retained within the material is the absorbance component.

The general equations for these relative units are:

$$\text{Reflectance} = \rho = \frac{\text{Radiation off of the surface of the material}}{\text{Radiation incident on the material}} \quad (11.1)$$

$$\text{Transmittance} = \tau = \frac{\text{Radiation through the material}}{\text{Radiation incident on the material}} \quad (11.2)$$

$$\text{Absorbance} = \alpha = \frac{\text{Radiation absorbed within the material}}{\text{Radiation incident on the material}} \qquad (11.3)$$

The relative units are calculated as ratios of the radiometric units measured. Commonly, one will ratio the irradiance or radiance, though any two equivalent units can be employed in the ratio.

Use of the above and other measurement and modeling approaches allows the evaluation of remote sensor experiments. These elements can be combined to study the response of materials to light energy, and to estimate the quantity of light that can be measured by remote sensors during an experiment.

Light is often measured or evaluated in the visible and near infrared using the units above. Light is also evaluated using absolute units of energy per unit time or rate, known as flux. It is also measured as a rate per unit area or irradiance, or as a rate per unit area per unit volume or radiance. These units and their application are discussed elsewhere, and they are beyond the scope of this effort.

11.4. DATA TYPES AND DATABASES

11.4.1. Data Types

Remote sensors and GIS technologies organize data according to two, general protocols. Remote sensors collect their data in grid cell or raster format, and GIS technologies process data in raster or vector form, or both, as needed.

11.4.2. Raster

GIS and remote sensor data are stored in files using one of two common methods, raster or vector. Raster storage is grid cell form and grid cells "build up" the image. Hence, a given area or polygon will be composed of a number of cells that have a certain area. The grid cells are referred to as "rasters," or as picture (pix...) "elements" (...els) or "pixels."

11.4.3. Vector

The other common way to store GIS data is in coordinate or "vector" form. A vector data set is stored as an x,y position for a point, two x,y positions and a line for a edge or curve vector, and/or a number of x,y points connected by line segments to form a polygon.

11.4.4. Attributes

An attribute is a characteristic or value of a variable stored in a GIS file. Depending upon the GIS software, one or several characteristics or attributes can be stored for a given point or area. Attributes can be almost anything. Examples for a given point or area could include: elevation; ownership; size or concentration; grid or map position; and/or a whole series of characteristics such as those that pertain to a given application.

11.4.5. Databases

A variety of data may be integrated and analyzed in the assessment of hydrological features and processes. This is truly a great capability of Geographic Information Systems and remote sensing. One often can access collections of data or databases, and obtain needed information. These databases are usually less expensive than the field work and data processing required to develop a custom database (Johnson and Goran, 1987). The use of existing data can expedite analyses, help focus field sampling activities, and generally ease the process and potentially lower costs of analyses.

An important thing to remember about data sources and databases is that they are inherently inexpensive. On a relative basis when compared to the costs of actual field sampling data, these sources can be acquired at lower relative cost. At a minimum, these sources can provide information for a "reconnaissance level" analysis before the more expensive and more detailed field evaluations are conducted (Lyon, 1993). The data eases the planning process, and allows the investigators to "come up to speed" quickly and at minimal, relative cost.

The implementation of existing databases is vital to most projects for the reasons above. In measure to their importance, the sources for these data and databases are provided in Appendix E. Also included are addresses where to write for the data of interest.

11.4.6. Aerial Photo Databases

Aerial photos of features of interest are acquired constantly to support science and engineering activities. These photographs are usually saved after the initial project, and may be later obtained from a variety of sources. Historical aerial photographs are particularly valuable as they are a record of a variety of conditions including general conditions of hydrology and soils (Lyon and Greene, 1992). They can be obtained and used in a variety of analyses and applications (Lyon and Drobney, 1983; Lyon, 1987; Williams and Lyon, 1991).

Aerial photographs can be obtained from a number of sources. They include: the USGS, the USDA, NOAA, local aerial photography firms and other valuable sources described in Appendix E. These photos have the added attraction of being relatively lower cost items. A paper, contact print of a nine by nine inch photo is between $3 and $8.

11.4.7. Digital Elevation Model (DEM)

DEMs are point elevation data stored in digital computer files. These data consist of x,y grid locations and point elevation data, or z variables. They are generated in a variety of ways and for a variety of map resolutions or scales by the USGS and others. The point elevation data are very useful as elevation data for input to a GIS. The data can be further processed to yield important derivative products, including digital maps of slope or aspect. With more detailed image processing, stream channels and sub-basin information can be derived. Stream channel and/or other network information can be the data input for more complex analyses (Argialas, Lyon, and Mintzer, 1988).

DEM data are available from the USGS, and local or regional sources can be located from the addresses provided in Appendix E.

A novel use of DEM data and satellite remote sensor data is the video "fly by" or flight simulation products. Examples include the popular "fly by" movies, such as "Los Angeles—the Movie" by the Jet Propulsion Laboratory.

11.4.8. Digital Terrain Model (DTM)

DTM data are an additional product created by a number of groups. DTM data provide digital topographic contours at a variety of scales.

DTM data are available from private sources, the USGS, and local or regional sources can be located from the addresses provided in Appendix E.

11.4.9. Digital Line Graph (DLG)

DLG data provide map-like presentations of road networks, stream courses, and other attributes for many USGS quadrangle areas. The data make very nice "overlays" of remote sensor or other map data, in addition to providing important details on the location of road and road types, and streams.

These DLG data are available from the USGS and from private sources, some of which can be located from addresses provided in Appendix E.

11.4.10. TIGER Data

Topologically Integrated Geographic Encoding and Referencing or TIGER data are demographic data available from the U.S. Department of Commerce, Bureau of the Census. These data are by census track and provide information on census-related issues. They are available in digital form, and can be very helpful when human and census-related data can assist in a hydrological analysis.

TIGER data are available from the Bureau of Census. TIGER data may also be obtained through vendors in either original or enhanced forms.

11.4.11. Satellite Data Sources

A number of satellite sensors acquire data over the earth's surface. One can learn about these sensors, their value and limitations, and make use of their data in hydrological evaluations.

A good way to become familiar with data types, and availability of data, is to "browse" data files of their characteristics. One such approach is the browsing of "metadata" files describing available satellite images.

The USGS and its EROS Data Center archive and maintain metadata information on satellite data. One can evaluate their holdings by a number of methods including use of the Internet World Wide Web (WWW) and a program such as Mosaic. The USGS Global Land Information System (GLIS) system stores descriptive characteristics and decompressed images of satellite sensor data sets. It also has information on improved data sets, such as those used in global change research program efforts.

One such program creates databases of moderate resolution satellite sensor data, and creates images of biomass during the year. The Advanced Very High Resolution Radiometer (AVHRR) satellite sensor of NOAA collects data in the red, near infrared, and thermal infrared portions of the spectrum. Use of a ratio of the red and infrared light make evident green biomass distribution of vegetation communities across the landscape. These products are created every two weeks from composite coverage during that period. Periodically, these products are made available through the internet and by ordering CD-ROM products from the EROS Data Center. These products may be browsed through GLIS.

North American Landscape Characterization (NALC) is another example of a value-added database of satellite sensor data.

11.4.12. North American Landscape Characterization (NALC)

NALC is a multiple year and agency project to develop current and historical Landsat Multispectral Scanner (MSS) data sets for evaluation of continental land cover and change in land cover (Lunetta, Sturdevant, and Lyon, 1993; Lunetta et al., 1993).

The goals of NALC include developing MSS data for the North American continent, correcting and packaging the data, analyzing the data sets for change in land cover, and using land cover to estimate standing stocks of carbon in biomass form.

These data and results of analyses will be used to address science issues identified by the U.S. Global Change Research Program (GCRP), and the work is being conducted by GCRP members including the USEPA, USGS and NASA.

The data sets are available as "triplicates" of MSS scenes from the USGS EROS Data Center. A given triplicate will contain three Landsat scenes from similar seasons during the epochs of 1970s (1973 $\pm$ one year), 1980s (1986 $\pm$ one year) and 1990s (1991 $\pm$ one year). The scenes are produced from the same ground area, and with geometric corrections and ground registration of coordinates. The original radiometric data are provided in the above form using standardized methods (Lunetta et al., 1993). Additional files include DEM data for the scene, and "housekeeping" files of pertinent information.

Information about NALC data products may be obtained from User Services, EROS Data Center, Sioux Falls, SD 57198 (Appendix E).

Details concerning image products supplied by the EROS Data Center may be examined or browsed using internet capabilities. Use of the EROS Data Center program GLIS, Global Land Information System, and internet programs such as World Wide Web (WWW) allow the user to evaluate data sets. The GLIS program describes many data sets in general terms or variables. The "meta" level of detail in data sets allows the user to: look at compressed versions of the original images; read information on data characteristics; obtain ordering information; and to collect appropriate documentation files from reading.

Addresses and documentation for GLIS may be obtained from the EROS Data Center.

11.4.13. Other Map Data

The power of computers allows one to capture data from any media and present it in computer format for processing. From a GIS perspective, digital entry allows for a whole variety of data sources to be integrated into the system for potential processing and analysis.

11.4.14. Digitized Data

There are several ways to enter data into the computer, and many of these are well known to the reader. The input of maps or map-like products can be simplified by "digitizing" the locations of points and the shapes of polygons on the map. In this manner, whole map details can be input to the computer in x,y format and "tagged" with attribute information as needed.

Digitizing is usually completed using a tablet, or digitizing instrument. The map is affixed to the tablet, and is registered absolutely. This is done by digitizing the location of corners or other located marks on the map in x,y digitizer coordinates, and recording the same in absolute coordinates such as State Plane Coordinates or Universal Transverse Mercator (UTM).

11.4.15. Scanned Data

Another successful method is to incorporate map or figure data into the computer is to "scan" the product with a video scanning instrument. These instruments capture all the details on the map or map-like product and stores them as a raster file. This method of data entry is particularly useful if the input product is a photo or image, or when the details presented on the input product are all required for the application project.

11.4.16. CD-ROM Data

CD-ROM is an acronym for Compact Disk-Read Only Memory. The compact disk is a storage device of great storage capacity. The Read Only Memory indicates that the stored information can only be read, and the medium does not allow the writing of additional data or other alterations of the CD.

A great many databases are now available on CD-ROM. This is a valuable method to distribute large data sets to many users at low cost. Examples include: USGS EROS Data Center AVHRR and biomass images for the U.S., the North American continent, and Alaska; NOAA-National Weather Service Airborne Hydrology Program (Minneapolis, MN) annual measurements of snow moisture equivalent by airborne gamma and snow course and SNOTEL measurements and AVHRR snow areal extent images; SSM/I passive microwave data and snow temperatures for North American from NOAA-World Data Center-A in Boulder, CO; Digital Chart of the World from the Defense Mapping Agency; and a number of others.

11.5. REMOTE SENSING CHARACTERISTICS OF WATER

Water has a variety of known characteristics and reactions to light that facilitate its evaluation. These characteristics are based on the portion of the spectrum used to measure the water body, and on the physical and chemical reactions of water to that portion of the spectrum.

Most materials or features including water react to light in one to three ways. For example, water is well known for absorbance of light energy and it is often much darker in the visible and near infrared portion of the spectrum due to the absorbance.

Water also reflects light from the surface (Figure 11.8) and from the water volume. Surface reflectance or sunglint can be seen near the water, or on aerial photos as a very bright or white area. Reflectance from the water volume upwells to the surface from the water itself or the bottom sediments, and it supplies the blue and blue-green color of the water.

Water also transmits light from the surface to the bottom and back. Transmittance of light serves to illuminate the water volume or column. In particularly clear waters, one can observe the bottom of a lake or ocean at 60 feet or more in depth due to the transmittance of light from the surface to the bottom and its reflectance back.

The First Law of Thermodynamics says that energy is conserved, and this is true for light energy. The incident quantity of light upon a feature such as water is portioned into reflectance, absorbance, and transmittance, and the sum of the three is equal to the total of the incident light (Equations 11.1–11.3). The same phenomena of absorbance, reflectance, and transmittance are found to a certain extent in water and water-borne materials, as well as other familiar materials like snow, concrete, and wood.

11.6. APPLICATIONS

11.6.1. General Characteristics of Applications

Applications of these technologies to hydrological problems has proven to be valuable. The "landscape-scale" of hydrology requires methods to gather spatially distributed information. The problems also require repeated sampling of the variables of interest to acquire information over large areas, and for the integration of all these data.

The costs and logistics of these actions can be high, and work is usually constrained by available resources. We need to develop more efficient technologies to gather spatially distributed information, hence the interest in remote sensor and GIS technologies (Lyon, 1993).

11.6.2. Planning

As one might think, the planning of scientific and engineered projects is a very appropriate application of remote sensor and GIS technologies. These technologies allow the collection of spatially distributed data, and these approaches can facilitate the analysis of image or spatially distributed data.

These approaches also allow one to engage in "scenarios" or in repetitive evaluations of project conditions. After the evaluations are made, the results can be produced in tabular form and in map or graphical forms. These capabilities can greatly facilitate any planning activity. The implementation of many of these technologies in business, government, and university arenas is an example of their widespread application and utility.

11.6.3. Site Determination

The identification and selection of a site for a project is a type of planning with special value to engineers who work to identify the optimal locations for a given construction project. Examples could include: siting a landfill; siting a raw materials processing plant in a good location between source materials and point of sale; optimizing the location of an office operation as compared to the adjacent market; and establishment of general site locations and addressing the information requirements of zoning and environmental permitting activities.

An important advantage of using GIS in siting are the visualization capabilities (Ervin, 1992; Garcia and Hecht, 1993). One can develop images that show the landscape, the engineered project, and related conditions (Green et al., 1993). The computer manipulation of the image and model allow the user to judge from visualization the "look" of the final project (Alber, Dobbins, and Buja, 1991). The repetition of different scenarios allows users to iteratively evaluate project and landscape conditions (Fels, 1992).

11.6.4. Management

To manage a facility or a resource requires information from monitoring activities. Water is constantly changing in quality and quantity, and day to day information as to its characteristics is vital (Lyon, 1993; Douglas, 1994; Verbyla, 1995). Hence, a variety of problems and a variety of applications may be addressed.

11.6.5. Best Management Practices

Best Management Practices (BMP) is the identification and implementation of management practices that minimize harm to the environment. BMP is a governmental term that encompasses a variety of activities that reduce and localize environmental problems. BMP has been applied to a number of commercial activities including changes in management practices that reduce erosion, pollution and other non-point sources of pollution from agricultural, mining, and manufacturing activities (Thenkbail, Ward, and Lyon, 1994).

Remote sensor and GIS technologies have a number valuable applications in the BMP arena. They can help to: identify the problems; monitor their influence and concentrations; map their locations; model the physical and chemical processes using GIS databases; and certify progress by determining the resulting impact of BMP as compared to past conditions.

11.6.6. Water Resource Applications

There is a huge variety of water resource applications of remote sensor and GIS technologies (Lyon, 1978; Lyon, 1979; Lyon, 1980; Lyon and Olson, 1983; Lyon et al., 1988; Lyon, 1989; Lyon and McCarthy, 1995). The limit on the application of these technologies is that of the capabilities of scientists and engineers. The key to successful and lower relative cost applications is knowledge of the problem at hand and of the capabilities of remote sensor and GIS tools.

The hydrology applications of remote sensing and GIS technologies are numerous and are constantly growing (e.g., Meyer et al., 1992; Mertz, 1993; Thenkbail, Ward, and Lyon, 1994; Lyon and McCarthy, 1995). This is not surprising because these are technologies with great applicability to spatial analysis and to the landscape scale of problem.

Many hydrological applications require land cover or water type information. This is because land cover is an important variable that influences many hydrological processes. These issues and processes include: erosion; storm runoff; ground water recharge; storm hydrograph characteristics; permeability; and others.

Analyses of hydrological characteristics also benefit from data being collected and summarized by watershed. A computer-based method allows data to be stored by an irregular boundary such as a watershed. It also allows the integration of hydrological characteristics for watershed or sub-watershed areas, which is very useful in analyses.

11.6.7. Quantity

Remote sensor and GIS technologies have been used to great effect in the measurement or prediction of water quantities. Usually these applications make use of land cover or topographic information. They may also utilize measurements of water or snow/ice directly via remote sensing.

The prediction of snowmelt runoff volumes has been improved by the use of remote sensor data. Several experimental and operational programs have utilized weather satellite data such as AVHRR and GOES-8 data, and airborne measurement to quantify the snow-moisture equivalent in target watershed snow packs (Carroll, 1988).

11.6.8. Quality

Water quality characteristics can be nicely measured from remote sensor technologies. The water is fairly transparent and absorbing, and as such it will exhibit characteristics of the materials that are dissolved or suspended in the water. This makes the identification and measurement of concentration of materials fairly simple (Hatchitt and Maddox 1993).

A number of variables in water have proven to be measured by remote sensor technologies. Some variables are measured directly and demonstrate a very good linear response of light reflectance or absorbance to concentration of the variable.

Surface-suspended sediment concentrations can be measured in a linear fashion over concentrations from 0–25 mg/l to as much as 600 mg/l and more. Chlorophyll "a" also shows good linear responses in concentrations of less than one mg/l to more than 25 mg/l and possibly upwards.

GIS technologies can greatly facilitate the storage and analysis of measurements. Many water quality and sediment data sets have been developed over the years and perhaps in a number of data collection "campaigns." The differences in sampling can be characterized with attribute entries in GIS files. The sampling points maintain their unique characteristics, yet they can then be evaluated with GIS technologies.

GIS technologies allow a number of products to be created and models to be run in support of hydrological applications (Lyon, Lunetta, and Williams, 1992). Products include digital files and hardcopy image maps of single variables, variable combinations, thematic maps of drainage and watershed variables, and similar products.

Complex products can be generated to present results of statistical or deterministic models. These products would include: visualizations of siting or management scenarios; land cover and land cover change maps; tabular results of statistical analyses; and results of model simulations.

11.6.9. Non-Point Sources

The addition of materials in water results in a reflectance from the water body that is a composite of the individual materials. Water that holds non-point materials such as suspended sediments and/or high concentrations of other pigments can be evaluated quantitatively using on-site sampling and remote sensor sampling (e.g., Lyon et al., 1988; Vieux and Needham, 1992; Mertz, 1993).

The problems associated with non-point sources of pollution can be addressed with remote sensor and GIS technologies. The utility of the data and technologies is due to: the size and distribution of the problem; the need for quantitative assessments or inventory of resources to manage; and the widespread sources of the problem.

11.6.10. Erosion Studies

Erosion is a widely distributed problem that has many sources. To identify sources, and to inventory, manage, and improve problem areas requires a variety of measurement technologies. Remote sensor technologies can supply data on locations of sources, and can be used to monitor conditions of sources and identify downstream impacts.

Sources of erosional products are usually bare earth areas. The location, extent, and temporal characteristics can be interpreted from aerial photographs or other appropriate-scale remote sensor products. Erosional areas can be identified by the light tone, or highly reflective, bare soil conditions (Lyon, 1987). These light areas are in contrast to relative dark toned areas of vegetation covered areas, and dark areas of clear water.

The downstream impact of erosional products can also be measured. Suspended sediment makes water highly reflective or lighter toned. High level concentrations of suspended sediments are the result of erosion, and are a non-point source of pollution. Suspended sediment conditions can be evaluated from a series of images, say from the Landsat satellite system. Quantitative assessments can be made if on-site measurements are taken in close temporal agreement with satellite data overpasses (Lyon et al., 1988).

11.6.11. Wetlands and Permitting

The wetland issue and development of property have resulted in controversy. This renewed focus is on the identification and avoidance of activities that may alter or harm wetlands. Now, wetlands have taken on a vital position in any activity that changes the landscape or in activities that seek to manage the landscape for various purposes (Lyon, 1993). It is necessary to identify, delineate, map, and store location and wetland type information in planning and management-oriented databases.

The wetland permitting process can be greatly assisted by the use of remote sensor and GIS technologies (Lee, 1991; Lyon, 1993). A key element of wetland delineations is the mapping and inventory of wetland areas (Lyon, Drobney, and Olson, 1983; Lyon, 1993; Mitsch and Gosselink, 1993). Remote sensor technologies such as aerial photographs, topographic maps, and engineering style maps are vital for documenting the size and location of wetlands (Lyon and Drobney, 1984; USACE, 1987; Lee, 1991; Williams and Lyon, 1991; Lyon and Green, 1992; Lyon, 1993; Lyon, 1995; Lee and Marsh, 1995). A combination of aerial photos or remote sensor data from many dates allows one to examine general hydrological conditions (Lyon, 1981; Lyon, 1993).

11.6.12. Reservoir Siting

The siting of a reservoir has much the same work requirements as siting any constructed project. However, remote sensor and GIS technologies can be particularly useful in the development of small or large reservoir impoundments. This is because the size of the watershed is important to calculations of runoff. The land cover in the watershed may also be important to the retention of runoff, or to the contribution of sediment from erosion and the filling of the reservoir.

11.6.13. Monitoring

The monitoring of a structure or management area is an important consideration. Information can assist in developing management scenarios, identifying problem conditions before a hazard exists, measuring the impact of management practices, and general inventory of resources. A potentially lower, relative cost method than traditional intensive field sampling is the use of remote sensor technologies (Lyon, McCarthy, and Heinen, 1986; Lyon et al., 1987; Lyon and Khumwaiter, 1989; Lyon, Lunetta, and Williams, 1992; Green et al., 1993; Ramsey and Jensen, 1995).

11.6.14. Hazardous Waste

A number of remote sensor and GIS technologies have been used in the identification, characterization, and management of hazardous waste sites (Lyon, 1987; James and Hewitt, 1992; Pickus and Hewitt, 1992). Often the use of remote sensor technologies helps to identify potential problems or educates the user as to the prevailing conditions of the feature or landscape (Lyon, 1987). These technological methods can also be generalized in their application to issues related to landfills and siting of these facilities. An example is the use of visualization technologies to help design the site (Garcia and Hecht, 1993).

To implement a remote sensor technology for monitoring requires knowledge of the problem, the characteristics that can be measured remotely, and a plan for collecting the requisite data. The budget available to support the effort is very important. It can be influential in the selection of the methods and instruments to be used.

11.6.15. Global Change Issues

Global change issues are at the forefront of new questions scientists and engineers must evaluate. To manage our cumulative activities such that global environments are maintained in the current or in an improved condition is very challenging. It is particularly difficult to deal with these complex issues. In the absence of data for analyses and in the absence of programs to collect such data, the scientist and engineer are at a disadvantage.

The global or regional scale of the problem requires thoughtful measurement and analysis technologies. The advent of remote sensing allows a variety of data to be collected over large areas of the earth's surface. GIS technologies offer the capability to organize huge quantities of measurements, and facilitate their statistical analysis and allow model simulations.

Hydrological issues can be appropriately addressed using remote sensor and GIS technologies. Water influences global change in the form of snow, runoff, precipitation, and storage of chemicals in forms of water. In particular, the ocean and large lakes are sources and/or sinks of carbon dioxide, methane, and nitrogen compounds.

Several projects now address global change issues. Examples include the Landsat and other Pathfinder series of projects. Their goals are to examine the value of future Earth Observation Systems (EOS) remote sensor products using contemporary products. An important example is the Landsat series of Pathfinders (Lunetta, Sturdevant, and Lyon, 1993) including North American Landscape Characterization (NALC).

These NALC triplicates form the basis for many potential analyses. Being uniform in execution of known characteristics, they can be applied in a number of hydrological applications. It is hoped that they can form an initial data set for GIS analyses, and foster the use of remote sensor data in global and regional change studies.

REFERENCES

Albers, B., J. Dobbins and K. Buja, 1991. Planning in paradise. Geo Info Systems, November–December, p.27–38.

Argialas, D., J. Lyon and O. Mintzer, 1988. Quantitative description and classification of drainage patterns. Photogrammetric Engineering and Remote Sensing, 54:505–509.

Bolstad, P. and J. Smith, 1992. Errors in GIS, assessing spatial data accuracy. Journal of Forestry, November, p.21–29.

Carroll, T., 1988. Airborne gamma snow and soil moisture equivalent measurements. National Weather Service, Minneapolis, MN.

Congalton, R. and K. Green, 1992. The ABCs of GIS, an introduction to geographic information systems. Journal of Forestry, November, p.13–20.

Douglas, W., 1994. Environmental GIS, Applications to Industrial Facilities. CRC/Lewis Publishers, Boca Raton, FL, 144 pp.

Ervin, S., 1992. Integrating visual and environmental analyses in site planning and design. GIS World, July, p.26–30.

Falkner, E., 1994. Aerial Mapping Methods and Applications. CRC/Lewis Publishers, Boca Raton, FL, 352 pp.

Fels, J., 1992. Viewshed simulation and analysis: an interactive approach. GIS World, July, p.54–59.

Garcia, J. and L. Hecht, 1993. GIS improves visualization, evaluation capabilities in superfund cleanup. GIS World, February, p.37–41.

Green, K., S. Bernath, L. Lackey, M. Brunengo and S. Smith, 1993. Analyzing the cumulative effects of forest practices: where do we start ? Geo Info Systems, February, p.31–41.

Hatchitt, J. and G. Maddox, 1993. Using DRASTIC methods to monitor the quality of Florida's groundwater. Geo Info Systems, January, p.42–46.

James, D. and M. Hewitt, 1992. To save a river, building a resource decision support system for the Blackfoot River drainage. Geo Info Systems, November–December, p.36–41.

Johnson, M. and W. Goran, 1987. Sources of digital spatial data for Geographic Information Systems. US Army Corps of Engineers, Construction Engineering Research Laboratory, Technical Report No. N-88-01, Champaign, IL, 33 pp.

Kessler, B., 1992. Glossary of GIS terms. Journal of Forestry, November, p.37–45.

Lee, C. and S. Marsh, 1995. The use of archival Landsat MSS and ancillary data in a GIS environment to map historical change in an urban riparian habitat. Photogrammetric Engineering and Remote Sensing, in press.

Lee, K., 1991. Wetland detection methods investigation. U.S. Environmental Protection Agency, Environmental Monitoring Systems Laboratory, Report No. 600/4-91/014, Las Vegas, NV, 73 pp.

Lunetta, R., J. Sturdevant and J. Lyon, 1993. North American landscape characterization (NALC), Research Brief. USEPA EPA/600/S-93/0005, Environmental Systems Monitoring Laboratory, Las Vegas, NV, 8 pp.

Lunetta, R., J. Lyon, J. Sturdevant, J. Dwyer, C. Elvidge, L. Fenstermaker, D. Yuan, S. Hoffer and R. Weerackoon, 1993. North American landscape characterization (NALC), Technical Research Plan. USEPA EPA/600/R-93/135, Environmental Systems Monitoring Laboratory, Las Vegas, NV, 417 pp.

Lyon, J., 1978. An analysis of vegetation communities in the Lower Columbia River Basin. Proceedings of the Pecora Symposium on Applications of Remote Sensing to Wildlife Management, Sioux Falls, SD, p. 321–327.

Lyon, J., 1979. Remote sensing of coastal wetlands and habitat quality of the St. Clair Flats, Michigan. Proceedings of the 13th International Symposium on Remote Sensing of Environment, Ann Arbor, MI, p. 1117–1129.

Lyon, J., 1980. Data sources for analyses of Great Lakes Wetlands. Proceedings of the Annual Meeting of the American Society for Photogrammetry, St. Louis, MO, p. 512–525.

Lyon, J., 1981. The influence of Lake Michigan water levels on wetland soils and distribution of plants in the Straits of Mackinac, Michigan. Doctoral Dissertation, University of Michigan, Ann Arbor, MI, 155 pp.

Lyon, J., 1983. Landsat-derived land cover classification for locating potential Kestrel nesting habitat. Photogrammetric Engineering and Remote Sensing, 49: 245-250.

Lyon, J., 1987. Maps, aerial photographs and remote sensor data for practical evaluations of hazardous waste sites. Photogrammetric Engineering and Remote Sensing, 53: 515-519.

Lyon, J., 1989. Remote sensing of suspended sediments and wetlands in Western Lake Erie. In Kreiger, K., (Ed.), Lake Erie estuarine systems: issues, resources, status, and management. NOAA Estuary-of-the Month Seminar Series, publication No. 14, U.S. Department of Commerce, Washington, D.C, p. 124–147.

Lyon, J., 1993. Practical handbook for wetland identification and delineation. Lewis Publishers, Chelsea, MI, 157 pp.

Lyon, J., 1995. Wetlands, how to avoid getting soaked. Professional Surveyor, 15:16–19.

Lyon, J. and R. Drobney, 1984. Lake level effects as measured from aerial photos. Journal of Surveying Engineering, 110: 103-111.

Lyon, J. and R. Greene, 1992. Lake Erie water level effects on wetlands as measured from aerial photographs. Photogrammetric Engineering and Remote Sensing, 58:1355–1360.

Lyon, J. and I. Khumwaiter, 1989. Cropland measurements using Thematic Mapper data and radiometric model. ASCE Journal of Aerospace Engineering, 2:130–140.

Lyon, J. and C. Olson, 1983. Inventory of coastal wetlands. Michigan Sea Grant Program Publication, University of Michigan, Ann Arbor, MI, 35 pp.

Lyon, J., R. Drobney and C. Olson, 1986. Effects of Lake Michigan water levels on wetland soil chemistry and distribution of plants in the Straits of Mackinac. Journal of Great Lakes Research, 12: 175-183.

Lyon, J., R. Lunetta and D. Williams, 1992. Airborne multispectral scanner data for evaluation of bottom types and water depths of the St. Mary's River, Michigan. Photogrammetric Engineering and Remote Sensing, 58:951-956.

Lyon, J. and J. McCarthy, 1995. Wetland and Environmental Applications of Geographic Information Systems. CRC/Lewis Publishers, Boca Raton, FL, 340 pp.

Lyon, J., J. McCarthy and J. Heinen, 1986. Video digitization of aerial photographs for measurement of wind erosion damage on converted rangeland. Photogrammetric Engineering and Remote Sensing, 52: 373-377.

Lyon, J., K. Bedford, J. Chien-Ching, D. Lee and D. Mark, 1988. Suspended sediment concentrations as measured from multidate Landsat and AVHRR data. Remote Sensing of Environment, 25: 107-115.

Lyon, J., J. Heinen, R. Mead and N. Roller, 1987. Spatial data for modeling wildlife habitat. ASCE Journal of Surveying Engineering, 113: 88-100.

Mertz, T., 1993. GIS targets agricultural nonpoint pollution. GIS World, April, p.41-46.

Meyer, S., T. Salem and J. Labadie, 1992. Geographic information systems in urban storm-water management. ASCE Journal of Water Resources Planning and Management, 119:206-228.

Mitsch, W. and J. Gosselink, 1993. Wetlands. Van Nostrand Reinhold, New York, NY, 539 pp.

Pickus, J. and M. Hewitt, 1992. Resource at risk: analyzing sensitivity of groundwater to pesticides. Geo Info Systems, November-December, p.50-56.

Ramsey, E. and J. Jensen, 1995. Remote sensing of mangroves: relating canopy spectra to site-specific data. Photogrammetric Engineering and Remote Sensing, in press.

Thenkbail, P., A. Ward and J. Lyon, 1994. Landsat-5 Thematic Mapper models of soybean and corn crop characteristics. International Journal of Remote Sensing, 15:49-61.

Verbyla, D., 1995. Satellite Remote Sensing of Natural Resources. CRC/Lewis Publishers, Boca Raton, FL, 224 pp.

Vieux, B. and S. Needham, 1992. Non-point-pollution model sensitivity to grid-cell size. ASCE Journal of Water Resources, Planning and Management, 119:141-157.

Williams, D. and J. Lyon, 1991. Use of a Geographical Information System data base to measure and evaluate wetland changes in the St. Mary's River, Michigan. Hydrobiologia, 219: 83-95.

U.S. Army Corps of Engineers, 1987. Corps of Engineers Wetlands Delineation Manual. Technical Report Y-87-1, Department of the Army, Washington, D.C.

U.S. Army Corps of Engineers, 1992. Photogrammetric mapping. Engineering Manual, Washington, D.C.

Problems

11.1 Select an application such as building a small impoundment for making an evaluation of non-point sources of pollution. Make a list of the various remote sensor and GIS technologies and products that could be used. Identify one or more variables that can be extracted from these technologies and/or products, and how variables can assist in the analysis of the given application.

11.2 Write to several of the sources for image and map data (Appendix E). Ask for information on their products and services. Send a copy of a map with a point location identified by latitude and longitude, and request information on the available images or maps for the location.

11.3 Using the relative portion of reflectance and transmittance (equations 11.1–11.3), calculate the absorbance of a material with reflectance of 0.40 and transmission of 0.02.

11.4 Conduct a literature search on remote sensor and GIS applications related to a specific issue of hydrology.

11.5 Make a list of sources of data one can obtain to supply different variable files or layers in a GIS.

CHAPTER 12

Practical Exercises on Conducting and Reporting Hydrologic Studies

Stanley W. Trimble and Andy D. Ward

12.1. INTRODUCTION

In the previous chapters we have provided technical information that should prove useful to scientists and engineers who will conduct studies relating to environmental hydrology. We have provided information on individual principles and procedures and, in most cases, much of the basic data needed to understand the examples and problems presented in each chapter. However, when solving real-world problems you will quickly discover that little or no information is initially available, each problem has unique aspects, and often the questions to be answered are not well defined. Also, many hydrologic studies require knowledge of several different topics to develop a comprehensive solution.

In this chapter we present eight practical exercises and information on how to approach conducting a hydrologic study. Several of the exercises can be combined to develop comprehensive solutions to hydrologic issues on a watershed of interest to you. These exercises are designed to enhance and enrich your hydrologic experience by introducing you to some sources and procedures used by practicing hydrologists, especially those in the surface water and environmental fields. Most sources will be available in good university libraries. Keep in mind that potential employers seek people who have practical experience.

Like many scientists and engineers (including the authors) you have probably focused your academic interests on subjects relating to science, engineering, and mathematics. While this is very appropriate, you need to recognize that much of your life will be spent communicating. Effective oral and written communication skills are vital to being a successful scientist or engineer. We have therefore included in this chapter a brief outline of how to prepare a technical report.

12.2. CONDUCTING A HYDROLOGIC STUDY

In order to conduct a hydrologic or environmental study it will be necessary to: (1) define the question; (2) conduct a preliminary investigation; (3) undertake detailed planning; (4) conduct the detailed study; and (5) report the results of the study. The first four aspects are discussed here and an outline of how to prepare a report is presented in Section 12.3.

12.2.1. Define the Question

Before undertaking any study the following questions need to be asked and answered:

- Who is the client?
- What is the purpose of the study?
- What is the motivation for the study?
- What issues are being addressed?
- What resources are available?
- What are the deadlines?
- Are there specific methods that must be used?
- Are there any legal constraints?

It is important that each of these questions be addressed regardless of whether you are a junior scientist in an agency or a senior partner in a consulting company. Failure to ask appropriate questions might result in the wrong issues being addressed, an inappropriate expenditure of resources (high or low), inadequate results, missing important deadlines, legal problems, the presentation of results at an inappropriate level of complexity, dissatisfaction by the client, and perhaps even a failure to be paid for your work or the loss of your job!

As an outcome of asking these questions you may need to prepare a proposal requesting resources to conduct the work or you might simply prepare a statement which better defines the question. Regardless of the end outcome, it is important that you carefully document all discussions, communications, and requests to conduct work. Make sure you document dates and participants in discussions and communications of substance. Be professional and always adhere to high ethical standards.

12.2.2. Conduct a Preliminary Investigation

Once the question has been defined and authorization has been provided to proceed, it will usually be necessary to conduct a preliminary investigation in order to provide sufficient information to: (1) design the study; (2) select analytical methods, models, procedures; and (3) prepare a budget. You will need to:

- Become familiar with the system which is being studied.
- Assemble and evaluate existing data.
- Obtain as much information as possible.
- Develop a scope of work.
- Identify data requirements; data deficiencies; and cost resource requirements.
- Identify system boundaries and boundary conditions.
- Prepare a preliminary diagram of the physical system.
- Identify issues which inhibit your ability to conduct the study and/or need further study.

A reconnaissance of the site should always be undertaken. Before visiting the site it is important that you define the purpose of the visit and prepare accordingly. Careful planning is critical to the success of any investigation. Obtain as much background information as possible. This should include reading reports on previous studies; obtaining topographic maps and aerial photographs (if available); and obtaining information on the climate, soils, geology, and land uses. Talk to people who are familiar with the location and communicate with local, state, and federal agencies to determine if they have any ongoing or recently completed studies in the region. Also, ascertain if there are climatic and stream gaging stations in the vicinity of the location.

Before leaving for a site visit (or any other field work) prepare a checklist of equipment and materials that will be needed. Obtain or prepare written descriptions of each task that will be performed. Make sure each of the people participating in the field reconnaissance is aware of the objectives and what they will be required to perform—this includes you! Make sure suitable data acquisition materials have been prepared to document all findings during the visit. Consider possible things that might go wrong and prepare contingency plans before going to the field. This is particularly important if electronic equipment is being used.

A common error when performing field work is to leave too much to memory. Take the time to get organized before you leave. Check that all items work, electronic instruments are fully charged and all items have been loaded into the vehicle(s) being used for the field work. Make careful field notes and take photographs or videotapes wherever possible.

12.2.3. Undertaking Detailed Planning

After the Preliminary Investigation, select an approach based on the objectives; the required accuracy; how the data will be used; time and resources needed to obtain the data; boundary conditions and assumptions; and data deficiencies and assumptions. Prepare a conceptual diagram of the system; identify data requirements; and identify sources and methods to obtain the data.

You then need to develop a detailed scope of work which specifies the administrative organization of the study, all resource requirements, operating procedures and methods, a schedule of activities, and deadlines. Before proceeding with the detailed study, the detailed scope of work needs to be approved by your supervisor or the client. In some cases it might serve as a proposal which will be submitted for funding consideration. Make sure you fully understand the status of the scope of work and have written confirmation of procedure before undertaking further work. Often the scope of work will be modified by your supervisor or the client. Request to be part of any negotiations relating to a scope of work which you have prepared. You will be in the best position to explain to the participants what is being proposed. Also, you will learn firsthand the rationale for any proposed changes.

12.2.4. Conducting the Detailed Study

Conducting the detailed study will require careful planning. Make sure you understand your role, the objectives, what resources are available, the expectations of the client, and the deadlines. A job file should be prepared for each project. This will normally be divided into sections such as communications, expenses, a number of data files or databases including raw data and field notes, files of calculations and work that was performed, drawings that were prepared, and reports. All materials need to be carefully cataloged and archived. Most organizations will have well established procedures on how studies should be conducted and documented.

Most studies relating to environmental systems will either be used as the basis for a legal decision, such as the issuing of a permit, or have the potential to be used in some future litigation. We cannot overemphasize the importance of documenting all data sources, procedures, assumptions, and disagreements (if any) on procedures which should be used. Also, it is important that the documentation be securely archived for many years after the completion of a project. Legal action associated with design failures, loss of property and/or loss of life could occur more than 50 years after a project was undertaken.

12.3. REPORTING A HYDROLOGIC STUDY

While you still have the opportunity we would strongly recommend taking additional courses on technical writing, english, word processing, and public speaking. Once you have completed your formal education the opportunity to obtain expert knowledge and assistance in these areas will greatly diminish, although on the job experience will help you become a more accomplished communicator.

This author (Dr. Ward) performed very poorly in English classes and I am sure that at times some of you were appalled at the grammar, incorrect use of punctuation, and lengthy sentences that can be found throughout the chapters which I prepared. However, in preparing this book, most students and reviewers commented more on the clarity of the statements and how easy it was to read and understand the concepts being presented. Reaching this level of acceptability of your writing style requires following a number of basic rules, feeding information to the reader in small doses, using simple words, and being as concise as possible.

In this section we have focused on the steps and rules that should be followed when preparing a technical report or paper. Modern word processing systems will help you overcome any spelling limitations you might have. Most word processing systems also incorporate grammar checking capabilities. However, it has been the author's experience that these capabilities often recommend inappropriate or incorrect changes, require a sophisticated knowledge of grammar by the user, and are generally not suitable for everyday use.

Many organizations have their own style guidelines. It is important that you become familiar with guidelines used by your organization. The guidelines presented below are based on the author's experience in preparing or reviewing many consulting reports, technical papers, theses, dissertations, proposals, book chapters, letters, and memos. It is also based in part on Moore, et al. (1990).

The following steps are critical to the timely completion of a well-written technically correct report: (1) define the audience of the report; (2) find published reports that could serve as models; (3) prepare a topical outline and have it reviewed; (4) and prepare an annotated outline and have it reviewed. Also, send a copy to cooperator(s) for review, if appropriate.

You should begin writing parts of the introduction of the report during early stages of the project. Write, revise and edit the first draft. It usually takes more than one draft to write a report. Some parts of this book have been rewritten more than six times. Have the report reviewed concurrently by within-office and out-of-office colleague reviewers.

The major objectives in having a report reviewed are to:

- Ensure that the report achieves the goals stated in the project description.
- Ensure the readability of the report.
- Ensure the technical quality of the report.
- Evaluate the suitability of proposed publication media.
- Evaluate the effectiveness of the presentation.
- Correct errors and other deficiencies that could embarrass the author or your organization.

Respond, in writing and in an appropriate manner to all reviewers. Have the report edited again, depending on the extent of revisions after colleague review.

Publish the report as quickly as possible after completing the study.

12.4. REPORT CONTENTS

Most reports and technical papers should be divided into distinct sections. This book has been divided into chapters, and then each chapter has been divided into several numbered sections and on occasions into subsections. It is not always necessary to number the sections but if the report is long, numbering will facilitate cross-referencing and finding materials which are located in different parts of the reports. Generally, sections should have titles which are highlighted (underlined and/or in bold, and use some upper case letters), centered on the page, or located on the left hand side above the section. We have presented below a brief description of the main sections which might appear in a report or technical paper. Please note that sometimes several of these sections will be combined and often the results section will be divided into several separate sections.

12.4.1. Title

The title is a concise description of the subject of the report and, as such, has to convey to the reader the content of the report.

12.4.2. Author(s) and Publication Date

List all authors and the publication date of the report. This information will usually be presented on a cover page which includes the name and address of the organization who prepared the report.

12.4.3. Table of Contents

The table of contents should be prepared before you begin writing the report as it will serve as a useful outline. It should include the main section headings and their page locations. It might also include the headings and pages of subsections. The table of contents will often be followed by lists of the figures and tables. However, in short reports, this additional information is sometimes omitted.

12.4.4. Executive Summary

This is an optional section or independent document. Consideration should be given to preparing an executive summary if the main report is longer than 20 pages. The executive summary should not exceed a length of 5 pages. It should include all the sections (but not the subsections) of the main report. An executive summary should be free of equations, references, and appendices. Summary tables and graphic materials can be used but should not contain excessive detail.

12.4.5. Abstract or Summary

A well-prepared abstract tells the reader the basic content of the report. It should include the purpose, scope, results, and conclusions of the study. The use of citations, abbreviations, and acronyms should be avoided.

12.4.6. Introduction

The introduction should state why and where the study was conducted. A summary of the purpose and scope of the report should also be presented.

The introduction might also provide information on the organization of the report.

12.4.7. Background and/or Justification

This section should include the problem addressed by the study; the objectives of the study; a statement on who commissioned the study and when; and a statement of cooperation, if applicable. (This section is sometimes called **Terms of Reference**.)

The introduction and background sections are often combined into a single section.

12.4.8. Purpose and Scope

This section describes the purpose and scope of the report, which may differ from that of the overall study. For example, the report might only address one aspect of a much larger study which was outlined in the background section.

This section is often omitted and the purpose and scope is presented in the introduction.

12.4.9. Description of Study Area

Briefly discuss the location and size of the study area, its climate, physiographic, geologic, hydrologic, and/or hydrogeologic setting. In short reports this section is also often omitted and the information is presented in the introduction, while in other reports, separate sections might be devoted to attributes of the system such as the climate, hydrology, or hydrogeology.

12.4.10. Methods and/or Procedures

Material under this heading pertains only to methods. No data should be included. Theory used in the study might be included in this section. However, be brief and only include key information which will allow the reader to better understand how the results were established. If an extensive literature review was conducted, it might be included in a separate section. If equations are presented, they should be numbered and cited in a sentence prior to the equations. Be sure that all terms in equations are defined.

12.4.11. Approach

The approach differs from the "methods" by presenting the rationale behind the study and the manner in which the study was performed. However, the methods and approach sections are often combined into a single section.

Avoid providing a "blow by blow" account of actions by you and other participants in the study. For example, in describing a field reconnaissance you might state that a team which included a hydrologist, soil scientist, geographer, environmental scientist, etc. participated in the reconnaissance. Their names might also be listed. However, avoid statements which say I/we did this and then we did that. Instead provide specific information on the observation and measurements which were made, how they were made, and why they were made.

12.4.12. Previous Studies

This section presents previous studies that have been conducted at the site, or are reported in the literature, which provide information and/or knowledge which were used in the study. Sometimes this information is presented in the background section. If no previous studies have been conducted, a statement to this effect should be presented earlier in the report.

12.4.13. Results

This section presents data that have been developed to address the problems/objectives stated in the introduction. This is the most important part of the report. The results will often be presented in several sections. This section is a statement of facts resulting from the study. Do not include information from other studies, theory, or opinions.

Each section should build on previous sections. In one of the previous sections, specific objectives of the study should have been presented. Wherever possible organize the results in the same order that the objectives were presented earlier and specifically tie the results to the objectives.

Tables and figures should be used to summarize and illustrate important findings. Each table or figure: (1) should have a number; (2) must have a caption; (3) should stand alone without the reader needing to refer to other parts of the report to understand its meaning; and (4) must be mentioned in the text of the report. Figures and tables should be located in the body of the text shortly after they are first mentioned. Their use is not restricted to the results section. For example, a map or layout drawing of the site might be presented in the introduction or background sections. Illustrations of procedures and concepts might be included in the methods and procedures sections.

12.4.14. Discussion of Results

This section presents an interpretation of the results. The discussion should be related back to the objectives. It will often compare the reported results with previous findings reported in the literature. The results will also be related back to the methods and approach.

Uncertainties and assumptions which influenced the results should be discussed. Limitations on the application or usefulness of the results should also be presented.

Many reports and papers combine a discussion of results with the results section. Try to avoid this practice as it is easy for facts based on the study to become cluttered with discussion on findings from other studies.

Sometimes additional work might be performed to evaluate a particular aspect of the results. If this work is not relevant to the specific objectives of the study, it should not be included in the report. However, if the work evaluates the validity of a result, it might be presented either in the results section or in the discussion of the results.

12.4.15. Summary and/or Conclusions

Conclusions summarize final results and interpretations of a study. The section answers to the objectives stated in the introduction and focuses on significant findings. A summary is a restatement of all the main ideas presented in the report beginning with the introduction. Unless required by your organization or the client, it is strongly recommended that only a single summary or abstract be placed at the beginning of the report.

12.4.16. Future Work

This is an optional section which identifies further work that should be performed.

12.4.17. Acknowledgments

This is an optional section which acknowledges extensive assistance by people other than the authors. It might be placed either near the beginning or end of the report.

12.4.18. References

A list of literature cited in the report should be presented at the end of the body of the report. In a very long report or a book the references for each section are sometimes located at the end of each section. References are usually listed alphabetically based on the surname of the first author. Sometimes they are numbered in the order that they appear in the text and then are listed according to this number. A third approach where the title is presented first is illustrated in the eight practical exercises which start in Section 12.6.

A bibliography which includes literature in addition to that cited in the report might be presented. However, if a bibliography is prepared, it is important that the reader be informed that not all documents listed were cited in the report.

The references which are listed at the end of each chapter in this book provide an example of how you might organize your references. In this section we indicated that this outline of how to organize a report was based in part on Moore et al. (1990). This reference is:

Moore, J.E., D.A. Aronson, J.H. Green, and C. Puente. 1990. Report Planning, Preparation, and Review Guide. U.S. Geological Survey, Open-File Report 89-275, Reston, Virginia.

12.4.19. Appendices

It is often desirable to include appendices providing further detail on the scope of work, pertinent background information, details on methods and procedures, data, and secondary results. Each appendix must be cited in the report and are usually sequentially identified by the letters A, B, C, etc. It is important that only materials which are not necessary for understanding the body of the report be included in the appendices.

12.5. GENERAL GUIDELINES TO PREPARING EXERCISE REPORTS

Following this section we have included eight practical exercises. Particular attention should be paid to specific directions which are provided by your instructor.

These might include preparing a report based on the guidelines presented earlier. As a general rule we would recommend that in all class reports you:

1. Include exercise directions.
2. Give complete citations for all documents used, including maps. Maps should include scale, agency and sheet name.
3. Include photocopies of all appropriate maps. Make tracings if necessary to avoid cluttering. Do a professional job. Show all measurements that you have made or show how you made them.
4. Do a professional write-up in complete sentences. Tell what you did in such a way that others can understand and that you can understand years from now.
5. Be a good scientist or engineer. For locations you select, for example, give Section number, Township and Range.
6. Most exercises require graphing. For those with the competency, a computer spreadsheet of some type will aid matters greatly. Not only will this be easier and neater, but the computer can do many of the required calculations. Most environmental agencies and consulting firms would demand the use of this software, so now is a good time to start.
7. All reports are to be prepared as if you were doing it for a paying client. You must satisfy the client to be paid.

The only difference between the guideline presented above and a report that you might prepare for your organization and/or a paying client is that some of these materials might form part of the job file or might only be presented in the appendices of the report.

12.6. EXERCISE 1: PRECIPITATION (see theory in Chapter 2)

Introduction:

Precipitation is the most important hydrologic input. This exercise is designed to familiarize you with procuring, processing, and understanding precipitation data.

Part I. Time Trends

A. Go to your library and find the *Climatic Summary for the United States,* (U.S. Weather Bureau), QC/983/U58CL Vols. 1 & 2 (Library of Congress subject headings), which cover up to 1930, and the 1931-52 supplements. The 1951-60 supplement and the annual supplements for individual years provide more recent data. Alternatively, procure the *Hydrodata* CD-Rom from EarthInfo, Inc. This contains hydrologic data, including precipitation records, for stations throughout the United States and is updated frequently.

B. Select any station and extract at least 20 continuous years' of annual precipitation values:

1. Plot the data against time (see Chapter 2).
2. Calculate the mean and standard deviation and indicate these as horizontal lines on the graph. Assuming an adequate sample and a normal distribution, it is probable that annual precipitation in the future should be within ___ & ___ inches 68% of the time, between ___ & ___ inches 95% of the time, and between ___ & ___ 99.7% of the time.

The standard deviation, SD, is determined as follows:

$$SD = [\Sigma(x\text{-}x)^2 / (n\text{-}1)]^{1/2}$$

where x is each entry in the distribution and n is the = number of entries in the distribution.

3. Divide the standard deviation by the mean (SD/x). This is the coefficient of variation.
4. Calculate and plot a 5-year moving average (Chapter 2 and Figure 12.1, this exercise). Is there a time trend?
5. Calculate a trend line for the raw data. Does it show a time trend? How does it compare with the mean and the 5-year moving average (see Figure 12.1)?

Part II. Frequency and Magnitude

A. **Annual Means.** Using data from the station above and the procedure in Chapter 2, construct a frequency table of annual precipitation magnitudes. Use the graph matrix or make your own.

B. **Individual Events.** This exercise introduces two of the most valuable documents available to hydrologists, geomorphologists, and planners of all stripes. They are:

"Rainfall Frequency Atlas of the U.S., "U.S. Weather Bureau *Tech Paper* No. 40, 1961. QC/851/U58t.

"Two to Ten Day Precipitation...2–100 years... U.S.," USWB *Tech Paper* No. 49. QC/851/ U58t.

Exercises:

1. **Frequency Magnitude of Events for a Location.** Using one inch ruled graph paper, set up the ordinate as 2 inches of precipitation/inch and the abscissa as 10 years/inch. Mark the 1, 5, 10, 25, 50 and 100 year points. These points may be sublabeled as 90+%, 20%, 10%, 4%, 2%, and 1% probability, respectively. From the maps of the contiguous U.S., select any point locality of interest to you and graph the probable 30 minute, 1 hour, 2 hour, 6 hour, 12 hour, 24 hour, 2 day, 4 day, and 10 day values

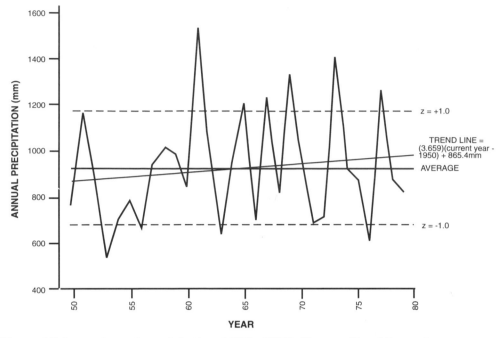

Figure 12.1.   Annual precipitation 1950–1979, Kansas City, Missouri.

with a continuous line (see Figure 12.2). Prepare the graph well enough to be presented in class and put your name on it. Write a short paragraph explaining the meteorological-climatological-landform conditions governing frequency and magnitude of precipitation events for your location.

2. **Seasonal Distribution.**   Inspect and understand Charts 52–54 of *Tech Paper* No. 40. For example, compare parts (Regions) 2 and 6. Why are they so different?

Part III. Spatial Distribution

A. The values for your graph are point values. Assume that you are working with a watershed of 140 mi^2. Using Figure 2.17 in Chapter 2, sketch a new 1-hour duration event curve on your graph for this area. Make this curve dashed to distinguish it from the point values (Figure 12.2).

B. Find U.S. Weather Bureau (now National Climate Center) Climatological Data QC/983/U58c. These reports give daily precipitation data for all U.S. Stations. While looking at these, find a fairly recent report and look at all the other good data included, such as class A pan evaporation.

1. Select a watershed with at least 5 weather stations (it might be well to be armed with a 1:250,000 topographic map of your area of interest). For an

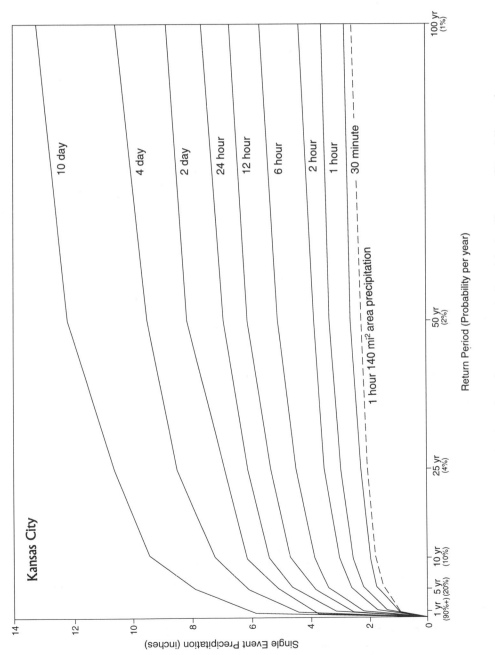

Figure 12.2.   Precipitation depth, duration, frequency relationship for Kansas City, Missouri.

excellent set of U.S. Basin maps at 1:500,000, see *River Basin Maps Showing Hydrologic Stations,* Interagency Comm. or Water Resources, Washington. Note on Hydrologic Activities No. 11,1961. Further, select some discrete storm event, i.e., one which is isolated in time and space. Compute the storm precipitation for the basin by (a) the arithmetic mean, (b) the Thiessen method, and (c) the isohyetal method.

For the last two methods, you will need to measure areas. This can be done by an overlay dot grid, or better by the use of a polar planimeter.

2. Draw decent diagrams of your endeavors. Write a paragraph comparing the three methods (for your case) telling which you think best and why.

12.7. EXERCISE 2: EVAPORATION (see theory in Chapter 4)

Part I. Annual Lake Evaporation

A. Find USGS *Water Supply Paper* 1838, "Reservoirs in the U.S." Select any U.S. reservoir of interest to you (the locational index is in a pocket in the back of the *WSP.* Please treat it with care). Find and record the surface area of the reservoir.

B. Find "Evaporation Maps for the U.S.," USWB *Technical Paper* 37, 1959. Read this paper; it is an important data source. From Plate 2, locate the average annual evaporation for your lake.

C. From Plate 1, determine the average annual class A pan evaporation at your reservoir location.

D. Estimate the annual lake evaporation from annual pan evaporation by use of the pan coefficient (Plate 3). Multiply the pan evaporation by the pan coefficient to get an estimate of the lake evaporation. Your number should agree with the figure from Plate 2.

E. What proportion of your total annual lake evaporation (in inches) should occur during the stress (hot) period between May and October (Plate 4)? You should be able to explain the distribution shown on Plate 4.

F. What is the standard deviation of pan evaporation (in inches) at your location (Plate 5)? Assuming this standard deviation is proportionately applicable to reservoirs, use the pan coefficient to estimate the standard deviation of annual evaporation from your reservoir in inches. Assuming you have a reasonably correct standard deviation of lake evaporation, future evaporation from your lake should be between ____ & ____ inches 68% of the time, between ____ & ____ inches 95% of the time, and between ____ & ____ inches 99.7% of the time.

G. If you were to map the coefficient of variation (SD/mean) for evaporation, how would the distribution vary over the U.S. (in very general terms)? Why?

Part II. Daily and Monthly Lake Evaporation

A. The nomograph on p. 2–4 of *Technical Paper* 37 allows the estimation of daily pan and lake evaporation for specified climatic elements. If you have weather data for a certain day and want evaporation, just plug in your data. You need average data if you want the average for a given month.

B. For average data, obtain the *Climatic Atlas of the U.S.* NOAA, 1974, ESSA, 1968 (QC/983/US7c). Data are on the following maps:

1. Mean daily air temperature (°F): pp. 1–24.
2. Solar radiation (Langleys/day): pp. 69–70.
3. Mean daily dew-point temperature (°F): pp. 57–58.
4. Wind movement (mph): pp. 73–74. You must multiply by 24 to get miles/day. Weather Service anemometers are located 22 ft above the ground and you need wind speeds for 2 (two) feet above the ground. Assume a roughness height of 0.1 ft.

C. The nomograph is entered at the upper left and you proceed in both directions to obtain daily evaporation. This figure is an average day of an average month. Use any month for this exercise, but July or January should yield more interesting figures.

Note 1: Your estimates of lake evaporation will probably be somewhat low if the lake is shallow and high if the lake is very deep.

Note 2: Actual ET rates are often quite similar to the lake values. This will depend, in large part, on vegetation type and cover and available soil moisture.

Part III. Estimating Actual ET Rates

A. The water balance equation is:

$$P = AET + Q \pm \Delta G \pm \Delta \theta$$

Where P is the precipitation depth, AET is actual ET, Q is runoff depth, ΔG is ground-water inflow or outflow, and $\Delta \theta$ is soil water change.

Transposing, we get:

$$AET = P - Q \pm \Delta G \pm \Delta \theta$$

Over long periods, ΔG and $\Delta \theta$ become negligible, so the equation reduces to:

$$AET = P - Q$$

This equation should properly be assigned to watersheds, but a reasonable estimate can be obtained for areas, especially in the eastern U.S.

B. Long term average precipitation can be handily obtained from *Climate and Man,* USDA Yearbook of Agriculture, 1941 from most state departments of natural resources. These data are also available on the EarthInfo CD-Rom. Find the average precipitation for the station nearest your reservoir.

C. Streamflow (runoff) can be obtained handily from "Annual Runoff in the Conterminous U.S.," USGS *Hydrologic Atlas* 212, 1965; or USGS *Hydrologic Atlas* 710, 1987. Extrapolate a value with reference more to the precipitation station than to the lake. What is the estimated AET? How does this compare with lake (fresh water surface evaporation) that you found?

D. Now, see Thornthwaite's map of PET Values (e.g., Strahler and Strahler, *Modern Physical Geography,* 4th ed., p. 169). This map has received wide, and often uncritical, acceptance and use since the '40s. It was compiled from the Thornthwaite PET equation which uses only temperature. Unlike the Blaney-Criddle equation, which you have used, the Thornthwaite equation does not consider vegetation, and like Blaney-Criddle, it does not consider humidity, radiation and wind. Keeping all this in mind, compare and contrast your estimated AET, Thornthwaite's PET, and lake evaporation. A short paragraph will suffice.

12.8. EXERCISE 3: RUNOFF (see theory in Chapter 5)

Introduction:

A data set is furnished for this exercise (Tables 12.1 and 12.2), or you can procure your own from either USGS *Water Supply Papers* or the CD-Rom *Hydrodata* from EarthInfo, Inc. The supplied data set consists of two months of daily streamflow records for two streams in Wisconsin, Coon Creek and Little La Crosse River, Table 12.1; and rainfall records for the same period of time, Table 12.2. Select one month for analysis.

A. On the furnished annotated graph paper, graph daily flow including storm peaks (Figure 12.3). Use the sheet for one stream for one month. Sketch an arbitrary line to separate baseflow from stormflow.

Note that baseflow varies day to day even when there is no precipitation. This is possibly attributable to (1) ET conditions, (2) barometric pressure fluctuations (3) upstream water withdrawals and (4) measurement errors. Despite these fluctuations, can you detect a baseflow recession curve line (line on this log paper) after precipitation events? Do you see that such a line would have to be a trend line because of the masking effect of the aforementioned fluctuations? According to John Hewlett, baseflow accounts for about 70% of total stream flow (water yield) in the Eastern U.S.A.

Table 12.1a.   Date Set, Exercise 3, Storm Discharge and Suspended Sediment, July 1938.

| Day | Coon Creek (77.2 sq. mi.) | | | | | Little LaCrosse River (77.1 sq. mi.) | | | | |
| | Mean Daily Discharge | Maximum Rate of Discharge | | | Suspended Matter | Mean Daily Discharge | Maximum Rate of Discharge | | | Suspended Matter |
	C.F.S.	C.F.S.	C.F.S. per sq. mi.	Time	Tons per day	C.F.S.	C.F.S.	C.F.S. per sq. mi.	Time	Tons per day
1	47				30.00	74	120	1.56	2:00	176.00
2	302	1,702	22.2	7:30	24,700.00	180	360	4.67	14:30	2,820.00
3	53				68.00	60				157.00
4	44				18.00	45				40.00
5	64	122	1.58	11:50	170.00	46				49.00
6	48				25.00	50	109	1.41	22:15	200.00
7	45				13.00	45				183.00
8	41				9.70	39				36.00
9	40				22.00	40				64.00
10	46	55	0.711	6:00	18.00	40				29.00
11	39				3.30	37				19.00
12	39				4.60	36				54.00
13	40	40	0.518	Constant	4.30	159	330	4.28	8:00	1,800.00
14	36				4.10	47				68.00
15	35				2.90	39				23.00
16	35				4.00	36				16.00
17	34				3.60	37				14.00
18	33				3.00	32				11.00
19	33				2.70	32				11.00
20	33				3.60	31				8.10

Table 12.1a. Continued.

| Day | Coon Creek (77.2 sq. mi.) | | | | | Little LaCrosse River (77.1 sq. mi.) | | | | |
| | Mean Daily Discharge | Maximum Rate of Discharge | | | Suspended Matter | Mean Daily Discharge | Maximum Rate of Discharge | | | Suspended Matter |
	C.F.S.	C.F.S.	C.F.S. per sq. mi.	Time	Tons per day	C.F.S.	C.F.S.	C.F.S. per sq. mi.	Time	Tons per day
21	101	400	5.18	17:00	2,050.00	38	90	1.17	21:30	203.00
22	55				128.00	38				122.00
23	38				13.00	34				17.00
24	36				6.40	32				11.00
25	34				2.70	32				9.80
26	37				4.40	32				8.50
27	37				6.50	36				16.00
28	37				5.30	38	49	0.635	4:45	39.00
29	33				6.00	32				12.00
30	33				3.70	31				9.10
31	33				3.70	32				8.70

Table 12.1b. Date set, exercise 3, storm discharge and suspended sediment, August 1938.

	Coon Creek (77.2 sq. mi.)					Little LaCrosse River (77.1 sq. mi.)				
	Mean Daily Discharge	Maximum Rate of Discharge			Suspended Matter	Mean Daily Discharge	Maximum Rate of Discharge			Suspended Matter
Day	C.F.S.	C.F.S.	C.F.S. per sq. mi.	Time	Tons per day	C.F.S.	C.F.S.	C.F.S. per sq. mi.	Time	Tons per day
1	32				1.60	31				6.50
2	31				5.70	30				8.50
3	31				2.30	30				8.30
4	29				1.90	29				6.80
5	91	480	6.21	11:00	2,210.00	52	120	1.56	15:30	212.00
6	37				20.00	36				28.00
7	33				7.50	44	102	1.32	18:40	160.00
8	34				7.40	52	117	1.52	18:30	272.00
9	39	52	0.673	7:15	13.00	40				54.00
10	33				3.30	34				20.00
11	32				4.30	31				27.00
12	29				3.20	30				11.00
13	29				2.60	29				8.70
14	29				4.20	28				7.80
15	36				6.10	30				9.70
16	225	860	11.1	14:30	2,550.00	252	730	9.46	0:30	1,380.00
17	62				41.00	205				450.00
18	42				6.80	45				23.00
19	38				4.90	40				14.00
20	35				5.00	38				10.00

Table 12.1b. Continued.

	Coon Creek (77.2 sq. mi.)					Little LaCrosse River (77.1 sq. mi.)				
	Mean Daily Discharge	Maximum Rate of Discharge			Suspended Matter	Mean Daily Discharge	Maximum Rate of Discharge			Suspended Matter
Day	C.F.S.	C.F.S.	C.F.S. per sq. mi.	Time	Tons per day	C.F.S.	C.F.S.	C.F.S. per sq. mi.	Time	Tons per day
21	32				4.20	37				7.30
22	33				2.60	35				6.80
23	47	65	0.84	9:00	12.00	200	385	5	11:00	1,640.00
24	34				2.00	51				47.00
25	34				1.60	39				16.00
26	34				2.60	36				10.00
27	32				1.60	34				7.70
28	32				2.20	32				6.70
29	32				1.60	32				6.40
30	32				1.90	32				6.60
31	31				2.00	32				6.30

Table 12.2a. Exercise 3, rainfall data (inches) for, July 1938.

Date	Coon Creek							Little LaCrosse River			
	Raingage Number										
	1	2	7	8	9	13	14	5	6	11	12
1	0.46	0.38	0.46	0.58	0.60	0.42	0.43	0.63	0.53	1.32	0.72
2	1.25	1.02	1.57	0.49	1.22	1.38	0.70	0.37	0.51	0.22	0.38
3											
4					0.01						0.01
5	0.47	0.48	0.30	0.05	0.55	0.30	0.30	0.42	0.25	0.19	0.11
6	0.02	0.02	0.20	0.25	0.32	0.05	0.05	0.10	0.01	0.04	0.05
7	0.09	0.37		0.14	0.11	0.05	0.22	0.06	0.04	0.02	0.03
8				0.01	0.01						
9	0.16	0.06	0.07	0.03	0.12	0.25	0.05	0.10	0.04	0.40	0.42
10	0.45	0.45	0.29	0.41	0.54	0.40	0.55	0.32	0.26	0.32	0.27
11											
12											
13	0.12	0.12	0.28	0.12	0.17	0.18	0.20	0.31	0.70	0.95	1.62
14	T			0.06	0.02		0.07		0.02	0.02	0.02
15	0.12										
16	T										
17											
18											
19											
20											

Table 12.2a. Continued.

Date	Coon Creek							Little LaCrosse River			
	Raingage Number										
	1	2	7	8	9	13	14	5	6	11	12
21											
22	1.21	1.51	0.60	0.90	1.50	0.95	1.45	1.14	0.41	0.27	0.27
23	1.21	0.02					0.03	0.03			0.01
24											
25		0.01						0.01			0.01
26											
27	0.50	0.34	0.44	0.17	0.38	0.47	0.25	0.48	0.34	0.49	0.29
28			0.29					0.04	0.04		
29											
30	0.05	0.04	0.01	0.02	0.06	0.52	0.10	0.04	0.03	0.01	0.03
31	0.50		0.01		0.04	T	T	0.07	0.01	0.06	0.03
Total	4.78	4.82	4.52	3.23	5.65	4.97	4.40	4.41	3.19	4.31	4.27

Table 12.2b. Exercise 3, rainfall data (inches) for, August 1938.

Date	Coon Creek							Little LaCrosse River			
	Raingage Number										
	1	2	7	8	9	13	14	5	6	11	12
1											
2											
3											
4											
5	1.47	0.64	0.76	0.64	0.49	1.28	0.85	0.91	0.79	0.38	0.61
6	0.10	0.06	0.20	0.28	0.18	0.07	0.05	0.10	0.19	0.10	0.29
7											0.21
8				0.34				0.05			0.82
9	0.70	0.57	0.26	0.72	0.16	0.45	0.30	0.55	0.20	0.49	0.56
10		0.02							0.49		
11											
12											
13											
14											
15	0.48	0.54	0.31	0.42	0.60	0.43	0.50	0.41	0.21	0.33	0.21
16	0.30	0.71	0.33	0.23	0.24	0.27	0.40	0.42	0.50	0.31	0.41
17	1.47	1.78	2.08	1.74	2.13	1.58	2.30	2.44	2.56	2.44	2.71
18											
19											
20											

Table 12.2b. Continued.

| | Coon Creek | | | | | | | Little LaCrosse River | | | |
| | Raingage Number | | | | | | | | | | |
Date	1	2	7	8	9	13	14	5	6	11	12
21											
22											
23	0.65	0.63	0.88	0.68	0.54	0.80	0.55	1.04	1.32	1.22	1.13
24											
25	0.65		0.02								
26		0.04						0.01			
27											
28											
29							T				
30											
31											
Total	5.17	4.99	4.84	5.05	4.34	4.88	4.95	5.93	6.26	5.27	6.95

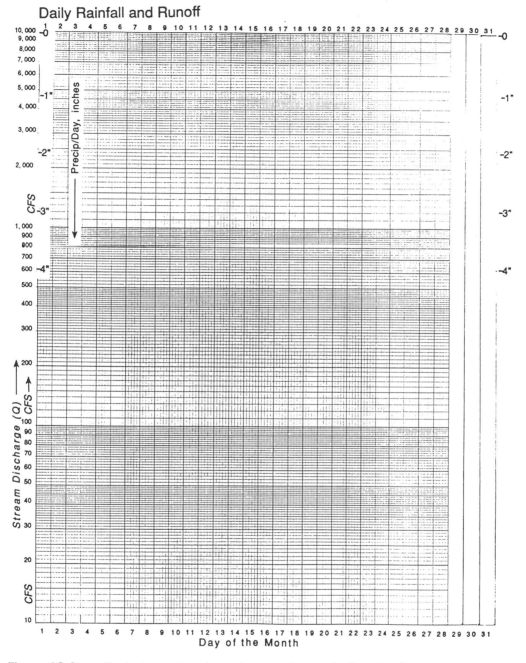

Figure 12.3.   Typical annotated graph paper for use in Exercise 3.

B. Make a hyetograph to accompany your hydrograph. From the precipitation data, take an arithmetic average for appropriate stations. The stations are well distributed and Thiessen weighting is not needed. The hyetograph bars should extend downward from the top according to the scale. Since the 24 hr. precipitation period ends at 8 a.m. of the noted day, the bar (0.25 in. wide) should be centered over 8 p.m. of the previous day. If the average is less than 0.01 in., put "T" for trace.

C. Note stream response to each storm (response = stormflow/precipitation).

Response varies within a basin depending on (a) precipitation amount, (b) precipitation intensity, (c) antecedent soil water [antecedent baseflow is used as a reasonable surrogate for soil water (θ) because drainage from unsaturated (often below FC) soil augments baseflow from deep groundwater], and (d) infiltration changes, especially from varying land use. Between basins, the list is longer; most important are (a) geology/soils, (b) slope, (c) drainage density, and (d) infiltration, especially regarding land use.

In your graph, response varies primarily as a function of antecedent θ and precipitation intensity. Note how response increases after a rainy period of a few days. According to Hewlett, average response for all eastern streams is about 10%, but varies (2–24%) according to physiographic province. Season or storm attributes and land use cause added variation. Additional variables must be considered for data sets in Winter and Spring: ground frost and snowmelt, effects of which often combine. This is very important in March and April.

D. Graph one storm event at a large arithmetic scale of your choice. Draw the hydrograph from the information given using the technique shown in Figure 12.4. Separate baseflow from stormflow as best you can. Baseflow should usually increase somewhat during the storm, depending on infiltration and duration.

Measure the area under the hydrograph above baseflow; this is total stormflow. This can be described by the expression:

$$Q_T \text{ (stormflow)} = \int_{t=0}^{t=n} Q(dt)$$

For those who desire, feel free to derive equations for the curves and solve this with integral calculus. Most will simply measure the area with a planimeter or dot grid and, using our graph scale, convert to total volume of stormflow (ft^3 or m^3 of water).

Now, divide total volume by the watershed area to get average watershed depth. Subtracting the stormflow depth from the precipitation depth, you will get an approximate idea of θ recharge.

Next, divide total stormflow (depth) by precipitation depth to get stream response in percent. Refer to your monthly hydrographs and see how the response of the stream to this storm compares with other storms. Important note: "stormflow" includes overland flow, thruflow, and channel precipitation, perhaps in different proportions for different storms. There is no reliable method of separating the

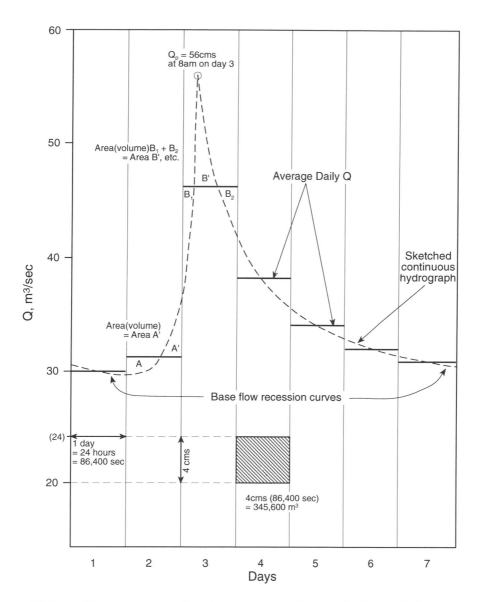

Figure 12.4.   Mass rate budgeting to create a continuous hydrograph from average daily Q and Peak Q values. From S.W. Trimble, *Am. Journal Science,* 277:876–87, 1983.

hydrograph into overland flow, thruflow, channel precipitation and baseflow!

E. **Sediment Yield (optional).**   Note that sediment transported past the gaging station is an exponential function of stream discharge, especially peak discharge (properly, the values should be instantaneous). The ambitious amongst you may want to graph sediment discharge against stream discharge. If so, use 3 x 5 cycle log-log paper (sediment discharge is the 5 cycle ordinate). Erosion and sedimentation will be discussed later.

12.9. EXERCISE 4: FLOW DURATION (see theory in Chapter 5)

Introduction:

One of the most important characteristics of a stream is flow duration. For example, does the stream have a very equitable flow regime or is it constantly up and down? Does it have permanent flow? What are the low flow and high flow characteristics (is it "flashy")? If the stream is to furnish power, transport sewage effluent away, or just dilute pollution, how much of the time can these jobs be accomplished? A flow duration curve can answer these questions, but you must know how to use it.

To extract enough flow data from USGS *Water Supply Papers* to construct a flow duration curve would take much time and effort. Instead, you are supplied with a data set in which the computer has done most of the work (Table 12.3). Daily flow for a stream has been tabulated and categorized and is ready for you to plot the flow duration curve. Those with the requisite computer skills might wish to obtain and process data from the *HydroData* CD-Rom from EarthInfo, Inc.

A. (Refer to Figures 12.5 and 12.6.) Plotting starts at the top left with the highest discharge (Q) values (lowest percent of time equaled or exceeded), and progresses to the lower Q values. Your set includes data for one stream during two time periods. Land use and land treatment have been improved considerably from the first period to the second, and thus, the stream regime might (but not necessarily) be more uniform (less flashy) for the second period.

B. After plotting the curve for each period on the same sheet (furnished), write a short paragraph pointing out differences in the two curves and giving the significance of these differences. Feel free to write on the graph. Has the stream been "tamed?" Keep in mind that the period is only a sample of the steam's behavior and the samples may be poor to good in terms of being representative. Generally, the quality of the sample, like precipitation, improves with number of years. Probably 15–20 years are required for a dependable curve, but this varies.

Some factors other than land use affecting flow duration are precipitation and other climatic considerations, geology and landforms, vegetation and other infiltration factors, and reservoirs. The latter may be the most important factor in some cases.

12.10. EXERCISE 5: STORM RUNOFF, TOTALS AND PEAKS
   (see theory in Chapter 5)

This exercise is designed to expose you to (a) a more sophisticated version of the SCS triangular hydrograph technique, (b) a source of much hydrologic information, the USDA *Soil Surveys* and (c) the possible effects of changing land use and land management on total and peak stormflow (direct runoff). You first need to obtain:

1. A modern (post 1955) soil survey (often catalogued as S591/AZ). Find a watershed of interest to you with *at least* two different soil mapping categories. Note that each soil mapping category or "Unit" usually includes (a) soil series, (b)

Table 12.3a. Cumulative flow duration for Zumbro River at Zumbro Falls, Minnesota, 1909–1945. Duration of Daily Values for Year Ending December 31.

C.F.S.	Σ%	C.F.S.	Σ%	C.F.S.	Σ%	C.F.S.	Σ%
0	100.00	180	65.50	960	8.50	5000	0.70
42	100.00	220	53.10	1200	6.40	6100	0.50
50	100.00	260	43.50	1400	5.00	7300	0.40
61	99.90	320	34.80	1700	3.80	8800	0.20
73	99.60	380	28.50	2000	3.00	11000	0.10
88	97.80	460	23.60	2400	2.30	13000	
110	92.00	550	18.70	2900	1.80	15000	
130	84.10	670	14.50	3500	1.30	18000	
150	77.70	800	11.20	4200	1.00		

95th percentile = 1400   75th percentile = 440   50th percentile = 230   10th percentile = 120
90th percentile = 870   70th percentile = 370   25th percentile = 160

Table 12.3b. Cumulative flow duration for Zumbro River at Zumbro Falls, Minnesota, 1954–1974. Duration of Daily Values for Year Ending December 31.

C.F.S.	Σ%	C.F.S.	Σ%	C.F.S.	Σ%	C.F.S.	Σ%
0	100.00	210	64.5	1100	8.60	6200	0.60
47	100.00	250	56.00	1400	6.20	7500	0.50
57	100.00	310	45.60	1700	4.60	9000	0.30
68	99.30	370	38.30	2000	3.60	1100	0.20
83	96.80	450	31.60	2400	2.90	13000	0.10
100	92.30	540	25.80	2900	2.10	16000	0.1
120	86.60	650	28.20	3500	1.60	19000	
150	78.50	790	17.30	4300	1.10	23000	
170	73.70	950	11.90	5100	0.90		

95th percentile = 1600   75th percentile = 560   50th percentile = 280
90th percentile = 1000   70th percentile = 470   25th percentile = 160

slope, and (c) degree of previous erosion. These designations are explained in the description for each soil. Average weighted slope should be less than 20%. Select a small watershed to make things easier and faster for yourself. Trace the watershed (with the soil mapping units) onto a sheet of tracing paper. To properly determine the watershed boundaries, you will probably need a topographic map (Figure 12.7). Include a photocopy of the topographic map with the basin outlined. For slopes, take the average for each category. For example, if a "B" slope is 2–6%, use 4%. The "length of water flow" is measured from the interfluve along the valley to the lower end of your basin. Soil Hydrologic Groups (A, B, C, D) for soil series are usually given in a table in your survey. While you are looking, consider all other good information (hydrologic and otherwise)

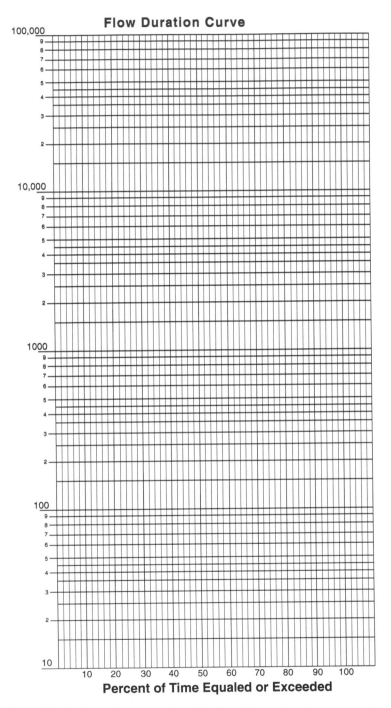

Figure 12.5. Typical graph paper for plotting flow duration exceedence relationships.

given in the tables. You will need a planimeter to obtain the area of the watershed in each soil mapping unit, or you may use a dot-grid overlay.

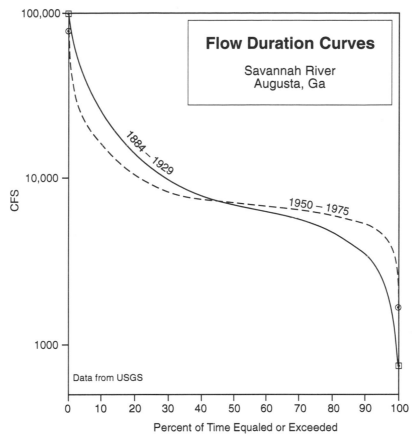

Figure 12.6.   Flow duration exceedence relationships for the Savannah River, Augusta, Georgia.

2. Obtain "A Method for Estimating Volume and Rate of Runoff in Small Watersheds;" (USDA-SCS TP 149) or see Chapter 5. This booklet gives complete details for the SCS triangular hydrography method. For land use and management, assume "before" and "after" conditions. That is, you will examine the effects of changing land use and management by simply assigning appropriate values from Table 2 for two time periods. Do whatever you wish, but keep it realistic. Assume any precipitation event that you wish, consistent with USWB *Technical Publications* 37 and 40 but hold it constant for both land use and conditions so that a comparison can be made. Calculate Q and Q_p for both land use conditions. Use Figure 3 and Table 2 in the SCS booklet, but use the techniques described in Chapter 5.

12.11. EXERCISE 6: EROSION AND ELEMENTARY SEDIMENT ROUTING (see theory in Chapter 6)

This exercise will familiarize you with the Universal Soil Loss Equation, sediment delivery ratios, and reservoir sediment trap efficiencies.

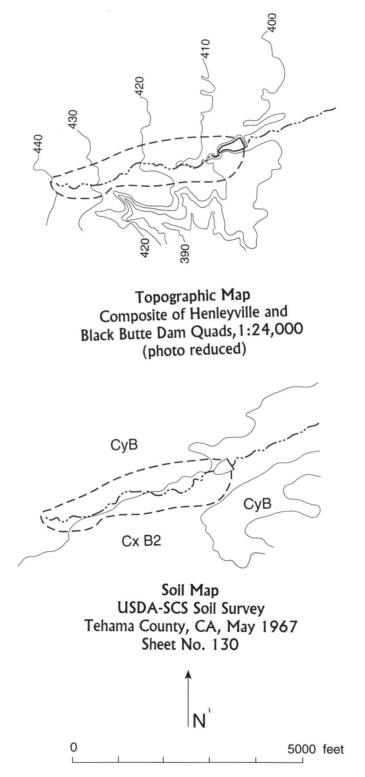

Topographic Map
Composite of Henleyville and
Black Butte Dam Quads, 1:24,000
(photo reduced)

Soil Map
USDA-SCS Soil Survey
Tehama County, CA, May 1967
Sheet No. 130

N

0   5000 feet

Figure 12.7.   Same-scale comparison of stream basin on soil and topographic maps
for Exercises 5 and 7.

<u>Materials Required</u>: Review of Chapter 6 or USDA-SCS *Technical Release* No. 51 (Rev.), "Procedure for Computing Sheet and Rill Erosion" and your watershed data from Exercise 5.

A. Calculate the total average annual soil loss for your watershed for one set of land conditions assumed in Exercise 5. Perhaps you will want to find soil losses for both sets of land conditions in Exercise 5. Soil erodibility, k, ranges from about 0.1 to 0.8; assume 0.3 for this exercise if values are not available.

B. **Sediment Delivery Ratios**. Soil loss is not the same as sediment yield. Eroded soil has many opportunities for deposition as colluvium or alluvium. Thus, not all eroded soil is delivered (at least in the short run) to some downstream point. That is, material is going into storage. This generalization has been embodied into a Sediment Delivery Curve (Figure 12.8). It shows how sediment delivery decreases as an exponential function of area. This curve is an average value. Basins with a high Relief Ratio and/or fine textured soil may have higher SDR's, and conversely for low RR and coarse texture.

Find the appropriate SDR for your basin size, multiply this by the total soil loss from your basin and you have estimated the sediment yield from your basin. By the way, this SDR cannot hold forever, eventually, the storage may become unsteady and be farther transported. For details, see Trimble, *Science,* 188, p. 1207–1208 and *Science,* 191, p. 871.

C. **Sediment Trap Efficiencies of Reservoirs**. Reservoirs make excellent sediment traps. Generally, the trap efficiency (TE) increases with the ratio of reservoir volume to average annual inflow. This is shown in Figures 12.8b and 12.8c. Where the sediment texture is very fine and/or the stream regime is highly variable, the TE is decreased. TE increases with coarse texture and equitable flow.

Assume that a reservoir of 100 acre-feet has been built at the mouth of your basin. What proportion of your sediment yield will be trapped and what proportion passes over the dam? Average annual runoff is available from USGS HA-212 or HA-710 (see Exercise 2).

12.12. EXERCISE 7: STREAMFLOW MEASUREMENT IN THE FIELD (see theory in Chapter 7)

Part I. Gaging streamflow by instrument

Using a current meter, measure the velocity of your stream in fairly short segments. Take the velocity about 1/3 of the depth below the surface (2/3 from the bottom). The total discharge is determined by the procedure in Figure 12.9.

Part II. Gaging streamflow by float method

Measure along your stream and measure the time it takes some floating object to pass the distance. An apple or orange works best. Multiply the average velocity

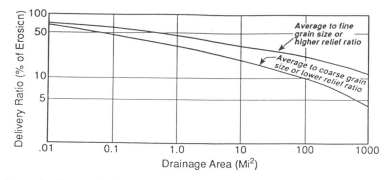

Figure 12.8a.   Sediment delivery ratio vs. size of drainage area (Roehl Curve).

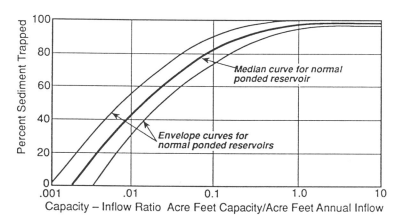

Figure 12.8b.   Trap efficiency of reservoirs as related to capacity - inflow ratio (Brune Curve).

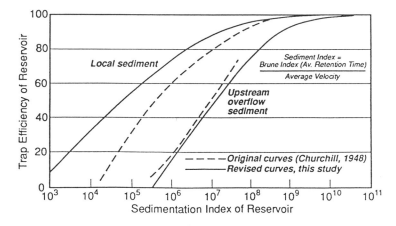

Figure 12.8c.   Trap efficiency vs. sedimentation index from Trimble & Bube. *The Environmental Professional* 12:255–72, 1990.

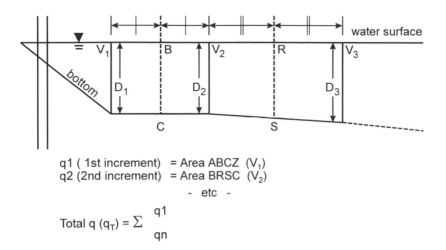

q1 (1st increment) = Area ABCZ (V_1)
q2 (2nd increment) = Area BRSC (V_2)

- etc -

$$\text{Total q } (q_T) = \Sigma \begin{array}{c} q1 \\ qn \end{array}$$

Figure 12.9.   Geometric relationships for determining streamflow based on current meter measurements.

times the cross-sectional area to get q. How does this compare with your measured q? Usually, a diminished coefficient is necessary. For relatively deep streams, 0.8 is a common coefficient but for shallow streams, it may fall below 0.7. What is yours?

Part III. Stream gaging exercise

The purpose of this exercise is to construct a stage-discharge relationship for a stream as if you intended to install a gauging station. Check out an automatic level, tripod, rod, tape, compass (or transit), and chaining pins. If possible, select a reach of stream which is reasonably straight and uniform with minimum vegetative growth. Select a fixed local point for a bench mark. Survey a profile across the stream at bankfull level or above. Make sure you describe the profile well including the azimuth. Then, carefully measure the slope of the floodplain. On a well-formed floodplain you might do this by measuring the difference in elevation between two points about 500–1000 feet apart. However, streams in many areas tend to have poorly formed floodplains, so you may have to survey a long profile along the floodplain and then fit a regression line to the points, which will give you a slope.

Use the Manning equation to construct a stage-discharge relationship from channel bottom (zero discharge) to bankfull. Let each member of the team calculate discharge for at least one stage between zero and bankfull but have at least 3 points on your curve. Manning's "n" factors are found in the Chapter 7, Table 7.4. Each stage will probably have a different "n" value. It is a good idea to assign an "n" value for each foresight taken in surveying the profile. Then, average the "n" values for each stage. Show your work for each stage and plot your stage-discharge relationship on semi-log graph paper. Give a short description of what you did so that you can do it later.

Your calculations should be done by hand to fully understand the procedure. However, most state highway departments have a computer program which will use your profile data to describe a stage-discharge relationship for whatever stage interval you specify.

12.13. EXERCISE 8: WATERSHED OR DRAINAGE BASIN MORPHOLOGY (see theory in Chapter 7)

Introduction:

The watershed is the focal point for surface water hydrology as well as fluvial geomorphology. Watershed morphology is greatly influenced by hydrologic processes or, in turn, watershed morphology greatly influences hydrologic processes. Over a long period of time, a generally efficient system of surface and subsurface drainage develops. A hydrologist should be familiar with the characteristics of watersheds and the interrelationships of these variables. This exercise is intended as a primer.

A. Using a 1:24,000 topographic map, select a watershed small enough to fit on an 8½ x 11" sheet of tracing paper or mylar. Select a fourth- or fifth-order watershed if possible.

B. Trace the drainage pattern of your watershed (Figure 12.10). Extend the stream network up valleys to the highest contour line which has been crenelated to indicate the stream valley. Streams may be classified by the period of time during which flow occurs:

1. Perennial - shown on the map by solid blue lines - exist during greater part of year, usually in well-formed channel.
2. Intermittent - shown by dashed blue line - flow only during wet season, usually less than half the time.
3. Ephemeral - not shown on map, flow occurs in (and formed) higher portions of valleys above intermittent streams. Channels are usually poor or non-existent.

Give a general description of your basin regarding the incidence of these three types.

C. **Stream Orders: Numbers, lengths, slopes (Horton Analysis).** On a separate sheet of paper, set up and complete Table 12.4 based on the following instructions:

1. The Strahler stream order system is most commonly used, but Shreve's system appears to be better correlated with hydrologic processes (see Figure 12.11). For this exercise, use the Strahler system. Indicate stream order on your stream network outline. This is most easily done by color, or by tick marks. Then count the number of segments, or links, for each order and enter them in the table.

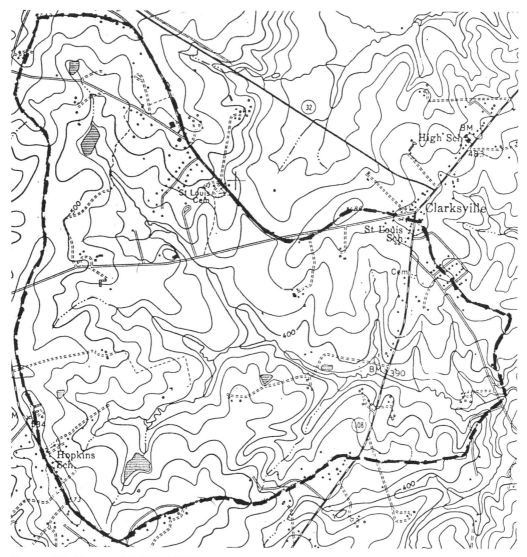

Figure 12.10.   1:24,000 USGS topographic map showing Clarksville, MD watershed (figure reduced 88%).

2. **Length:** just measure valley length instead of stream length. Stream length is always greater than valley length, the degree depending on sinuosity, but there won't be too much difference in lower order streams such as your basin. To measure, lay a piece of paper alongside the valley, mark off the distance on the paper, repeating the process with each stream valley segment. Compute the total length and average length per segment.
3. **Average basin size by order:** select a few typical basins for each order and planimeter them. Average the sizes and enter them in the table.
4. **Average slope by order:** select a few typical streams for each order and measure the slopes (drop in elevation ÷ valley length, expressed as a dimensionless decimal fraction, e.g., 0.0042).

Table 12.4. Bifurcation ratios and stream orders, numbers, lengths, and slopes.

Stream Order	Segments (#)	Total Length per order (ft)	Average Length (ft)	Average Subwatershed Area (acres)	Average Slope (ft/ft)	Bifurcation Ratio
1						
2						
3						
4						

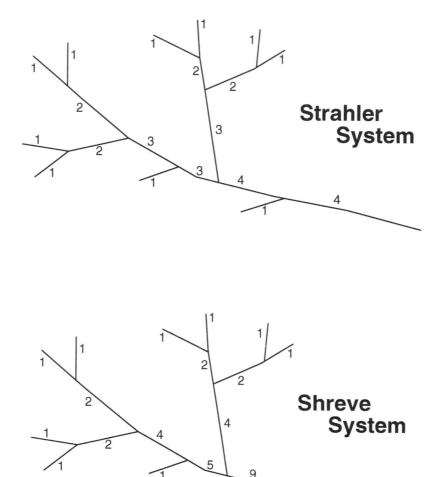

Figure 12.11.   Comparison of the Strahler and Shreve stream ordering systems.

5. **Bifurcation Ratio:** indicates the number of streams of one order feeding into streams of the next higher order (number of first-order streams ÷ number of second-order streams). BR is determined from data already calculated. BR tends to be constant for each order so an average is usually given for the whole basin.

Horton Analysis:

Stream Order	Number of Segments	Total Length Per Order, mi	Average Length, mi	Average Basin Size, mi²	Average Slope	Bifurcation Ratio
1	81	14	.17	.1	.044	5.4
2	15	4.76	.32	.25	.025	5.0
3	3	1.52	.51	.7	.017	3
4	1	1.90	1.9	3.05	.009	

4.5 Average

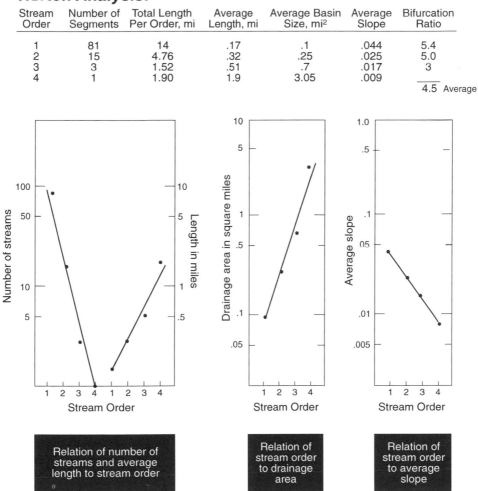

Figure 12.12.   Stream ordering for Clarksville, MD watershed.

D. **Graph.**   To see interrelationships among the above characteristics, graph each against stream order.   With stream order as the abscissa, plot number of segments, average length, (basin size), and average slope.   Fashion your graphs after those in Figure 12.12.   In fact, you can just photocopy those matrices and use them, but you may have to revise the ordinate numbering.   Be sure to use semi-log paper.

E. **Profile of main channel (valley), and relief ratio (RR).** Draw a topographic profile of the main stream valley (the longest possible stream) from interfluve to the lowest end of the watershed (Figure 12.13).   Use at least five (5) points in this profile.   Calculate the overall slope.   This is the RR.   RR is significant in predicting storm discharge rapidity and magnitude as well as sediment yield.

F. **Drainage Density (DD).** DD is the total length of stream channels per unit area. DD equals total stream length ÷ total basin area. DD is indicative of the surface flow from an area and is thus highly significant. DD is related to precipitation (frequency and magnitude), other climatic conditions and to factors of infiltration and percolation (e.g., soils, geology, and land use). DD can be changed by varying land use and land management. DD is often called texture, with low DD being coarse and high DD being fine. Some examples of DD (in mi/mi^2) are:

Humid forested areas in Eastern U.S.	8–16
Humid forested Rocky Mountains	8–16
California coast ranges, igneous rocks	20–30
California coast ranges, weak sediments	30–40
Drier areas of Rocky mountains	50–100
Badlands, SD	200–400

G. **Constant of Channel Maintenance.** This constant is the reciprocal of DD. It is an interesting and significant variable because it tells how much area (e.g., ft^2) is required to maintain one unit (e.g., foot) of drainage channel.

H. **Basin shape.** Our interest here is the **Elongation Ratio** (Diameter of circle with same area as basin divided by basin length). For basin length, use length of main valley from above. The diameter of a circle is:

$$2 \cdot (A/\pi)^{1/2}$$

where A = is the area of your basin. Elongation of a basin is important in regulating the rapidity and magnitude of flood peaks. Elongation is often related to bifurcation ratio; an elongated basin may have a high BR.

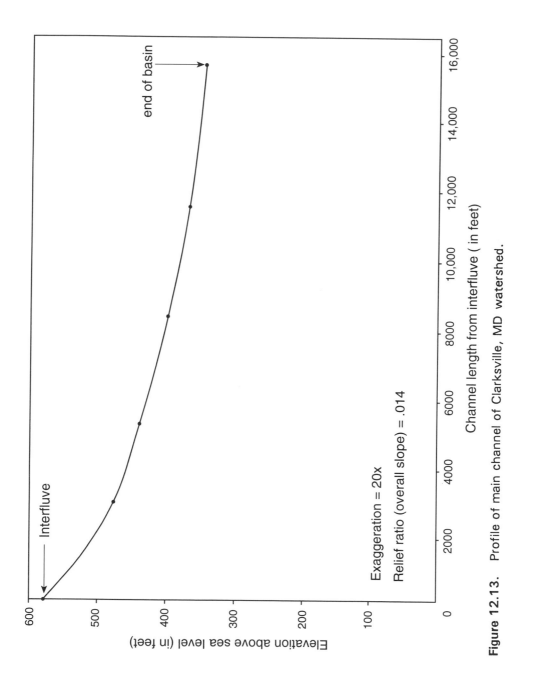

Figure 12.13. Profile of main channel of Clarksville, MD watershed.

APPENDIX A

Unit Conversion Factors

Multiply the U.S. customary unit		By	To obtain the SI unit	
Name	Symbol		Symbol	Name
Acceleration				
feet per second squared	ft/s²	0.3048	m/s²	meters per second squared
inches per second squared	in./s²	0.0254	m/s²	meters per second squared
Area				
acre	acre	0.4047	ha	hectare
acre	acre	4.0469×10^{-3}	km²	square kilometer
square foot	ft²	9.2903×10^{-2}	m²	square meter
square inch	in.²	6.4516	cm²	square centimeter
square mile	mi²	2.5900	km²	square kilometer
square yard	yd²	0.8361	m²	square meter
Energy				
British thermal unit	Btu	1.0551	kJ	joule
foot-pound (force)	ft • lb$_f$	1.3558	J	joule
horsepower-hour	hp • h	2.6845	MJ	megajoule
kilowatt-hour	kW • h	3600	kJ	kilojoule
kilowatt-hour	kW • h	3.600×10^{6}	J	joule
watt-hour	W • h	3.600	kJ	kilojoule
watt-second	W • s	1.000	J	joule
Force				
pound force	lb$_f$	4.4482	N	newton
Flow rate				
cubic foot per second	ft³/s	2.8317×10^{-2}	m³/s	cubic meters per second
gallons per day	gal/d	4.3813×10^{-2}	L/s	liters per second
gallons per day	gal/d	3.7854×10^{-3}	m³/d	cubic meters per day
gallons per minute	gal/min	6.3090×10^{-5}	m³/s	cubic meters per second
gallons per minute	gal/min	6.3090×10^{-2}	L/s	liters per second
million gallons per day	Mgal/d	43.8126	L/s	liters per second
million gallons per day	Mgal/d	3.7854×10^{3}	m³/d	cubic meters per day
million gallons per day	Mgal/d	4.3813×10^{-2}	m³/s	cubic meters per second
Length				
foot	ft	0.3048	m	meter
inch	in.	2.54	cm	centimeter
inch	in.	0.0254	m	meter
inch	in.	25.4	mm	millimeter
mile	mi	1.6093	km	kilometer
yard	yd	0.9144	m	meter
Mass				
ounce	oz	28.3495	g	gram
pound	lb	4.5359×10^{2}	g	gram
pound	lb	0.4536	kg	kilogram
ton (short: 2000 lb)	ton	0.9072	Mg (metric ton)	megagram (10^3 kilogram)
ton (long: 2240 lb)	ton	1.0160	Mg (metric ton)	megagram (10^3 kilogram)

Multiply the U.S. customary unit		By	To obtain the SI unit	
Name	Symbol		Symbol	Name
Power				
British thermal units per second	Btu/s	1.0551	kW	kilowatt
foot-pounds (force) per second	ft • lb$_f$/s	1.3558	W	watt
horsepower	hp	0.7457	kW	kilowatt
Pressure (force/area)				
atmosphere (standard)	atm	1.0133×10^2	kPa (kN/m²)	kilopascal (kilonewtons per square meter)
inches of mercury (60°F)	in. Hg	3.3768×10^3	Pa (N/m²)	pascal (newtons per square meter)
inches of water (60°F)	(60°F) in. H₂O	2.4884×10^2	Pa (N/m²)	pascal (newtons per square meter)
pounds (force) per square foot	(60°F) lb$_f$/ft²	47.8803	Pa (N/m²)	pascal (newtons per square meter)
pounds (force) per square inch	lb$_f$/in.²	6.8948×10^3	Pa (N/m²)	pascal (newtons per square meter)
pounds (force) per square inch	lb$_f$/in.²	6.8948	kPa (kN/m²)	kilopascal (kilonewtons per square meter)
Temperature				
degrees Fahrenheit	°F	0.555(°F - 32)	°C	degrees Celsius (centigrade)
degrees Fahrenheit	°F	0.555 (°F + 459.67)	°K	degrees kelvin
Velocity				
feet per second	ft/s	0.3048	m/s	meters per second
miles per hour	mi/h	4.4704×10^{-1}	km/s	kilometers per second
Volume				
acre-foot	acre-ft	1.2335×10^3	m³	cubic meter
cubic foot	ft³	28.3168	L	liter
cubic foot	ft³	2.8317×10^{-2}	m³	cubic meter
cubic inch	in.³	16.3871	cm³	cubic centimeter
cubic yard	yd³	0.7646	m³	cubic meter
gallon	gal	3.7854×10^{-3}	m³	cubic meter
gallon	gal	3.7854	L	liter
ounce (U.S. Fluid)	oz (U.S. fluid)	2.9573×10^{-2}	L	liter

APPENDIX B

Glossary of Terms

Most of the definitions are based on ASAE Standard: ASAE 5256 Soil and Water Engineering Terminology, American Society of Agricultural Engineers, St. Joseph, Michigan.

Acid mine drainage: Water draining from areas that have been mined for coal or other mineral ores. The drainage water is acidic, sometimes having a pH less than 2.0 because of its contact with sulfur-bearing material.

Acid rain: Precipitation that has a low pH (less than 5.6 which is normal for "natural" precipitation). The precipitation becomes acidic when moisture in the air reacts with sulfur and nitrogen pollutants in the atmosphere. Acid rain has a harmful effect on some plants, aquatic organisms, soils, and buildings.

Actual evaporation: Evaporation depends on climatic conditions and conditions at the surface from which the water will evaporate. Actual evaporation is a measure of the amount of water actually evaporated from a surface and accounts for the surface conditions as well as climatic conditions. Potential evaporation only accounts for climatic conditions.

Actual evapotranspiration: Evapotranspiration depends on climatic conditions and conditions of the plants and soil surface. Actual evapotranspiration is a measure of the amount of water actually evapotranspired from the plants and soil system and takes into account the conditions of the plant/soil system.

Actual vapor pressure: The pressure a gas exerts on the liquid it is contacting is its vapor pressure. In the context of this book, the gas is water vapor and the liquid is water. The actual vapor pressure is the amount of pressure the water vapor in the air exerts on the surface it contacts.

Advection: The process by which solutes are transported by the motion of flowing groundwater.

Aerodynamic roughness: A measure of the roughness of a surface in regards to the disturbance that surface would cause for wind moving over it. A measure of the ability of the surface and its roughness elements to create a sufficiently large turbulent boundary layer as to reach the surface.

Albedo: A measure of the light reflectance properties of a soil and crop surface; the ratio of shortwave electromagnetic radiation reflected from a soil and crop surface to the amount incident upon that surface.

Alkali soil: Soil containing sufficient exchangeable sodium to interfere with water penetration and the growth of most crops. The exchangeable sodium content is greater than 15% (preferred term is Saline-sodic soil).

Anaerobic decomposition: The decay of organic matter by microorganisms in the absence of oxygen.

Anisotropic soils: Soils not having the same physical properties when the direction of measurement is changed. Commonly used in reference to permeability changes with direction of measurement.

Application rate: Rate that water is applied to a given area. Usually expressed in units of depth per time.

Aquiclude: Underground geologic formation that neither yields nor allows the passage of an appreciable quantity of water, although it may be saturated with water itself.

Aquifer: A geologic formation that holds and yields useable amounts of water. Aquifers can be classified as confined or unconfined.

Aquitard: Underground geologic formation that is slightly permeable but yields inappreciable amounts of water when compared to an aquifer.

Arid climates: Climate characterized by low rainfall and high evaporation potential. A region is usually considered as arid when precipitation averages less than 250 mm (10 in.) per yr.

Artesian aquifer: Aquifer that contains water under pressure as a result of hydrostatic head. For artesian conditions to exist, an aquifer must be overlain by a confining material or aquiclude and receive a supply of water. The free water surface stands at a higher elevation than the top confining layer.

Available plant water: The portion of water in a soil that can be readily absorbed by plant roots. It is the amount of water released between in situ field capacity and the permanent wilting point.

Bank storage: Water leaving a stream channel during rising stages of stream flow, during falling stages.(See Floodplain storage.)

Base flow: Sustained low flow of a stream which is often due to groundwater inflow to the stream channel. Often written as a single word.

Bed load: Coarse sediment or material moving on or near the bottom of a flowing channel by rolling, sliding, or bouncing.

Berm: Strip or area of land, usually level, between the upper edge of a spoil bank and the edge of a ditch or canal.

Best management practice, BMP: Structural, nonstructural, and managerial techniques recognized to be the most effective and practical means to reduce surface and ground-water contamination while still allowing the productive use of resources.

Biodegradation: Breaking down of natural or synthetic organic materials by microorganisms in soils, natural bodies of water, or wastewater treatment systems.

Broad-crested weir: Weir of water measurement having a rounded or wide crest in the direction of the stream.

Bulk density: (Soil) The mass of dry soil per unit bulk volume. The bulk volume is determined before drying to constant weight at 105°C (220°F).

Canopy: Vegetative cover over the land surface of a catchment area.

Capillary fringe: A zone in the soil just above the water table that remains saturated or almost saturated. The extent depends upon the size-distribution of pores.

Capillary pressure head: Height water will rise by surface tension above a free water surface in the soil, expressed as length unit of water. Sometimes called "capillary rise."

Capillary soil moisture: (Preferred term is Soil-water potential.)

Capture zone, steady-state: The region surrounding a well that contributes flow to the well and which extends upgradient to the groundwater divide of the drainage basin.

Capture zone, traveltime-related: The region surrounding a well that contributes flow to the well within a specified period of time.

Catchment: See Watershed.

Cation exchange capacity: A measure of the quantity of cations a given mass of soil can hold. It is related to clay content and type, and organic matter content.

Channel capacity: Flow rate in a ditch, canal, or natural channel when flowing full or at design flow.

Channel improvement: Increasing the cross section, straightening or clearing vegetation from a channel to change its hydraulic characteristics, increase its flow capacity, and reduce flooding.

Channel stabilization: Erosion prevention and stabilization of channel by use of vegetation, jetties, drops, revetments, or other measures.

Channel storage: 1) (Hydrology) Water temporarily stored in channels while en-route to an outlet. 2) (Drainage) The volume of water that can be stored above the start-pumping level in ditches or floodways without flooding cropland.

Chlorinated hydrocarbon: Synthetic compound that contains chlorine, hydrogen, and carbon; a main ingredient in some pesticides.

Clay: A soil separate consisting of particles less than 2 μm in equivalent diameter.

Concentration gradient: A concentration gradient exists when there is more of a substance in one place than another and the two places are in contact, either directly or by a material through which this substance can flow.

Cone of depression or influence: The water table or piezometric surface, roughly conical in shape, produced by the extraction of water from a well.

Confined aquifer: An aquifer whose upper, and perhaps lower, boundary is defined by a layer of natural material that does not transmit water readily.

Confining layer: A body of material of low hydraulic conductivity that overlies or underlies an aquifer.

Conservation tillage: A tillage practice that leaves plant residues on the soil surface for erosion control and water conservation.

Consumptive use: The total amount of water taken up by vegetation for transpiration or building of plant tissue, plus the unavoidable evaporation of soil moisture, snow, and intercepted precipitation associated with vegetal growth.

Conventional tillage: The traditional tillage practice that involves inverting the tillage layer, burying most of the plant residues, and leaving the soil bare.

Cover crop: Close-growing crop that provides soil protection, seeding protection, and soil improvement between periods of normal crop production, or between trees in orchards and vines in vineyards. When plowed under and incorporated into the soil, cover crops may be referred to as green manure crops.

Crop residue: Portion of a plant or crop left in the field after harvest.

Crop rotation: A system of farming in which a succession of different crops are planted on the same land area, as opposed to growing the same crop time after time (monoculture).

Curve number: An index of the runoff potential which is related to the soil and vegetation conditions of the site. Used in SCS Runoff Equations.

Darcy's law: A concept formulated by Henry Darcy in 1856 to describe the rate of flow of water through porous media. The rate of flow of water in porous media is proportional to, and in the direction of, the hydraulic gradient and inversely proportional to the thickness of the bed.

Deep percolation: Water that moves downward through the soil profile below the root zone and cannot be used by plants.

Design runoff rate: Maximum runoff rate expected for a given design return period storm.

Detention storage: Water in excess of depression storage which is temporarily stored in the watershed while enroute to streams. Most eventually becomes surface runoff, but some may infiltrate or evaporate.

Dewpoint temperature: Air typically contains water vapor. The amount of water vapor a given parcel of air can hold depends on the temperature of the air. Warmer air can hold more water than cooler air. The dewpoint temperature is the temperature to which a given parcel of air must be cooled (at constant pressure and water vapor content) in order for the air to be saturated with water.

Diffuse: The process of a substance moving from an area of higher concentration to an area of lower concentration of that substance.

Diffusion coefficient: A measure of the ease with which a particular substance can diffuse in a given system.

Discharge: Rate of water movement.

Discharge curve: Rating curve showing the relation between stage and flow rate of a stream, channel, or conduit.

Drain: Any closed conduit (perforated tubing or tile) or open channel, used for removal of surplus ground or surface water.

Drainage: Process of removing surface or subsurface water from a soil or area.

Drainage basin: See Watershed.

Drain tile: Short length of pipe made of burned clay, concrete, or similar material. Usually laid with open joints, to collect and remove subsurface water.

Emissivity: The ratio of the emittance of a given surface at a specified wavelength and temperature to the emittance of an ideal black body at the same wavelength and temperature.

Ephemeral gully: Small channels eroded by runoff which can be easily filled and removed by normal tillage, only to reform again in the same location.

Ephemeral stream: A stream that is dry most of the year and only contains water during and immediately after a rainfall event.

Equipotential line: a contour line of a potentiometric surface along which the hydraulic head of the groundwater flow system is the same for all points on the line.

Erosion: The wearing away of the land surface by running water, wind, ice, or other geological agents, including such processes as gravitational creep. The following terms are used to describe different types of water erosion:

    Accelerated erosion: Erosion much more rapid than normal, natural, or geological erosion, primarily as a result of the influence of the activities of humans or, in some cases, of animals.

    Geological erosion: The normal or natural erosion caused by geological processes acting over long geological periods (Synonymous with Natural erosion.)

    Gully erosion: The erosion process whereby water accumulates in narrow channels and, over short periods, removes the soil from this narrow area to considerable depths ranging from 0.5 m (1.6 ft) to as much as 30 m (97 ft.)

    Interrill erosion: The removal of a fairly uniform layer of soil on a multitude of relatively small areas due to raindrop impact and by shallow surface flow.

    Natural erosion: Wearing away of the earth's surface by water, ice, or other natural agents under natural environmental conditions of climate, vegetation, etc. undisturbed by humans (See Geological erosion.)

    Normal erosion: The graduate erosion of land used by society which does not greatly exceed natural erosion. (See Natural erosion.)

    Rill erosion: An erosion process by concentrated overland flow in which numerous small channels of only several centimeters in depth are formed; occurs mainly on recently cultivated soils (See Rill.)

    Sheet erosion: The removal of soil from the land surface by rainfall and surface runoff. Often interpreted to include rill and interrill erosion.

    Splash erosion: The detachment and airborne movement of small soil particles caused by the impact of raindrops on soils.

Evaporation: The physical process by which a liquid is transformed to a gaseous state.

Evaporation pan: A pan, typically of specific materials and dimensions, which is filled with water and left open to the environment. Evaporation from the pan is measured, and this evaporation can be related to evapotranspiration from a nearby crop.

Evapotranspiration: The combination of water transpired from vegetation and evaporated from the soil and plant surfaces.

Exchangeable cation: A positively charged ion held on or near the surface of a solid particle by a negative surface charge of a colloid and which may be replaced by other positively charged ions in the soil solution.

Exchangeable sodium percentage: The fraction of the cation exchange capacity of a soil occupied by sodium ions.

Field capacity: Amount of water remaining in a soil when the downward water flow due to gravity becomes negligible.

Floodplain storage: Volume of water that spreads out and is temporarily stored in a flood plain.

Flood routing: Process of determining stage height, storage volume, and outflow from a reservoir or reach of a stream for a given hydrograph of inflow.

Flood spillway: An auxiliary channel to carry a flood flow that exceeds a given design rate to the channel downstream. (Preferred term is Emergency spillway.)

Flowline: A line indicating the instantaneous direction of groundwater flow throughout a flow system, at all times in a steady-state flow system or at a specific time in a transient flow system. In an isotropic medium, flowlines are drawn perpendicular to equipotential lines.

Flow rate: Rate of water movement. Often written as a single word and expressed in cfs or m^3/s.

Flume: 1) Open conduit for conveying water across obstructions. 2) An entire canal elevated above natural ground. An aqueduct. 3) A specifically calibrated structure for measuring open channel flows.

Flux density: The rate of flow of any quantity, such as water vapor, through a unit area of specified surface.

Freeboard: Vertical distance between the maximum water surface elevation anticipated in design and the top of retaining banks, pipeline vents, or other structures, provided to prevent overtopping because of unforeseen conditions.

Friction slope: Friction head loss per unit length of conduit.

Gabion: Rectangular or cylindrical wire mesh cage filled with rock for protecting aprons, streambanks, shorelines, etc., against erosion.

Gaining stream: Stream or part of a stream that has an increase in flow because of inflow from ground water.

Gage height: 1) (Surveying) The vertical distance from the sight bar, batter board or receiver to the bottom of the finished cut. 2) (Hydraulics) Elevation of a water surface measured by a gauge.

Gaging station: Section in a stream channel equipped with a gage or facilities for obtaining stream flow data.

Geographic Information System, GIS: Computer data base management system for spatially distributed attributes.

Grade: (noun) Degree of slope to a road, channel, or ground surface. (verb) To finish the surface of a canal bed, roadbed, top of embankment, or bottom of excavation.

Gradually varied flow: Steady, nonuniform open-channel flow in which the changes in depth and velocity from section to section are gradual enough such that accelerative forces are negligible.

Grassed waterway: Natural or constructed channel covered with an erosion-resistant grass, that transports surface runoff to a suitable discharge point at a nonerosive rate.

Gravitational water: Soil water which moves into, through, or out of the soil under the influence of gravity. (Preferred term is Soil-water potential.)

Groundwater: Water occurring in the zone of saturation in an aquifer or soil.

Groundwater divide: A ridge in the water table or potentiometric surface from which groundwater moves away in both directions.

Ground water flow: Flow of water in an aquifer or soil. That portion of the discharge of a stream that is derived from ground water.

Growing season: The period, often the frost-free period, during which the climate is such that crops can be produced.

Guard cells: Specialized cells in a leaf which surround the stomates. The guard cells can expand and contract to control the loss of water vapor through stomates.

Gully: Eroded channel where runoff concentrates, usually so large that it cannot be obliterated by normal tillage operations.

Gully head advance: Upstream migration of the upper end of a gully.

Gypsum block: An electrical resistance block in which the absorbent material is gypsum.

Head: The height to which water can raise itself above a known datum (commonly sea level) exerting a pressure on a given area, at a given point. It is synonymous with hydraulic head.

Head loss: Energy loss in fluid flow.

Heterogeneous: Pertaining to a nonuniform geologic material having different characteristics and hydraulic properties at different locations.

Homogeneous: Pertaining to a uniform geologic material having identical characteristics and hydraulic properties everywhere.

Humid climates: Climate characterized by high rainfall and low evaporation potential. A region is usually considered as humid when precipitation averages more than 500 mm (20 in.) per yr.

Hydraulic conductivity: The ability of a porous medium to transmit a specific fluid under a unit hydraulic gradient; a function of both the characteristics of the medium and the properties of the fluid being transmitted. Usually a laboratory measurement corrected to a standard temperature and expressed in units of length/time. Although the term hydraulic conductivity is sometimes used interchangeably with the term permeability, the user should be aware of the differences.

Hydraulic gradient: Change in the hydraulic head per unit distance.

Hydraulic head: The total mechanical energy per unit weight of water that is equal to the sum of the elevation head, pressure head, and the velocity head at a given point in the flow system.

Hydraulic length: Longest flow path on a watershed.

Hydraulic resistance: Friction along the wetted boundary of a channel or conduit that causes a loss in head.

Hydrograph: Graphical or tabular representation of the flow rate of a stream with respect to time.

Hydrologic cycle: Term used to describe the movement of water in and on the earth and atmosphere. Numerous processes such as precipitation, evaporation, condensation, and runoff comprise the hydrologic cycle.

Hyetograph: A plot of rainfall intensity with respect to time.

Impeller meter: A rotating mechanical device for measuring flow rate in a pipe or open channel.

Impermeable layer: (Soil) Layer of soil resistant to penetration by water, air, or roots.

Infiltration: The downward entry of water through the soil surface into the soil.

Infiltration rate: The quantity of water that enters the soil surface in a specified time interval. Often expressed in volume of water per unit of soil surface area per unit of time.

Influent stream: Stream or portion of stream that contributes water to the groundwater supply. (See Losing stream.)

Initial storage: That portion of precipitation required to satisfy canopy interception, the wetting of the soil surface, and depression storage. Sometimes called "initial abstraction."

Interception: That portion of precipitation caught by vegetation and prevented form reaching the soil surface.

Interflow: Water that infiltrates into the soil and moves laterally through the upper soil horizons until it returns to the surface, often in a stream channel.

Intermittent stream: Natural channel in which water does not flow continuously.

Intrinsic permeability: The property of a porous material that expresses the ease with which gases or liquids flow through it. (See Permeability.)

Isotropic: (Soil) The condition of a soil or other porous media when physical properties, particularly hydraulic conductivity, are equal in all directions.

Kinematic wave: A method of mathematical analysis of unsteady open channel flow in which the dynamic terms are omitted because they are small and assumed to be negligible.

Lag time: 1) (Hydrology) The interval between the time when one half of the equivalent uniform excess rain (runoff) has fallen and the time when the peak of the runoff hydrograph occurs. 2) (Irrigation) The interval after water is turned off at the upper end of a field, until it recedes (disappears from that point).

Laminar flow: Flow in which there are no cross currents or eddies and where the fluid elements move in approximately parallel directions. Flow through granular materials is usually laminar. Sometimes called "streamline" or "viscous flow."

Langley: A unit of energy per unit area commonly used in radiation measurements. One langley is equal to 1 gram calorie per square centimeter.

Latent heat of vaporization: The heat released or absorbed per unit mass of water when evaporation occurs (1 kg of water at 20°C requires 2.45 MJ of heat to vaporize the water).

Leachate: Water that moves downward through some porous media and contains dissolved substances removed from the media.

Leaching: Removal of soluble material from soil or other permeable material by the passage of water through it.

Losing stream: A channel that loses water into the bed or banks. (See Influent stream.)

Lysimeter: An isolated block of soil usually undisturbed and in situ, for measuring the quantity, quality, or rate of water movement through or from the soil.

Mass flow: Movement of a substance occurring when force is exerted on the substance by some outside influence such as pressure or gravity such that all the molecules of the substance tend to move in the same direction.

No-tillage or no-till: A tillage system in which the soil is not tilled except during planting when a small slit is made in the soil for seed and agrichemical placement. Pest control is achieved through the use of pesticides, crop rotation, and biological control rather than tillage. Sometimes called "zero tillage."

Nonpoint source pollution, NPS: Pollution originating from diffuse areas (land surface or atmosphere) having no well-defined source.

Nonsaline-alkali soil: Soil containing sufficient exchangeable sodium to interfere with the growth of most crops   (preferred term is Sodic soil).

Normal depth: Depth of flow in an open channel during uniform flow for the given conditions.

Observation well: Hole bored to a desired depth below the ground surface for observing the water table level.

Orographic storm: A weather pattern in which precipitation is caused by the rising and cooling of air masses as they are forced upward by topography.

Overland flow: Surface runoff occurring at relatively shallow depths across the land surface prior to concentration in drainage ways. May cause sheet and rill erosion.

Particle-size analysis: Determination of the various amounts of different separates in a soil sample, usually by sedimentation, sieving or micrometry.

Perched water table: A water table, usually of limited area, maintained above large ground-water bodies by the presence of an intervening, relatively impervious confining stratum.

Percolating water: Subsurface water that flows through the soil or rocks. (See Seepage.)

Percolation: 1) Downward movement of water through porous media such as soil. 2) Intake rate used for designing wastewater absorption systems.

Perennial stream: A stream which flows throughout the year.

Permanent wilting point: Soil water content below which plants cannot readily obtain water and permanently wilt. Sometimes called "permanent wilting percentage."

Permeability: 1) (Qualitative) The ease with which gases, liquids or plant roots penetrate or pass through a layer of soil or porous media. 2) (Quantitative) The specific soil property designating the rate at which gases and liquids can flow through the soil or porous media.

Permeameter: Device for containing a soil sample and subjecting it to fluid flow in order to measure permeability or hydraulic conductivity.

Permissible velocity: Highest water velocity in a channel or conduit that does not cause erosion.

Pipe drain: Any circular subsurface drain, including corrugated plastic tubing and concrete or clay tile.

Pipe spillway: A pipe drain for transporting water through an embankment. Sometimes called a "culvert."

Porosity: 1) (Soil) The volume of pores in a soil sample, divided by the combined volume of the pores and the soil of the sample. 2) (Aquifer) The sum of the specific yield and the specific retention.

Porosity, effective: The volume of the void spaces through which water or other fluids can travel in a rock or sediment, divided by the total volume of the rock or sediment.

Potential evaporation: Evaporation from a surface when all surface-atmosphere interfaces are wet so there is no restriction on the rate of evaporation from the surface.

Potential evapotranspiration: Rate at which water, if available, would be removed from soil and plan surfaces.

Potentiometric surface: An imaginary surface representing the static head of ground water and defined by the level to which water will rise in a well.

Precipitation intensity: Rate of precipitation, generally expressed in units of depth per time. (See Rainfall intensity.)

Preferential flow: Flow into and through porous media or soil by way of cracks, root holes, and other paths of low resistance rather than uniformly through the whole media.

Raindrop erosion: Soil detachment resulting from the impact of raindrops on the soil. (See Erosion.)

Rainfall erosivity: A measure of rainfall's ability to detach and transport soil particles.

Rainfall frequency: Frequency of occurrence of a rainfall event whose intensity and duration can be expected to be equalled or exceeded. (Preferred term is Return period.)

Rainfall intensity: Rate of rainfall for any given time interval, usually expressed in units of depth per time.

Rating curve: Graphic or tabular presentation of the discharge of or flow through a structure or channel section as a function of water stage or depth of flow. Sometimes called a "rating table."

Reach: A length of a stream or channel with relatively constant characteristics.

Receiving waters: Distinct bodies of water, such as streams, lakes, or estuaries, that receive runoff or wastewater discharges.

Recession curve: Descending portion of a stream flow or hydrograph.

Recharge: Process by which water is added to the zone of saturation to replenish an aquifer.

Recharge area: Land area over which water infiltrates and percolates downward to replenish an aquifer. For unconfined aquifers, the area is essentially the entire land surface overlaying the aquifer and for confined aquifers, the recharge area may be a part of or unrelated to the overlaying area.

Reference crop evapotranspiration: The evapotranspiration predicted from a specific crop, arbitrarily called a reference crop, under a given climatic condition assuming water is available to the crop.

Regime: Condition of a stream with respect to its rate of flow.

Relative humidity: Ratio of the amount of water present in the air to the amount required for saturation of the air at the same dry bulb temperature and barometric pressure, expressed as a percentage.

Reservoir: Body of water, such as a natural or constructed lake, in which water is collected and stored for use.

Retention: Precipitation on an area that does not escape as runoff; the difference between total precipitation and total runoff.

Return period: The frequency of occurrence of a hydrologic event whose intensity and duration can be expected to be equalled or exceeded usually expressed in years.

Rill: Small channels eroded into the soil surface by runoff which can be filled easily and removed by normal tillage.

Riparian: 1) Pertaining to the banks of a body of water, a riparian owner is one who owns the banks. 2) A riparian water right is the right to use and control water by virtue of ownership of the banks.

Root zone: Depth of soil that plant roots readily penetrate and in which the predominant root activity occurs.

Row grade: The slope in the direction of crop rows.

Runoff: The portion of precipitation, snow melt, or irrigation that flows over and through the soil, eventually making its way to surface water supplies.

Runoff coefficient: Ratio of peak runoff rate to rainfall intensity.

Runoff duration: Elapsed time between the beginning and end of a runoff event.

Safe well yield: Amount of ground water that can be withdrawn from an aquifer without degrading quality or reducing pumping level.

Saline-sodic soil: Soil containing sufficient exchangeable sodium to interfere with the growth of most crops and containing appreciable quantities of soluble salts. The exchangeable-sodium-percentage is greater than 15, and the electrical conductivity of the saturation extract is greater than 4 mS/cm (0.01 mho/in.).

Sand: Soil particles ranging from 50 to 200 μm in diameter. Soil material containing 85% or more particles in this size range.

Saturated flow: Flow of water through a porous material under saturated conditions.

Saturation vapor pressure: The vapor pressure at which a liquid-vapor system is in a state of dynamic equilibrium where the number of molecules escaping from the liquid equals the number of molecules leaving the vapor and recaptured into the liquid. The saturation vapor pressure increases exponentially with temperature.

Sedimentation: Deposition of waterborne or wind borne particles resulting from a decrease in transport capacity.

Sediment basin: Pond at the upper end of a conveyance or reservoir for detaining particle-laden water for a sufficient length of time for deposition to occur.

Sediment load: Amount of sediment carried by running water or wind.

Seepage: The movement of water into and through the soil from unlined canals, ditches, and water storage facilities.

Semiarid climate: Climate characterized as neither entirely arid nor humid, but intermediate between the two conditions. A region is usually considered as semiarid when precipitation averages between 250 mm (10 in.) and 500 mm (20 in.) per year.

Sheet flow: Water, usually storm runoff, flowing in a thin layer over the soil or other smooth surface.

Silt: 1) A soil separate consisting of particles between 2 and 50 μm in diameter. 2) (Colloquial) Deposits of sediment which may contain soil particles of all sizes.

Silt bar: A deposition of sediment in a channel.

Sinuosity: Ratio of the stream length to the valley length.

Sludge: The solids which are removed from raw water or wastewater during water treatment.

Snow course: A designated line along which the snow is sampled at appropriate times to determine its depth and density (water content) for forecasting water supplies.

Snow density: Water content of snow expressed as a percentage by volume.

Sodic soil: A nonsaline soil containing sufficient exchangeable sodium to adversely affect crop production and soil structure. The exchangeable-sodium-percentage is greater than 15 and the electrical conductivity of the saturation extract is less than 4 mS/cm (0.01 mho/in).

Sodium adsorption ratio, SAR: The proportion of soluble sodium ions in relation to the soluble calcium and magnesium ions in the soil water extract. (Can be used to predict the exchangeable sodium percentage.)

Sodium percentage: Percentage of total cations that is sodium in water or soil solution.

Soil: The unconsolidated minerals and material on the immediate surface of the earth that serves as a natural medium for the growth of plants.

Soil aeration: Process by which air and other gases enter the soil or are exchanged.

Soil and water conservation district, SWCD: A local governmental entity within a defined water or soil protection area that provides assistance to residents in conserving natural resources, especially soil and water.

Soil compaction: Consolidation, reduction in porosity, and collapse of the structure of soil when subjected to surface loads.

Soil conservation: Protection of soil against physical loss by erosion and chemical deterioration by the application of management and land-use methods that safeguard the soil against all natural and human-induced factors.

Soil erodibility: A measure of the soil's susceptibility to erosional processes.

Soil erosion: Detachment and movement of soil from the land surface by wind or water. (See Erosion.)

Soil structure: The combination or arrangement of primary soil particles, into secondary particles, units, or peds that make up the soil mass. These secondary units may be, but usually are not, arranged in the profile in such a manner as to give a distinctive characteristic pattern. The principal types of soil structure are platy, prismatic, columnar, blocky, and granular.

Soil texture: Classification of soil by the relative proportions of sand, silt, and clay present in the soil.

Soil water: All water stored in the soil.

Soil-water characteristic curve: Soil-specific relationship between the soil-water matric potential and soil-water content.

Soil-water deficit or depletion: Amount of water required to raise the soil water content of root zone to field capacity.

Specific heat of water: The amount of heat required to raise the temperature of one gram of water one degree Celsius.

Specific retention: Amount of water that a unit volume of porous media or soil, after being saturated, will retain against the force of gravity. (Compare to Specific yield.)

Specific yield: Amount of water that a unit volume of porous media or soil, after being saturated will yield when drained by gravity. (Compare to Specific retention.)

Spillway: Conduit through or around a dam for the passage of excess water. May have controls.

Staff gauge: Graduated scale, generally vertical, from which the water surface elevation may be read.

Stage: Elevation of a water surface above or below an established datum; gage height.

Static lift: Vertical distance between source and discharge water levels in a pump installation.

Steady flow: Open-channel flow in which the rate and cross-sectional area remain constant with time at a given station.

Steady-state flow: A condition of groundwater flow where there is no change in head with time, which occurs when, at any point in a flow field, the magnitude and direction of the flow velocity are constant with time.

Stefan-Boltzman constant: A universal constant used in the equation relating the rate of emission of radiant energy from the surface of a body to the emissivity of the surface and the temperature of the body.

Stem flow: 1) Precipitation intercepted by vegetation that reaches the ground by flowing down the stems or trunks of vegetation. 2) Flow in the xylem of plants.

Stomata: Pores on the leaf surface which lead to the intracellular spaces within the leaves. It is through the stomata that the water vapor which is transpired exits the leaf.

Storage, specific: The volume of water released from or taken into storage per unit volume of a porous medium per unit change in head.

Storativity: The volume of water an aquifer releases from or takes into storage per unit surface area of the aquifer per unit change in head. It is equal to the product of specified storage and aquifer thickness. In an unconfined aquifer, the storativity is equivalent to the specific yield.

Stratified soils: Soils that are composed of layers, usually varying in permeability and texture.

Stream: A stream is a flow of running water that runs along a channel and has a surface open to the atmosphere. If it flows under the ground it is called a subterranean stream.

Stream bank stabilization: Vegetative or mechanical control of erodible stream banks, including measures to prevent stream banks from caving or sloughing such as lining banks with riprap, or matting and constructing jetties or revetments, as necessary, for permanent protection.

Stream-channel erosion: Scouring of soil and the cutting of channel banks or beds by running water. Sometimes called "streambed erosion" or "stream bank erosion."

Streamflow: The rate of water movement in a stream. Often written as two words.

Subgrade: Earth material beneath a subsurface drain or foundation.

Subsoiling: Tillage operation to loosen the soil below the tillage zone without inversion and with a minimum of mixing with the tilled zone.

Subsurface drain: Subsurface conduits used primarily to remove subsurface water from soil. Classifications of subsurface drains include pipe drains, tile drains, and blind drains.

Surface drainage: The diversion or orderly removal of excess water from the surface of land by means of improved natural or constructed channels, supplemented when necessary by shaping and grading of land surfaces to such channels.

Surface inlet: Structure for diverting surface water into an open ditch, subsurface drain, or pipeline.

Surface irrigation: Broad class of irrigation methods in which water is distributed over the soil surface by gravity flow.

Surface retention: That portion of precipitation required to satisfy interception, the wetting of the soil surface, and depression storage. (See Initial storage.)

Surface roughness: A measure of the impact of surface vegetation on wind speed. Surface roughness is equal to 0.123 times the height of the vegetation.

Surface runoff: Precipitation, snow melt, or irrigation in excess of what can infiltrate or be stored in small surface depressions.

Surface sealing: Reorienting and packing of dispersed soil particles in the immediate surface layer of soil and clogging of surface pores, resulting in reduced infiltration.

Surface storage: Sum of detention and channel storage excluding depression storage, represents at any given moment, the total water enroute to an outlet from an area or watershed.

Surface water: Water flowing or stored on the earth's surface.

Suspended sediment: Material moving in suspension in a fluid, due to the upward components of the turbulent currents or by colloidal suspension. Sometimes called "suspended load."

Tensiometer: Instrument, consisting of a porous cup filled with water and connected to a manometer or vacuum gage, used for measuring the soil-water matric potential.

Terrace: 1) A broad channel, bench, or embankment constructed across the slope to intercept runoff and detain or channel it to protected outlets. 2) A level plain, usually with a steep front, bordering a river, lake, or sea.

Time of concentration: The time it takes water to travel along the hydraulic length.

Transient flow: Unsteady flow that occurs when, at any point in a flow field, the magnitude or direction of the flow velocity changes with time.

Transmissivity: The rate at which water of a prevailing density and viscosity is transmitted through a unit width of an aquifer or confining bed under a unit hydraulic gradient. It is a function of properties of the liquid, the porous media, and the thickness of the porous media.

Transpiration: The process by which water in plants is transferred to the atmosphere as water vapor.

Transpiration ratio: The ratio of weight of water transpired to the weight of dry matter contained in the plant.

Trapezoidal weir: A sharp-crested weir of trapezoidal shape.

Unconfined aquifer: An aquifer whose upper boundary consists of relatively porous natural material which transmits water readily and does not confine water. The water level in the aquifer is the water table.

Uniform flow: Flow in which the velocity and depth are the same at each cross section.

Unsaturated flow: Movement of water in soil in which the pores are not completely filled with water.

Unsaturated zone: The part of the soil profile in which the voids are not completely filled with water.

Vapor pressure: The pressure a gas exerts on the liquid with which it is in contact. In the context of this book, the gas is water vapor, and the liquid is water.

Vapor pressure deficit: Difference between the existing vapor pressure and that of a saturated atmospheric vapor pressure at the same temperature.

Wash load: That part of the sediment load of a stream that is composed of suspended clay and silt particles.

Wastewater: Water of reduced quality that has been used for some purpose and discarded.

Water holding capacity: Amount of soil water available to plants. (See Available soil water.)

Water rights: Legal rights to use water supplies derived from common law, court decisions, or statutory enactments.

Watershed: Land area that contributes runoff (drains) to a given point in a stream or river. Synonymous with catchment and drainage or river basin.

Watershed gradient: The average slope in a watershed measured along a path of water flow from a given point in the stream channel to the most remote point in the watershed.

Water table: The upper surface of saturated zone below the soil surface where the water is at atmospheric pressure.

Weir: 1) Structure across a stream to control or divert the flow. 2) Device for measuring the flow of water. Classification includes sharp-crested or broad-crested with rectangular, trapezoidal, or triangular cross sections.

Well casing: Pipe installed within a borehole to prevent collapse of sidewall material, to receive and protect pump and pump column, and to allow water flow from the aquifer to pump intake.

Well screen: That part of the well casing which has openings through which water enters.

Well test: Determination of the well yield versus drawdown relationship with time.

Well yield: Discharge that can be sustained from a well through some specified period of time. (See Safe well yield.)

Wetlands: Area of wet soil that is inundated or saturated under normal circumstances and would support a prevalence of hydrophytic plants.

Wetted perimeter: Length of the wetted contact between a conveyed liquid and the open channel or closed conduit conveying it, measured in a plane at right angles to the direction of flow.

Wind erosion: Detachment, transportation, and deposition of soil by the action of wind. The removal and redeposition may be in more or less uniform layers or as localized blowouts and dunes.

Xylem: The woody vascular tissue of a plant that conducts water and mineral salts in the stems, roots, and leaves and gives support to the softer tissues.

APPENDIX C

Extreme Event Information for the United States
(Provided by the USDA-NRCS)

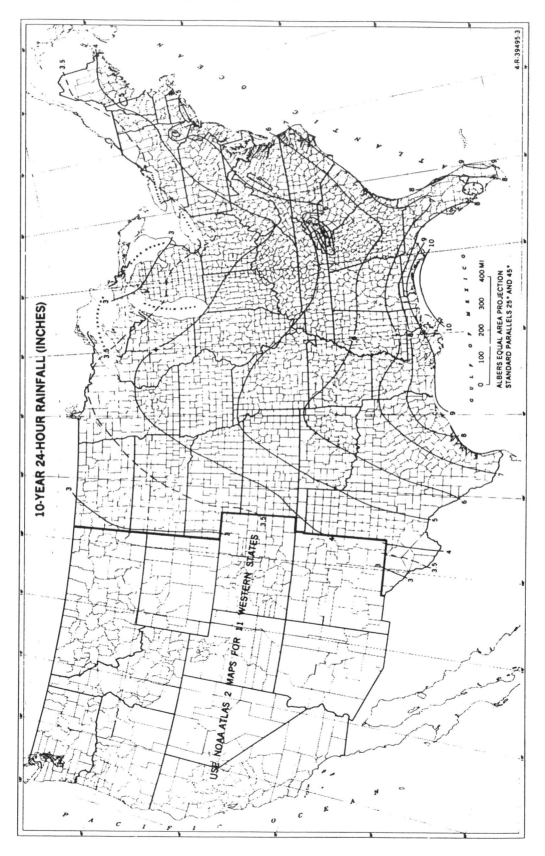

10-YEAR 24-HOUR RAINFALL (INCHES)

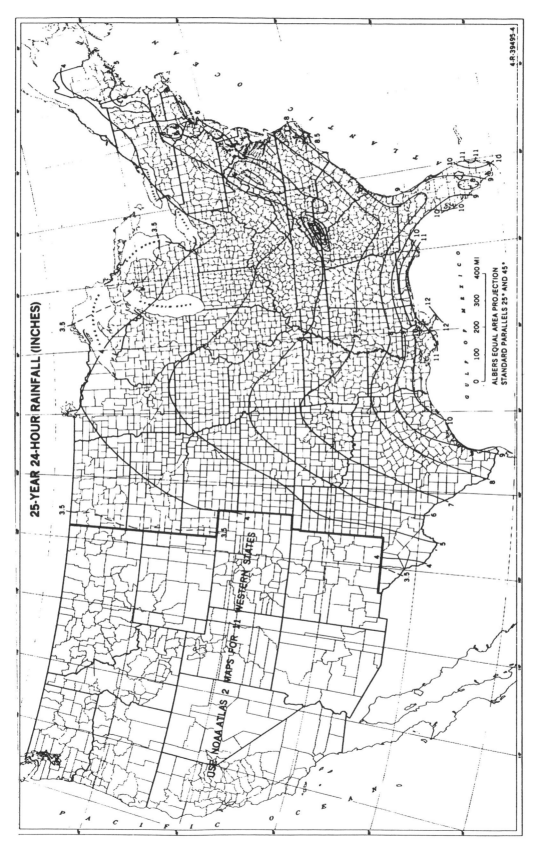

25-YEAR 24-HOUR RAINFALL (INCHES)

USE NOAA ATLAS 2 MAPS FOR 11 WESTERN STATES

ALBERS EQUAL AREA PROJECTION
STANDARD PARALLELS 25° AND 45°

0   100   200   300   400 MI

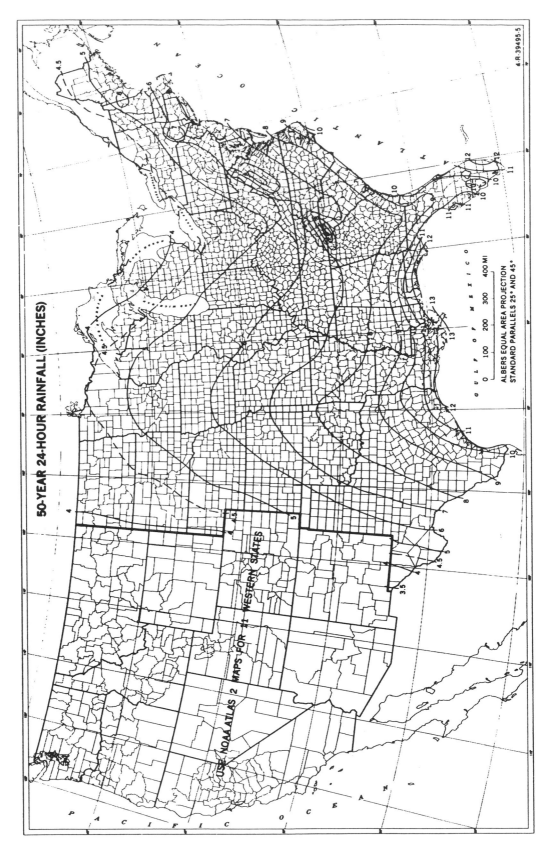

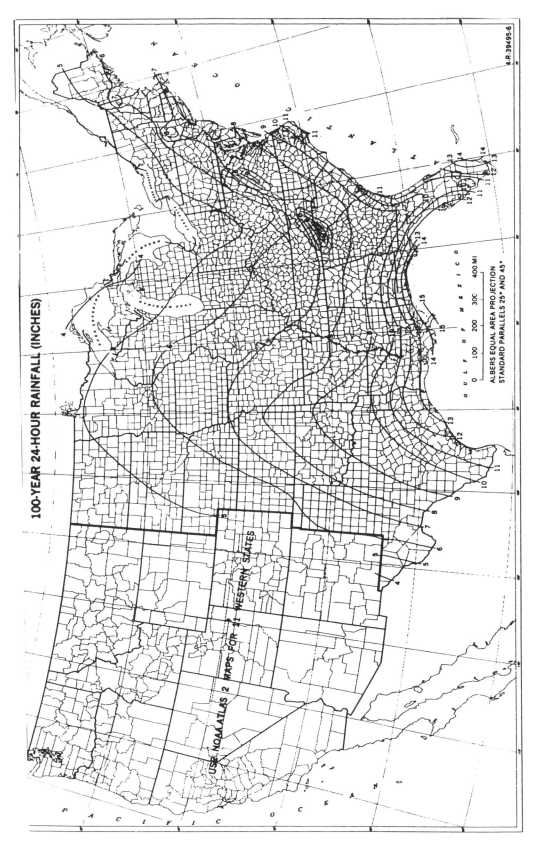

100-YEAR 24-HOUR RAINFALL (INCHES)

USE NOAA ATLAS 2 MAPS FOR 11 WESTERN STATES

GULF OF MEXICO

0 100 200 300 400 MI

ALBERS EQUAL AREA PROJECTION
STANDARD PARALLELS 25° AND 45°

4-R-39495-6

APPENDIX D

Hydrologic Soil Group and Erodibility Factors for the Most Common Soils in Each State
(Provided by the USDA-NRCS)

Soil Series	Group Type	K Values	Soil Series	Group Type	K Values
Alabama					
Troup	A	.10 - .10	Almontevallo	D	.28 - .28
Luverne	C	.24 - .24	Algorgas	D	.17 - .20
Smithdale	B	.17 - .17	Townley	C	.28 - .37
Nauvoo	B	.28 - .28	Bibb	D	.20 - .20
Lorangeburg	B	.10 - .10	Dothan	B	.15 - .15
Alaska					
Tolstoi	D	.37 - .37	Akina	D	.05 - .05
Salamatof	D	.05 - .05	Mosman	D	.15 - .15
Maybeso	D	.05 - .05	Godstream	D	.05 - .05
Kupreanof	B C	.24 - .24	Strandline	B	.37 - .37
McGilvery	D	.05 - .05	Kushneahin	D	.05 - .05
Arizona					
Winona	D	.15 - .32	Sheppard	A	.10 - .20
Thunderbird	D	.05 - .24	Mohall	B	.20 - .43
Barkerville	C	.20 - .20	Gilman	B D	.28 - .55
Springerville	D	.05 - .28	Denure	B	.10 - .55
Clovis	B	.17 - .28	Gunsight	B	.02 - .15
Arkansas					
Enders	C	.32 - .32	Nella	B	.15 - .15
Mountainburg	D	.17 - .24	Leadvale	C	.43 - .43
Linker	B	.24 - .28	Sacul	C	.28 - .28
Sharkey	D	.43 - .43	Guyton	D	.43 - .43
Carnasaw	C	.24 - .32	Perry	D	.37 - .37
California					
Cajon	A/B	.10 - .24	Gaviota	D	.24 - .24
Cieneba	C	.20 - .20	Auburn	D	.15 - .32
Hanford	B	.20 - .32	Millsholm	D	.20 - .37
San Joaquin	D	.32 - .32	Rositas	A C	.10 - .43
Maymen	D	.20 - .24	Los Osos	C	.32 - .32
Colorado					
Valent	A	.17 - .24	Olney	B	.15 - .28
Wiley	B	.32 - .32	Platner	C	.20 - .28
Ascalon	B	.15 - .24	Baca	B C	.20 - .37
Vona	B	.20 - .28	Manvel	B C	.28 - .43
Weld	C	.20 - .37	Stoneham	B	.37 - .37
Connecticut					
Charlton	B	.20 - .24	Canton	B	.20 - .24
Hollis	C/D	.20 - .20	Sutton	B	.20 - .24
Paxton	C	.20 - .24	Ridgebury	C	.20 - .24
Woodbridge	C	.20 - .24	Leicester	C	.24 - .28
Hinckley	A	.20 - .20	Wethersfield	C	.24 - .28

Soil Series	Group Type	K Values	Soil Series	Group Type	K Values
Delaware					
Fallsington	B/D	.24 - .24	Matapeake	B	.49 - .49
Sassafras	B	.28 - .28	Rumford	B	.17 - .17
Evesboro	A	.17 - .17	Chincoteague	D	.32 - .32
Pocomoke	B/D	.20 - .20	Johnston	D	.17 - .17
Woodstown	C	.24 - .24	Keyport	C	.43 - .43
Florida					
Myakka	D B/D	.10 - .10	Blanton	A B	.10 - .10
Candler	A	.10 - .10	Riviera	D B/D C/D	.10 - .10
Immokalee	D B/D	.10 - .10	Basinger	D B/D	.10 - .10
Lakeland	A	.10 - .10	Beaugallie	D B/D	.10 - .10
Smyrna	D B/D	.10 - .10	Pineda	D B/D	.10 - .15
Georgia					
Tifton	B	.10 - .10	Pelham	B/D	.10 - .10
Cecil	B	.28 - .28	Orangeburg	B	.10 - .10
Madison	B	.24 - .24	Dothan	B	.15 - .15
Fuquay	B	.15 - .15	Leefield	C	.10 - .10
Pacolet	B	.15 - .20	Cowarts	C	.15 - .15
Hawaii					
Akaka	A	.05 - .05	Kahaluu	D	.02 - .02
Kekake	D	.02 - .02	Mawae	A	.02 - .02
Kiloa	A	.02 - .02	Guam	D	.05 - .05
Keei	D	.02 - .02	Hanipoe	B C	.10 - .17
Puu Pa	A	.10 - .20	Honokaa	A	.05 - .05
Idaho					
Portneuf	B	.49 - .49	Power	B	.43 - .43
Purdam	C	.32 - .43	Rexburg	B	.32 - .49
Pancheri	B	.43 - .49	Arbidge	C	.20 - .24
Ririe	B	.43 - .49	Chilcott	C D	.24 - .49
Wickahoney	D	.20 - .20	Declo	B	.37 - .49
Illinois					
Drummer	B/D	.28 - .28	Rozetta	B	.37 - .43
Tama	B	.28 - .43	Sable	B/D	.28 - .28
Ipava	B	.28 - .28	Bluford	C	.43 - .43
Hickory	C	.37 - .37	Flanagan	B	.28 - .28
Fayette	B	.32 - .43	Hosmer	C	.43 - .43
Indiana					
Crosby	C	.37 - .43	Wellston	B	.37 - .37
Blount	C	.43 - .43	Glynwood	C	.43 - .43
Pewamo	C/D	.24 - .24	Fincastle	C	.37 - .37
Miami	B	.37 - .37	Crider	B	.32 - .32
Brookston	B/D	.28 - .28	Cincinnati	C	.43 - .43
Iowa					
Clarion	B	.24 - .28	Canisteo	B/D	.24 - .24
Fayette	B	.32 - .43	Webster	B/D	.24 - .24
Colo	B B/D	.28 - .37	Marshall	B	.28 - .43
Galva	B	.28 - .32	Tama	B	.28 - .43
Nicollet	B	.24 - .24	Monona	B	.28 - .43

Soil Series	Group Type	K Values	Soil Series	Group Type	K Values
\multicolumn{6}{c}{**Kansas**}					
Harney	B	.32 - .32	Holdrege	B	.32 - .32
Ulysses	B	.32 - .32	Uly	B	.32 - .32
Richfield	B	.32 - .32	Irwin	D	.32 - .37
Keith	B	.32 - .32	Colby	B	.43 - .43
Crete	C	.37 - .37	Clime	C	.20 - .28
\multicolumn{6}{c}{**Kentucky**}					
Shelocta	B	.32 - .32	Zanesville	C	.43 - .43
Eden	C	.17 - .43	Faywood	C	.32 - .37
Latham	D	.43 - .43	Crider	B	.32 - .32
Lowell	C	.37 - .37	Caneyville	C	.28 - .43
Loring	C	.49 - .49	Baxter	B	.28 - .28
\multicolumn{6}{c}{**Louisiana**}					
Sharkey	D	.20 - .43	Allemands	D	.32 - .32
Guyton	D	.43 - .43	Fausse	D	.20 - .20
Sacul	C	.28 - .28	Barbary	D	
Commerce	C	.37 - .37	Moreland	D	.43 - .43
Ruston	B	.15 - .15	Clovelly	D	
\multicolumn{6}{c}{**Maine**}					
Lyman	C/D	.20 - .28	Plaisted	C	.20 - .20
Monarda	D	.15 - .28	Burnham	D	.24 - .24
Marlow	C	.20 - .24	Colonel	C	.17 - .20
Thorndike	C/D	.17 - .20	Hermon	A	.10 - .17
Telos	C	.15 - .28	Dixfield	C	.17 - .20
\multicolumn{6}{c}{**Maryland**}					
Sassafras	B	.20 - .28	Beltsville	C	.43 - .43
Manor	B	.32 - .37	Mattapex	C	.37 - .37
Glenelg	B	.24 - .32	Matapeake	B	.43 - .49
Othello	D C/D	.24 - .37	Elkton	D C/D	.24 - .43
Fallsington	B/D	.24 - .24	Woodstown	C	.24 - .28
\multicolumn{6}{c}{**Massachusetts**}					
Paxton	C	.20 - .24	Woodbridge	C	.20 - .24
Hinckley	A	.17 - .20	Freetown	D	
Lyman	C/D	.20 - .28	Canton	B	.20 - .24
Carver	A	.10 - .10	Hollis	C/D	.20 - .24
Merrimac	A	.24 - .24	Charlton	B	.20 - .24
\multicolumn{6}{c}{**Michigan**}					
Kalkaska	A	.15 - .15	Capac	C	.32 - .32
Marlette	B	.32 - .32	Parkhill	B/C	.17 - .24
Spinks	A	.15 - .15	Graycalm	A	.10 - .10
Rubicon	A	.10 - .15	Houghton	D A/D	
Oshtemo	B	.24 - .24	Boyer	B	.17 - .24
\multicolumn{6}{c}{**Minnesota**}					
Canisteo	D B/D	.16 - .24	Hamerly	C	.28 - .28
Clarion	B	.24 - .28	Glencoe	D B/D	.28 - .28
Webster	B/D	.24 - .24	Ves	B	.17 - .24
Barnes	B	.20 - .28	Nicollet	B	.24 - .24
Lester	B	.28 - .28	Seelyeville	D A/D	.10 - .10
\multicolumn{6}{c}{**Mississippi**}					
Smithdale	B	.17 - .17	Sweatman	C	.37 - .37
Sharkey	D	.20 - .43	Alligator	D	.37 - .37
Providence	C	.49 - .49	Ruston	B	.15 - .15
Memphis	B	.49 - .49	Ora	C	.28 - .28
Loring	C	.49 - .49	Savannah	C	.37 - .37

Soil Series	Group Type	K Values	Soil Series	Group Type	K Values
Missouri					
Clarksville	B	.28 - .28	Keswick	C	.32 - .37
Goss	B	.24 - .24	Shelby	B	.28 - .37
Armstrong	C	.32 - .37	Lamoni	C	.37 - .37
Mexico	D	.43 - .43	Gara	C	.28 - .37
Menfro	B	.37 - .37	Lagonda	C	.37 - .37
Montana					
Cabbart	D	.37 - .37	Phillips	C	.43 - .43
Yawdim	D	.32 - .32	Delpoint	C	.20 - .37
Williams	B	.20 - .43	Neldore	D	.32 - .32
Cabba	D	.17 - .24	Scobey	C	.20 - .43
Zahill	C	.28 - .37	Cambert	C	.37 - .37
Nebraska					
Valentine	A	.15 - .17	Uly	B	.32 - .32
Valent	A	.17 - .24	Nora	B	.32 - .32
Coly	B	.43 - .43	Crete	C	.37 - .37
Holdrege	B	.17 - .32	Moody	B	.32 - .32
Hastings	B	.32 - .32	Hobbs	B	.32 - .32
Nevada					
Stewval	D	.10 - .10	Theon	D	.10 - .20
Palinor	D	.24 - .24	Mazuma	B C	.28 - .55
Cleavage	D	.10 - .20	Orovada	B	.15 - .49
Unsel	B	.10 - .24	Downeyville	D	.05 - .24
Chiara	D	.20 - .55	Sumine	C	.17 - .24
New Hampshire					
Marlow	C	.20 - .24	Canton	B	.20 - .24
Becket	C	.17 - .20	Tunbridge	C	.20 - .24
Monadnock	B	.17 - .28	Berkshire	B	.20 - .24
Lyman	C/D	.20 - .28	Peru	C	.20 - .24
Hermon	A	.10 - .17	Colton	A	.15 - .15
New Jersey					
Downer	B	.20 - .20	Rockaway	C	.17 - .24
Sassafras	B	.20 - .28	Lakewood	A	.10 - .10
Atsion	D C/D	.17 - .17	Manahawkin	D	.05 - .05
Evesboro	A	.17 - .17	Aura	B	.43 - .43
Lakehurst	A	.17 - .17	Freehold	B	.28 - .28
New Mexico					
Deama	D	.05 - .20	Pastura	D	.37 - .37
Ector	D	.10 - .15	Travessilla	D	.10 - .55
Kimbrough	D	.37 - .37	Upton	C	.15 - .15
Clovis	B	.28 - .28	Lozier	D	.15 - .15
Amarillo	B	.24 - .24	Berino	B	.17 - .17
New York					
Volusia	C	.24 - .37	Bath	C	.24 - .24
Mardin	C	.24 - .32	Honeoye	B	.24 - .32
Lordstown	C	.20 - .20	Ontario	B	.24 - .32
Arnot	C/D	.24 - .24	Oquaga	C	.20 - .28
Nassau	C	.32 - .32	Howard	A	.32 - .32
North Carolina					
Pacolet	B	.15 - .20	Georgeville	B	.24 - .43
Cecil	B	.24 - .28	Goldsboro	B	.20 - .20
Norfolk	B	.20 - .20	Appling	B	.24 - .24
Rains	B/D	.15 - .15	Chewacla	C	.24 - .24
Evard	B	.15 - .24	Badin	B	.15 - .32

Soil Series	Group Type	K Values	Soil Series	Group Type	K Values
North Dakota					
Barnes	B	.20 - .28	Buse	B	.20 - .28
Williams	B	.15 - .28	Fargo	D	.32 - .32
Svea	B	.28 - .28	Cabba	D	.20 - .24
Hamerly	C	.28 - .28	Parnell	C/D	.28 - .28
Zahl	B	.28 - .28	Tonka	C/D	.32 - .32
Ohio					
Blount	C	.43 - .43	Westmoreland	B	.28 - .37
Hoytville	C/D	.24 - .28	Bennington	C	.43 - .43
Pewamo	C/D	.24 - .24	Miamian	C	.37 - .37
Gilpin	C	.24 - .32	Upshur	D	.37 - .43
Crosby	C	.43 - .43	Mahoning	D	.43 - .43
Oklahoma					
Dennis	C	.43 - .43	Hector	D	.10 - .15
Stephenville	B	.17 - .20	Port	B	.37 - .37
Richfield	B	.32 - .32	Darnell	C	.20 - .20
Quinlan	C	.37 - .37	Clarksville	B	.28 - .28
Carnasaw	C	.32 - .32	Woodward	B	.37 - .37
Oregon					
Lickskillet	C D	.05 - .24	Condon	C	.43 - .43
Bohannon	C	.10 - .15	Bakeoven	D	.05 - .10
Ritzville	B	.49 - .49	Klickitat	B	.20 - .20
Walla Walla	B	.43 - .43	Peavine	C	.28 - .28
Preacher	B	.17 - .17	Simas	C	.17 - .37
Pennsylvania					
Hazleton	B	.15 - .17	Berks	C	.17 - .17
Gilpin	C	.24 - .32	Cookport	C	.24 - .32
Dekalb	A C	.17 - .17	Ernest	C	.32 - .43
Weikert	B/D	.20 - .28	Wellsboro	C	.24 - .32
Oquaga	C	.20 - .28	Buchanan	C	.24 - .24
Rhode Island					
Canton	B	.20 - .24	Paxton	C	.20 - .24
Charlton	B	.20 - .24	Newport	C	.24 - .28
Hinckley	A	.20 - .20	Bridgehampton	B	.43 - .49
Woodbridge	C	.20 - .24	Ridgebury	C	.20 - .24
Merrimac	A	.24 - .24	Sutton	B	.20 - .24
South Carolina					
Cecil	B	.28 - .28	Coxville	D	.24 - .24
Pacolet	B	.20 - .20	Rains	B/D	.15 - .15
Lakeland	A	.10 - .10	Wilkes	C	.24 - .24
Lynchburg	C	.15 - .15	Madison	B	.24 - .24
Goldsboro	B	.20 - .20	Johnston	D	.17 - .17
South Dakota					
Sansarc	D	.37 - .37	Highmore	B	.32 - .32
Opal	D	.37 - .37	Williams	B	.15 - .28
Clarno	B	.20 - .20	Houdek	B	.20 - .28
Pierre	D	.37 - .37	Promise	D	.37 - .37
Samsil	D	.37 - .37	Lakoma	D	.37 - .37
Tennessee					
Bodine	B	.28 - .28	Lexington	B	.49 - .49
Memphis	B	.49 - .49	Talbott	C	.32 - .37
Smithdale	B	.17 - .17	Grenada	C	.49 - .49
Loring	C	.49 - .49	Ramsey	D	.17 - .20
Baxter	B	.28 - .37	Mimosa	C	.28 - .37

Soil Series	Group Type	K Values	Soil Series	Group Type	K Values
Texas					
Tarrant	D	.10 - .20	Miles	B	.24 - .24
Pullman	D	.32 - .32	Houston Black	D	.32 - .32
Ector	D	.15 - .15	Olton	C	.32 - .32
Amarillo	B	.24 - .24	Crockett	D	.43 - .43
Reagan	B	.32 - .37	Sherman	D	.32 - .32
Utah					
Rizno	D	.28 - .32	Chipeta	D	.43 - .43
Skumpah	B D	.43 - .55	Tooele	B	.17 - .37
Saltair	D	.49 - .49	Hiko Peak	B	.10 - .20
Moenkopie	D	.10 - .15	Sheppard	A	.20 - .20
Amtoft	D	.10 - .24	Begay	B	.49 - .49
Vermont					
Tunbridge	C	.20 - .24	Vergennes	C	.49 - .49
Berkshire	B	.20 - .24	Rawsonville	C	.43 - .49
Lyman	C/D	.20 - .28	Cabot	D	.28 - .32
Peru	C	.20 - .24	Houghtonville	C	.43 - .49
Marlow	C	.20 - .24	Woodstock	D	.24 - .24
Virginia					
Cecil	B	.28 - .28	Madison	B	.24 - .24
Appling	B	.24 - .24	Hayesville	B C	.15 - .24
Frederick	B	.28 - .32	Nason	B C	.24 - .43
Berks	C	.17 - .17	Weikert	B/D	.20 - .28
Emporia	C	.28 - .28	Tatum	B	.20 - .37
Washington					
Ritzville	B	.49 - .49	Walla Walla	B	.43 - .43
Alderwood	C	.15 - .15	Palouse	B	.32 - .32
Shano	B	.55 - .55	Warden	B	.55 - .55
Athena	B	.37 - .37	Newbell	B	.24 - .28
Quincy	A	.15 - .32	Aits	B	.24 - .37
West Virginia					
Gilpin	C	.24 - .32	Weikert	B/D	.20 - .28
Dekalb	A C	.17 - .17	Pineville	B	.20 - .24
Berks	C	.17 - .17	Westmoreland	B	.37 - .37
Upshur	D	.37 - .43	Calvin	C	.15 - .37
Muskingum	C	.24 - .37	Cateache	C	.28 - .32
Wisconsin					
Pence	B	.24 - .24	Kewaunee	C	.17 - .37
Plainfield	A	.15 - .17	Magnor	C	.37 - .37
Menahga	A	.15 - .15	Fayette	B	.32 - .43
Padus	B	.24 - .24	Seaton	B	.37 - .37
Newglarus	B C	.37 - .37	Valton	B C	.32 - .32
Wyoming					
Shingle	D	.02 - .37			
Hiland	B	.20 - .37			
Forkwood	B C	.32 - .43			
Kishona	B C	.28 - .43			
Theedle	C	.32 - .32			
American Samoa					
Aua	B	.17 - .17	Puapua	D	.10 - .10
Pavaiai	C	.10 - .10	Sogi	C	.10 - .10
Ofu	B	.10 - .10	Leafu	C	.17 - .17
Oloava	B	.10 - .10	Tafuna	A	.02 - .02
Iliili	D	.05 - .05	Fagasa	C	.10 - .10

Soil Series	Group Type	K Values	Soil Series	Group Type	K Values
Fed. Sts. Micronesia					
Tolonier	B	.05 - .05	Umpump	B	.15 - .15
Dolen	B	.05 - .05	Rumung	C	.10 - .10
Fomseng	C	.10 - .10	Weloy	C	.10 - .10
Naniak	D	.05 - .05	Yap	B	.10 - .10
Dolekei	B	.10 - .10	Ilachetomel	D	.05 - .05
Guam					
Guam	D	.05 - .05	Inarajan	C	.17 - .24
Akina	B	.20 - .20	Ylig	C	.24 - .24
Pulantat	C	.24 - .24	Togcha	B	.15 - .15
Agfayan	D	.20 - .20	Atate	B	.15 - .15
Ritidian	D	.02 - .02	Shioya	A	.15 - .15
Marshall Islands					
Ngedebus	A	.05 - .10	Majuro	A	.02 - .02
North Mariana Islands					
Chinen	D	.10 - .15	Laolao	B	.15 - .15
Luta	D	.10 - .10	Kagman	C	.05 - .15
Takpochao	D	.10 - .10	Shioya	A	.15 - .15
Dandan	C	.15 - .15	Banaderu	D	.20 - .20
Saipan	B	.02 - .15	Akina	B	.20 - .20
Palau					
Aimeliik	B	.10 - .10	Ngardmau	B	.05 - .05
Palau	B	.10 - .10	Dechel	D	.15 - .15
Ilachetomel	D	.05 - .05	Wollei	D	.10 - .10
Ngardok	B	.15 - .15	Peleliu	D	.05 - .05
Babelthuap	B	.05 - .05	Tabecheding	C	.17 - .17
Puerto Rico					
Mucara	D	.10 - .10	Descalabrado	D	.24 - .24
Caguabo	D	.24 - .24	Pandura	D	.17 - .17
Humatas	C	.02 - .02	Soller	D	.17 - .17
Consumo	B	.10 - .10	Naranjito	C	.10 - .10
Los Guineos	C	.10 - .10	Callabo	C	.10 - .10

APPENDIX E

SOURCES OF DATA AND IMAGES
by John Grimson Lyon

Sources of Aerial Photographs and Sensor Images

U.S. Geological Survey (USGS)

The EROS Data Center archives and produces copies of aerial photographs and other imagery acquired by Department of the Interior agencies. These acquisitions include USGS mapping photographs, products from high altitude aerial photography programs such as NAPP and NHAP, and products which are of different emulsion types such as color infrared, color, and black and white infrared, and aerial radar images. The EROS Data Center also archives Landsat satellite data and NASA space and aerial images. A computerized data base of products can be accessed by latitude and longitude locators, or by USGS quadrangle or other reference location. Computer printouts and microfiche of high altitude aerial photographs are available at no charge. Copies of products can be ordered from EDC.

For assistance please contact User Services, EROS Data Center, Sioux Falls, SD 57198, (605) 594-6151.

U.S. Department of Agriculture

The former USDA-ASCS holds photographs acquired by agencies of the Department of Agriculture. These groups include the U.S. Forest Service, U.S. Natural Resource Conservation Service (formerly the Soil Conservation Service, SCS), the former ASCS and other agencies. Listings of holdings are supplied by computer printout and are available at no cost. One can access their holdings by latitude and longitude of the site, or by county name.

For information contact the Aerial Photography Field Office, USDA, Sales Branch, 2222 West 2300 South, Salt Lake City, UT 84125, or POB 30010, Salt Lake City, UT 84130-0010, (801) 524-5856.

National Archives and Record Service (NARS)

This agency archives a variety of photo and map data and, in particular, archives aerial photographs flown by the government previous to 1945. Generally, one or two dates of aerial photography coverage are available for counties in the U.S. They publish a pamphlet that describes available coverage and provides ordering details. For more information, please contact the National Archives and Records Service, Cartographic Branch, Washington, D.C., 20408, (703)765-6700.

National Oceanic and Atmospheric Administration (NOAA)

National Climatic Data Center

The Satellite Data Service Division of the National Climatic Data Center provides a variety of image and digital data products from weather and environmental satellites. These products from the geostationary weather satellites (GOES), include the moderate resolution satellite data used to make biomass maps of regional and continental areas (Advanced Very High Resolution Radiometer, AVHRR), and other atmospheric and weather satellite data. Technical and ordering information are available from Satellite Data Service Division (NOAA-NCDC), Room 100, Princeton Executive Square, Washington, D.C. 20233.

National Ocean Service (NOAA-NOS)

The Photogrammetry Branch has acquired data over coastal and offshore areas in support of its mapping mandates. For more information contact the Photogrammetry Branch, National Oceanic and Atmospheric Administration (NOAA), 6001 Executive Blvd,, Rockville, MD 20852, (301)443-8601.

Other Sources of Data or Information

Research and engineering efforts often involve federal lands themselves, or land in-holdings found within federal management and ownership boundaries. In these cases, appropriate aerial photographs may be available locally or regionally from the following agencies.

U.S. Forest Service (USFS)

The USFS contracts and acquires aerial photographic coverage over National Forest lands and other areas related to their mandates. A variety of scales and film types have been employed. Most of these photographs are available through the USDA Aerial Photography Field Office in Salt Lake City.

The Forest Service has nine regions with regional headquarters where regional aerial photographic coverage can often be viewed and ordered. Consult the government section of your telephone book for details.

Inquiries may also be referred to the Division of Engineering, U.S. Forest Service, Washington, D.C., 20250, or to the Public Affairs Office, USDA-U.S. Forest Service, POB 96090, Washington, D.C. 20090-6090.

U.S. Bureau of Land Management (BLM)

The BLM has obtain coverage of the lands the manage. Local coverage is available through the USDA-ASCS Aerial Photography Field Office in Salt Lake City, UT.

Inquiries may also be made to U.S. BLM, Office of Public Affairs, Washington, D.C. 20240.

U.S. National Park Service (NPS)

NPS contracts for aerial photography over and about U.S. National Park land and other areas such as National Monuments and National Recreational Areas. Their holdings are available through the EROS Data Center in Sioux Falls, SD. Additional information may be available from the National Park Service, Office of Public Inquiries, Room 1013, Washington, D.C. 20240 or NPS, Denver Service Center, 655 Parfet Street, POB 25287, Denver, CO, 80225, (303)234-5132.

Sources of Maps

Small scale topographic and other maps are available from the U.S. Geological Survey. These maps may be purchased locally from vendors such as map store, climbing and outdoor shops, and hunting and fishing stores. Naturally, these maps can be obtained from the USGS. One should be aware of the time frame necessary for delivery of products. It may be faster to obtain maps locally, or to borrow maps available in an archive such as a library or state natural resource agency.

Map orders may be placed through the USGS Map Sales, Box 25286, Denver, CO 80225.

The public can obtain information and purchase some maps on a "walk up" basis at USGS's Earth Science Information Centers (ESIC). The USGS operates the larger ESIC's, and they include:

Anchorage-ESIC
4230 University Drive, Room 101
Anchorage, AK 99508-4664

Anchorage-ESIC
Room G-84
605 West 4th Avenue
Anchorage, AK 99501

Denver-ESIC
169 Federal Building
1961 Stout Street
Denver, CO 80294

Lakewood-ESIC
Box 25046, Federal Center, MS 504
Building 25, Room 1813
Denver, CO 80225-0046

Menlo Park-ESIC
Building 3, MS 532, Room 3128
345 Middlefield Road
Menlo Park, CA 94025

Reston-ESIC
USGS
507 National Center
Reston, VA 22092

Rolla-ESIC
1400 Independence Road
Rolla, MO 65401

Salt Lake City-ESIC
8105 Federal Building
125 South State Street
Salt Lake City, UT 84138

San Francisco-ESIC
504 Custom House
555 Battery Street
San Francisco, CA 94111

Sioux Falls-ESIC
EROS Data Center
Sioux Falls, SD 57198

Spokane-ESIC
678 U.S. Courthouse
West 920 Riverside Avenue
Spokane, WA 99201

Stennis Space Center-ESIC
Building 3101
Stennis Space Center, MS 39529

U.S. Department of Interior
1849 C Street, NW, Room 2650
Washington, D.C. 20240

In all the states there exists a local ESIC contact. These units are based at state natural resource agencies, state universities, and other agencies that have an interest in helping the public obtain maps and aerial photographs. Commonly, the local ESIC will maintain aerial photographs of the state, and they have access to USGS databases of existing aerial photos. Hence, they can be very valuable in identifying local aerial coverage.

The local ESIC offices often are very knowledgeable concerning aerial photographs held by state agencies and private companies. It may also be possible to access these local sources of photos by interaction with a regional or national data base. Locate the local or regional ESIC office in your state by directory assistance.

The U.S. Geological Survey also vends a variety of digital cartographic and geographic data. These computer compatible files allow the user to access data digitally, and conduct computer processing and graphical display exercises. The most interesting data are the Digital Elevation Map (DEM) products, which are files of point elevations, and the Digital Line Graph (DLG) products which display cultural or planimetric details such as roads and resource data such as drainage systems.

Should the application require these sorts of information, DEM and DLG data sets can be of great assistance in regional analyses of hydrological variables. You may wish to identify other sources of digital data.

Navigation Maps

Conventional navigation maps and bathymetric maps are available from Mapping and Charting, National Oceanic and Atmospheric Administration (NOAA), 6001 Executive Blvd,, Rockville, MD 20852, or for orders contact the Distribution Branch, National Ocean Service, NOAA, Riverdale, MD 20737, (301)436-6990.

National Wetland Inventory Maps

National Wetland Inventory Maps (NWI) and information are available from the Earth Science Information Center, USGS, 507 National Center, Reston, VA 22092, (703)648-5920 or (800)USA-MAPS.

These wetland inventory maps are also available in digital form to facilitate computer processing and display. A very good use of such data would be in a Geographic Information System (GIS) database. A sample of the utility of these data is provided by the 45-year study of wetlands along the St. Mary's River area in Michigan (see Lyon, 1993). These digital data are available from the National Wetlands Inventory Digital Cartographic and Geographic Data, US Geological Survey, Earth Science Information Center (ESIC), 507 National Center, Reston, VA 22092. They are also available via Internet.

Floodplain Maps

Floodplain maps are available from the Federal Emergency Management Agency (FEMA) through its local, state and regional offices. These offices can be located through the government section of the telephone directory. These maps and information can also be obtained from FEMA, 500 C Street, Washington, D.C. 20472.

Other Data Sources

The National Weather Service of the Department of Commerce can be contacted for a variety of weather records. Products which record the weather conditions at a neighboring station are available on a daily basis. Summary statistics by location, region, and state are also available. Many times these publications and general records are deposited in libraries and at universities, where they may be accessed quickly and at no charge.

For weather records contact the NOAA National Climatic Data Center, Federal Building, 37 Battery Park Ave., Ashville, NC 28801-2733, (704)259-0682.

For airborne measurements of hydrological variables such as airborne-measured snow moisture equivalents, maps of snow cover during the winter, and other synoptic hydrological data, contact the National Weather Service (NOAA-NWS), National Operational Hydrologic Remote Sensing Center, 6301 - 34th Ave. South, Minneapolis, MN 55450-2985, (612)725-3039.

Regional Hydrological Information

The U.S. Geological Survey maintains offices in most states, and many of these offices are responsible for collecting hydrological information. The information may include water gaging station data for major and minor rivers and streams, and water quality sampling for selected locations.

The local USGS offices concerned with hydrology can be accessed locally through the government pages of the telephone directory. One can also contact the USGS directly at the larger centers listed under the map section, and through the USGS national clearinghouses of information. One can also contact the Hydrologic Information Unit, U.S. Geological Survey, 419 National Center, Reston, VA 22092

Regional Soils Information

The USDA and the Natural Resource Conservation Service (formerly SCS) has created several databases of soil boundaries and soil attributes for the US. These include programs on local soils (SSURGO), regional soils (SSURGO, STATSGO) and national soils (STATSGO, NATSGO). These databases are very amenable to GIS studies. Information can be obtained from: National Cartographic and Geographic Information Systems Center, USDA, POB 6567, Fort Worth, TX 76115, (817)334-5559.

A Hydrologic Units Geographic Database is also available from the Center above, and should be available in DLG-like format.

Additional Sources

The National Technical Information Service archives government and governmental contracted reports of science and engineering activities. It can be a great source of very detailed and technical information on a number of subjects. Write to the National Technical Information Service, U.S. Department of Commerce, 5285 Port Royal Road, Springfield, VA 22161.

The U.S. Government Printing Office is a good source of current governmental publications. They operate a number of bookstores throughout the nation, and they are commonly found in federal buildings in larger cities. Locate the local bookstores through the government section of the telephone directory. The Office may also be contacted at the Superintendent of Documents, U.S. Government Printing Office, North Capitol and H Streets, NW, Washington, D.C. 20402.

Bureau of the Census

The Department of Commerce, Bureau of the Census creates a number of useful products. In particular, the Topologically Integrated Geographic Encoding and Referencing (TIGER) system databases are used to form GIS variables for analyses. TIGER line files and other extract type products provide administrative boundaries, water and coastal boundaries, and other geographic information including attribute characteristics. Information is available from Customer Services, Bureau of the Census, Washington Plaza, Room 315, Washington, D.C. 20233.

U.S. Army Corps of Engineers Commands

When corresponding with U.S. Army Corps of Engineers (USACE) Commands, it is best to locate the District level office that has regional jurisdiction in the area of interest. The District jurisdictions are often based on the boundaries of watersheds, and may not be easy to ascertain. Use the local telephone directory and the government pages to locate the appropriate Command. The USACE Districts and Divisions are also listed in Lyon (1993).

The field activities of USACE are organized under Divisions and the activities are conducted by the Divisions and Districts. The initial point of contact for hydrological or wetland related issues is the Hydraulics and Hydrology Branch, or the Regulatory Functions Branch of the District.

References

Lyon, J. 1993.   Practical Handbook for Wetland Identification and Delineation. Lewis/CRC Press, Boca Raton, FL, 157 pp.